T0137409

Advanced Structured Materials

Volume 120

Series Editors

Andreas Öchsner, Faculty of Mechanical Engineering, Esslingen University of
Applied Sciences, Esslingen, Germany

Lucas F. M. da Silva, Department of Mechanical Engineering, Faculty of
Engineering, University of Porto, Porto, Portugal

Holm Altenbach, Faculty of Mechanical Engineering,
Otto-von-Guericke-Universität Magdeburg, Magdeburg, Sachsen-Anhalt, Germany

Common engineering materials reach in many applications their limits and new developments are required to fulfil increasing demands on engineering materials. The performance of materials can be increased by combining different materials to achieve better properties than a single constituent or by shaping the material or constituents in a specific structure. The interaction between material and structure may arise on different length scales, such as micro-, meso- or macroscale, and offers possible applications in quite diverse fields.

This book series addresses the fundamental relationship between materials and their structure on the overall properties (e.g. mechanical, thermal, chemical or magnetic etc) and applications.

The topics of *Advanced Structured Materials* include but are not limited to

- classical fibre-reinforced composites (e.g. glass, carbon or Aramid reinforced plastics)
- metal matrix composites (MMCs)
- micro porous composites
- micro channel materials
- multilayered materials
- cellular materials (e.g., metallic or polymer foams, sponges, hollow sphere structures)
- porous materials
- truss structures
- nanocomposite materials
- biomaterials
- nanoporous metals
- concrete
- coated materials
- smart materials

Advanced Structured Materials is indexed in Google Scholar and Scopus.

More information about this series at http://www.springer.com/series/8611

Holm Altenbach · Wolfgang H. Müller ·
Bilen Emek Abali

Editors

Higher Gradient Materials
and Related Generalized
Continua

 Springer

Editors
Holm Altenbach
Institut für Mechanik
Otto-von-Guericke-Universität Magdeburg
Magdeburg, Germany

Wolfgang H. Müller
Institut für Mechanik
Technische Universität Berlin
Berlin, Germany

Bilen Emek Abali
Institut für Mechanik
Technische Universität Berlin
Berlin, Germany

ISSN 1869-8433   ISSN 1869-8441   (electronic)
Advanced Structured Materials
ISBN 978-3-030-30408-9   ISBN 978-3-030-30406-5   (eBook)
https://doi.org/10.1007/978-3-030-30406-5

This Springer imprint is published by the registered company Springer Nature Switzerland AG
The registered company address is: Gewerbestrasse 11, 6330 Cham, Switzerland

Preface

The idea for this volume of the Advanced Structured Materials Series was born during two seminars, namely "New Developments in Micropolar Theory," and "Advanced Seminar: Generalized Continua in Engineering—Theory, Experiments, and Applications," which were held on November 6-7, 2017 and September 3-5, 2018, respectively, both at the Technische Universität Berlin. The first seminar was organized by Wolfgang H. Müller (Berlin) and Elena Vilchevskaya (St. Petersburg) and the second by Wolfgang H. Müller (Berlin) & Holm Altenbach (Magdeburg) and attended by many scientists from Germany, Russia, Italy, USA, Sweden, Georgia, France, Estonia, and Finland. The organizers were assisted by B. Emek Abali (Berlin).

Generalized Continua have recently seen a formidable renaissance: the Cosserat brothers gave a first summary in 1909. Their ideas stayed dormant for a while and were picked up after World War II by Ericksen & Truesdell resulting in a continuous stream of theoretical papers until today. Most recently, the theory was complemented by applications and embedded in experiments focusing on how to determine the various new material parameters required for making the theory applicable.

During the last decade various colloquia held in Paris (2009)[1], Wittenberg (2010, 2012)[2,3], and Magdeburg (2015)[4], as well as the CISM Course "Generalized Continua—from the Theory to the Engineering Applications" (Udine, 2011)[5] helped to promote this type of research. During this new "Advanced Seminar," attention

---

[1] Maugin, G.A., Metrikine, A.V. (eds) Mechanics of Generalized Continua: One Hundred Years After the Cosserats. Springer, New York, 2010

[2] Altenbach, H., Maugin, G.A., Erofeev, V. (eds) Mechanics of Generalized Continua. Advanced Structured Materials, vol. 7. Springer, Berlin, Heidelberg, 2011

[3] Altenbach, H., Forest, S., Krivtsov, A. (eds) Generalized Continua as Models for Materials. Advanced Structured Materials, vol. 22. Springer, Berlin, Heidelberg

[4] Altenbach, H., Forest, S. (eds) Generalized Continua as Models for Classical and Advanced Materials. Advanced Structured Materials, vol. 42. Springer, Cham

[5] Altenbach H., Eremeyev V.A. (eds) Generalized Continua from the Theory to Engineering Applications. CISM International Centre for Mechanical Sciences (Courses and Lectures), vol 541. Springer, Vienna, 2013

was paid on the most recent research items, e.g., new generalized models, materials with significant microstructure, multi-field loadings or identification of constitutive equations. Finally yet importantly, a comparison with discrete modeling approaches and experiments was discussed.

As editors, we intend to thank all authors for their crucial contributions as well as all reviewers for their invaluable time and effort. We delightedly acknowledge Dr. Christoph Baumann (Springer Publisher) for initiating the book project. In addition, we have to thank Dr. Mayra Castro (Senior Editor Applied Sciences; Materials Science; Materials Engineering; Nanotechnology and Nanomedicine) and Mr. Ashok Arumairaj (Production Administrator) giving the final support. Last but not least, the first editor has to acknowledge the Fundacja na rzecz Nauki Polskiej (Fundation for Polish Science) allowing to finalize this book at the Politechnika Lubelska (host: Prof. dr.hab.inż. Tomasz Sadowski, dr.h.c.) with the help of the Alexander von Humboldt Polish Honorary Research Felloship.

Berlin, Magdeburg   *Holm Altenbach*
July 2019   *Wolfgang H. Müller*
                                                                                 Bilen Emek Abali

Contents

11   Theoretical Estimation of the Strength of Thin-film Coatings 221
Sergey N. Romashin, Victoria Yu. Presnetsova, Larisa Yu. Frolenkova,
Vladimir S. Shorkin, and Svetlana I. Yakushina

List of Contributors

Bilen Emek Abali
Chair of Continuum Mechanics and Constitutive Theory, Institute of Mechanics,
Technische Universität Berlin, Einsteinufer 5, 10587 Berlin, Germany
e-mail: bilenemek@abali.org

Faris Alzahrani
NAAM Research Group, Department of Mathematics, King Abdulaziz University,
Jeddah 21589, Saudi Arabia
e-mail: faris.kau@hotmail.com

Ugo Andreaus
Department of Structural and Geotechnical Engineering, Università di Roma La
Sapienza, 18 Via Eudossiana, Rome, Italy
e-mail: ugo.andreaus@uniroma1.it

Emilio Barchiesi
International Research Center for the Mathematics and Mechanics of Complex
Systems - M&MoCS, Università dell'Aquila, L'Aquila
Department of Structural and Geotechnical Engineering, Università di Roma La
Sapienza, 18 Via Eudossiana, Rome, Italy
e-mail: barchiesiemilio@gmail.com

Antonio Battista
International Research Center for the Mathematics and Mechanics of Complex
Systems - M&MoCS, Università dell'Aquila, L'Aquila, Italy
LaSIE, Université de La Rochelle, La Rochelle, France
e-mail: antonio.battista@uni-lr.fr

Petr Belov
Institute of Applied Mechanics of Russian Academy of Sciences, Moscow, Russia
e-mail: belovpa@yandex.ru

Francesco dell'Isola
International Research Center for the Mathematics and Mechanics of Complex
Systems - M&MoCS, Università dell'Aquila, L'Aquila
Department of Structural and Geotechnical Engineering, Università di Roma La
Sapienza, 18 Via Eudossiana, Rome, Italy
Research Institute for Mechanics, Nizhny Novgorod Lobachevsky State University,
23, Gagarin av. 603950 Nizhny Novgorod, Russia
e-mail: francesco.dellisola.me@gmail.com

Alessandro Della Corte
International Research Center for the Mathematics and Mechanics of Complex
Systems - M&MoCS, Università dell'Aquila, L'Aquila
Department of Structural and Geotechnical Engineering, Università di Roma La
Sapienza, 18 Via Eudossiana, Rome, Italy
e-mail: alessandro.dellacorte.memocs@gmail.com

Victor A. Eremeyev
Faculty of Civil and Environmental Engineering, Gdańsk University of Technology,
ul. Gabriela Narutowicza 11/12, 80-233 Gdańsk, Poland
Research Institute for Mechanics, Nizhny Novgorod Lobachevsky State University,
23, Gagarin av. 603950 Nizhny Novgorod, Russia
e-mail: eremeyev.victor@gmail.com

Vladimir I. Erofeev
Mechanical Engineering Research Institute of Russian Academy of Sciences
Research Institute for Mechanics, Nizhny Novgorod Lobachevsky State University,
23, Gagarin av. 603950 Nizhny Novgorod, Russia
e-mail: erof.vi@yandex.ru

Larisa Yu. Frolenkova
Orel State University named after. I.S. Turgenev, 29 Naugorskoe Shosse, 302020
Orel, Russia
e-mail: Larafrolenkova@yandex.ru

Ivan Giorgio
Department of Structural and Geotechnical Engineering, Università di Roma La
Sapienza, 18 Via Eudossiana, Rome
International Research Center for the Mathematics and Mechanics of Complex
Systems - M&MoCS, Università dell'Aquila, L'Aquila, Italy
e-mail: ivan.giorgio@uniroma1.it

Rainer Glüge
Institute of Mechanics, Otto-von-Guericke–University Magdeburg, Universitätsplatz 2, 39106 Magdeburg, Germany
e-mail: gluege@ovgu.de

Tasawar Hayat
Department of Mathematics, Quaid-I-Azam University, Islamabad, Pakistan
NAAM Research Group, Department of Mathematics, King Abdulaziz University,
Jeddah 21589, Saudi Arabia
e-mail: pensy_t@yahoo.com

Leonid A. Igumnov
Research Institute for Mechanics, Nizhny Novgorod Lobachevsky State University,
23, Gagarin av. 603950 Nizhny Novgorod, Russia
e-mail: Igumnov@mech.unn.ru

Aleksandr A. Ipatov
Research Institute for Mechanics, Nizhny Novgorod Lobachevsky State University,
23, Gagarin av. 603950 Nizhny Novgorod, Russia
e-mail: ipatov@mech.unn.ru

Tomasz Lekszycki
Warsaw University of Technology, Warsaw, Poland
e-mail: t.lekszycki@wip.pw.edu.pl

Anna V. Leontyeva
Mechanical Engineering Research Institute of Russian Academy of Sciences,
Research Institute for Mechanics, Nizhny Novgorod Lobachevsky State University,
23, Gagarin av. 603950 Nizhny Novgorod, Russia
e-mail: aleonav@mail.ru

Svetlana Yu. Litvinchuk
Research Institute for Mechanics, Nizhny Novgorod Lobachevsky State University,
23, Gagarin av. 603950 Nizhny Novgorod, Russia
e-mail: litvinchuk@mech.unn.ru

Sergey Lurie
Institute of Applied Mechanics of Russian Academy of Sciences, Moscow, Russia
e-mail: salurie@mail.ru

Alexey O. Malkhanov
Research Institute for Mechanics, Nizhny Novgorod Lobachevsky State University,
23, Gagarin av. 603950 Nizhny Novgorod, Russia
e-mail: alexey.malkhanov@gmail.com

Mikhail Nikabadze
Lomonosov Moscow State University
Bauman Moscow State Technical University, Moscow, Russia
e-mail: nikabadze@mail.ru

Panayiotis Papadopoulos
Department of Mechanical Engineering, 6131 Etcheverry Hall, University of
California, Berkeley, CA, 94720-1740, USA
e-mail: panos@me.berkeley.edu

Igor S. Pavlov
Mechanical Engineering Research Institute of Russian Academy of Sciences,
Research Institute for Mechanics, Nizhny Novgorod Lobachevsky State University,
23, Gagarin av. 603950 Nizhny Novgorod, Russia
e-mail: ispavlov@mail.ru

Andrey N. Petrov
Research Institute for Mechanics, Nizhny Novgorod Lobachevsky State University,
23, Gagarin av. 603950 Nizhny Novgorod, Russia
e-mail: andrey.petrov@mech.unn.ru

Aron Pfaff
Fraunhofer Institute for High-Speed Dynamics, Ernst-Mach-Institut, Ernst-
Zermelo-Str. 4, Freiburg, 79104, Germany
e-mail: Aron.Pfaff@emi.fraunhofer.de

Victoria Yu. Presnetsova
Orel State University named after. I.S. Turgenev, 29 Naugorskoe Shosse, 302020
Orel, Russia
e-mail: alluvian@mail.ru

Sergey N. Romashin
Orel State University named after. I.S. Turgenev, 29 Naugorskoe Shosse, 302020
Orel, Russia
e-mail: sromashin@yandex.ru

Vladimir S. Shorkin
Orel State University named after. I.S. Turgenev, 29 Naugorskoe Shosse, 302020
Orel, Russia
e-mail: vshorkin@yandex.ru

Armine Ulukhanyan
Bauman Moscow State Technical University, Moscow, Russia
e-mail: armine_msu@mail.ru

Hua Yang
Chair of Continuum Mechanics and Constitutive Theory, Institute of Mechanics,
Technische Universität Berlin, Einsteinufer 5, 10587 Berlin, Germany
e-mail: hua.yang@campus.tu-berlin.de

Svetlana I. Yakushina
Orel State University named after. I.S. Turgenev, 29 Naugorskoe Shosse, 302020
Orel, Russia
e-mail: jakushina@rambler.ru

Chapter 1
A Computational Approach for Determination of Parameters in Generalized Mechanics

Bilen Emek Abali, Hua Yang, and Panayiotis Papadopoulos

Abstract Metamaterials are functionalized by specifying a structure at a micro-scopic length scale such that they provide a tailored deformation response at a macroscopic length scale. Their modeling at the macroscale is attained by using the generalized mechanics that incorporates higher gradients of the displacement leading to additional parameters effected by the "inner" structure at the microscale. As these additional parameters are a consequence of the inner structure, we propose a general methodology for determining them by using a computational approach. The inner structure is given and the presented strategy achieves numerical values of all homogenized parameters to be used in the generalized mechanics for modeling a structure at the macroscale.

Keywords: Parameter determination · Material modeling · Inverse analysis · Finite element method · Generalized mechanics

1.1 Introduction

Homogenization is an approach to include the effects in another time or length scale. We concentrate at the homogenization in space, where at least two different length scales are existing. At atomistic length scale, discrete atoms are modeled as *points* and their interaction with scalar potentials. In micrometer length scale, the amount of such *points* is so high, more conveniently we use continuum mechanics resulting

B. E. Abali · H. Yang
Technische Universität Berlin, Institute of Mechanics, Einsteinufer 5, 10587 Berlin, Germany
e-mail: bilenemek@abali.org; hua.yang@campus.tu-berlin.de

P. Papadopoulos
University of California, Berkeley, Department of Mechanical Engineering, 6131 Etcheverry Hall, Berkeley, CA 94720, USA
e-mail: panos@me.berkeley.edu

© Springer Nature Switzerland AG 2019
H. Altenbach et al. (eds.), *Higher Gradient Materials and Related Generalized Continua*, Advanced Structured Materials 120,
https://doi.org/10.1007/978-3-030-30406-5_1

1

from a homogenization approach. Homogenization links two different physics and bridges discrete and continuum mechanical models. Such an approach can be based on a statistical method as introduced in Irving and Kirkwood (1950). An analogous attempt is used for composite materials, a heterogeneous material at the microscale can be modeled as a homogeneous material at the macroscale. In this case the scale separation is motivated by the material properties, at both scales, continuum mechanics models the underlying system. Such an approach uses energy equivalence at both scales as proposed in Hill (1972). For a composite material, at least two different materials, with known material models and parameters, generate a homogenized material modeled with a predetermined constitutive equation. Determination of material parameters of the homogenized material is a challenging task. Analytical solutions can be obtained by using homogenization approaches leading to estimates of the homogenized parameters (Voigt, 1889; Reuss, 1929; Hashin and Shtrikman, 1962), for a more detailed explanation of these methods, we refer to Dormieux et al (2006); Zohdi and Wriggers (2008); Nemat-Nasser and Hori (2013); Kachanov and Sevostianov (2013); Zohdi (2004). Analogous ideas are used to obtain analytical or semi-analytical solutions for parameter determination in the case of special types of inclusions—voids in the bulk material at the microscale—leading to different structure related materials response at the macroscale. These homogenization methods use several solution methods as in Eshelby (1957); Mori and Tanaka (1973); Levin (1976); Willis (1977); Kanaun and Kudryavtseva (1986); Hashin (1991); Nazarenko (1996); Shafiro and Kachanov (2000); Castañeda and Tiberio (2000); Zheng and Du (2001); Sevostianov and Kachanov (2014); Berryman (2005); Sburlati et al (2018); Nazarenko et al (2018).

In addition to these (semi-)analytical methods, computational approaches allow one to apply the same type of strategy for every microstructure as in Kushnevsky et al (1998); Kouznetsova et al (2001); Miehe et al (2002); Lebensohn et al (2004); Sevostianov and Kachanov (2006); Nazarenko et al (2009); Ladeveze et al (2010); Nemat-Nasser and Srivastava (2011); Mercer et al (2015); Nazarenko et al (2016). Often, the idea in Hill (1972) is employed such that a representative volume element (RVE) is postulated. The main assumption (an RVE relies on) is the axiomatic start that there exists an RVE with periodic boundary conditions accurately describing the deformation response of the whole geometry. The general consensus is the applicability of this postulate, at least for classical mechanics with an energy definition depending on the first gradient of displacement. For the generalized mechanics, higher gradients of displacement are incorporated in the energy definition as introduced in different versions in Eringen and Suhubi (1964); Mindlin (1964); Eringen (1968); Steinmann (1994); Eremeyev et al (2012); Polizzotto (2013a,b); Ivanova and Vilchevskaya (2016); Abali (2019). Although different concepts are used for motivating a generalized mechanics, all versions consider using a second space gradient in displacement. We doubt that the same postulate is reasonable and refrain ourselves from taking for granted that an RVE exists in the case of generalized mechanics.

At the microscale, the formulation is sufficient with first gradients in the energy formulation. By abusing the wording, we call the formulation with first gradients "classical mechanics." For describing the material response accurately, we may use classical mechanics (also called CAUCHY continuum or BOLTZMANN continuum) and incorporate a detailed description of the inner structure (substructure). Consider a simple plate of 100 mm length manufactured by a 3D printer (additively) with gaps or pores of 10 μm length. As we want to model this structure by the classical mechanics, we need to use an extensive computational effort. Instead, we may use a generalized mechanics formulation with energy definition incorporating higher gradients. The additional parameters lead naturally as a consequence of the homogenization procedure between micro- and macroscales, we refer to Pideri and Seppecher (1997); Bigoni and Drugan (2007); Seppecher et al (2011); Abdoul-Anziz and Seppecher (2018); Mandadapu et al (2018). Therefore, we start off with the *a priori* knowledge that two approaches are identical: the detailed modeling at the microscale by classical mechanics and an efficient modeling at the macroscale by generalized mechanics. For the microscale modeling, we need solely the known material parameters—they are directly measurable—as well as a huge computational effort. For the macroscale modeling by generalized mechanics, we need additional (effective) parameters—they are unknown. Since both modeling approaches are identical, we can extract the unknown parameters by a comparison. In this work, we intend to describe a general approach for this comparison. For the sake of clarity, we use an isotropic material and a quadratic energy formulation; however, the method is applicable for every symmetry class and nonlinear formulation.

Parameter determination in generalized mechanics has been studied before, for example see Forest et al (1999); Pietraszkiewicz and Eremeyev (2009). An often used technique is the asymptotic analysis for scale separation as applied in mechanics by Bensoussan et al (1978); Hollister and Kikuchi (1992); Chung et al (2001); Temizer (2012) as well as in generalized mechanics by Forest et al (2001); Li (2011); Eremeyev (2016); Barboura and Li (2018). Often an RVE is used, in rare cases the same critical inspection is applied about the existence of an RVE (Rahali et al, 2015). Beyond the difficulty of an RVE, the length scale is of importance when it comes to the question if generalized mechanics is adequate. For example the so-called "size-effect" phenomenon is demonstrated in experiments various times, see for example, Namazu et al (2000); Lam et al (2003); Chen et al (2010). Basically, this effect starts dominating by increasing the bending stiffness of a beam whenever the length scale in the macroscale divided by the length scale of the microscale decreases and approaches to unity. This effect is precisely captured by strain gradient theory based computations in Abali et al (2017), where the same generalized mechanics formulation is used for simulating geometries in different sizes. We emphasize that the formulation remains the same; but the response varies depending on the ratio of length scales. In other words, materials response is not the same for a large sample and for a small sample. Therefore, we conclude that the additional parameters are structure (geometry) related as well as we fail to define the appropriate length scale ratio. Moreover, we fail to exploit an RVE (a small sample) differing in response regarding the whole body (a large sample). Hence, we avoid on purpose introducing an RVE with the cell length tantamount to the length scale ratio.

In this work we concentrate on an innovative method for determining all additional parameters emerging in a generalized mechanics formulation. The novelty of this method relies on:

- circumventing the use of an RVE,
- being applicable for any anisotropic structure (even effected by the substructure),
- the capability of determining parameters for geometric and material nonlinearities as well.

For providing a clear demonstration of the algorithm, we will show results for the simplest case: a linear, isotropic material at the microscale with a given substructure and known *two* material parameters. This configuration is assumed to result in a linear, isotropic, centro-symmetric strain gradient material with *seven* parameters. The outcome is determining these seven parameters. The computational characterization of the homogenized material is based on simulations obtained by the finite element method. Algorithm is explained in detail and the code for the computation is made public in order to encourage a transparent and efficient scientific exchange.

1.2 Homogenization Between Micro- and Macroscales

By following Abali and Müller (2016) we begin with the axiom that there is a LaGRANGEan density, $\mathcal{L}$, depending on any number of coordinates, x_μ, leading to an action formulation:

$$\mathcal{A} = \int_\Omega \mathcal{L} \, d\Sigma \,, \quad d\Sigma = dx_1 \, dx_2 \dots dx_m \,, \tag{1.1}$$

for the domain of interest Ω in this m-dimensional space. The definition of the LaGRANGEan is the key aspect. For classical mechanics, it depends on so-called "primitive variables" and their first derivatives, we use herein a primitive variable as a function simply existing axiomatically, in the following, it will be simply the displacement for isothermal deformation without phase change or any other chemical or electrodynamics interactions, etc. For generalized mechanics, such as a *second gradient* theory, it depends on primitive variables, their first and *second* derivatives. The coordinates are space, $\boldsymbol{X}$, and time, t, such that $x_\mu = \{\boldsymbol{X}, t\}$ defines a material frame where the particles rest and this configuration is the reference frame without strain energy. In the case of neglecting gravitational as well as inertial terms in the energy formulation, we end up with the LaGRANGEan being equal to (minus of) the stored energy density, $\mathcal{L} = -w$. For this energy we distinguish between microscale energy density, $^{\mathrm{m}}w$, and the macroscale energy density, $^{\mathrm{M}}w$. The action obtained at the microscale by using the substructure is equal to the action obtained at the macroscale. This assumption is conceptually different to the HILL–MANDEL condition, which states that the mean product of microscopic stress times strain over the region equals the product of the mean stress times mean strain. Herein, we skip introducing stress or its averaging over the volume and work with energies at both scales. We solely consider a system defined by the action and equivalence of this action in micro- and macroscales that is possible by using different definitions of the microscale energy density, ^{m}w, and the macroscale energy density, ^{M}w.

Consider an elastic material with the quadratic energy formulation in strains $^{m}\varepsilon_{ij}$ by using the stiffness tensor $^{m}C_{ijkl}$ expressed in Cartesian coordinates as follows:

$$^{m}w = \frac{1}{2}\,^{m}\varepsilon_{ij}\,^{m}C_{ijkl}\,^{m}\varepsilon_{kl} , \tag{1.2}$$

where for the sake of clarity, we assume a linear strain measure as well as constant material parameters in the stiffness tensor. The same formulation would be applicable for any other definition of the stored energy density as well. At the microscale, the material parameters are known, i.e., $^{m}C_{ijkl}$ is defined in each point of the continuum body. In the case of the strain gradient theory, the macroscale energy density is given by

$$^{M}w = \frac{1}{2}\,^{M}\varepsilon_{ij}\,^{M}C_{ijkl}\,^{M}\varepsilon_{kl} + \,^{M}\varepsilon_{ij}G_{ijklm}\,^{M}\varepsilon_{kl,m} + \frac{1}{2}\,^{M}\varepsilon_{ij,k}D_{ijklmn}\,^{M}\varepsilon_{lm,n} . \tag{1.3}$$

We are searching for the material parameters generating the homogenized stiffness tensor $^{M}C_{ijkl}$ as well as structure related material tensors G_{ijklm} and D_{ijklmn}. All three are unknowns; however, they are constants (in space) at the macroscale by simplifying our methodology to homogeneous materials. As we have excluded inertial terms as well as time-dependent material behavior, the continuum body, $\mathcal{B}$, deforms immediately to the steady state and the equivalence of action in multiscales reduces to

$$\int_{\mathcal{B}} {}^{m}w \, dv = \int_{\mathcal{B}} {}^{M}w \, dv ,$$

$$\int_{\mathcal{B}} {}^{m}\varepsilon_{ij}\,^{m}C_{ijkl}\,^{m}\varepsilon_{kl} \, dv = {}^{M}C_{ijkl} \int_{\mathcal{B}} {}^{M}\varepsilon_{ij}\,^{M}\varepsilon_{kl} \, dv +$$

$$+ 2G_{ijklm} \int_{\mathcal{B}} {}^{M}\varepsilon_{ij}\,^{M}\varepsilon_{kl,m} \, dv +$$

$$+ D_{ijklmn} \int_{\mathcal{B}} {}^{M}\varepsilon_{ij,k}\,^{M}\varepsilon_{lm,n} \, dv , \tag{1.4}$$

since the unknown material tensors are constant in space. Consider a continuum body at the reference frame and a particle of it with coordinates X_i as depicted in Fig. 1.1 The same massive particle is moved to $^{m}x_i$ as a consequence of the deformation. As this deformation is described at the microscale, the substructure is visualized as well. For the simplicity, an example is used as in composite materials with the orange inclusion (fiber) embedded in the blue material (matrix). For the homogenized case, the same particle moves to $^{M}x_i$ expressed at the macroscale without the substructure. We emphasize that micro- and macroscales are both expressed in the same coordinate system. There is no scale separation or different coordinate

Fig. 1.1 Homogenization procedure. Left: continuum body in the reference frame. Right top: deformation at the microscale. Right bottom: corresponding deformation at the macroscale.

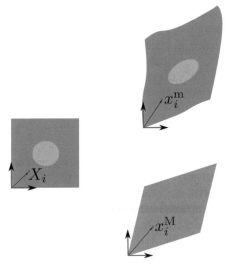

systems in this investigation. Two different cases are demonstrated: heterogeneous case at the microscale with known material properties versus homogeneous case at the macroscale with sought after parameters. The displacement at both scales are different

$$
\begin{aligned}
{}^{\mathrm{m}}u_i &= {}^{\mathrm{m}}x_i - X_i \ , \\
{}^{\mathrm{M}}u_i &= {}^{\mathrm{M}}x_i - X_i \ ,
\end{aligned}
\tag{1.5}
$$

and we intend to determine parameters fulfilling Eq. (1.4) with the same strain measure at both scales, for example we may use linearized strains

$$
\begin{aligned}
{}^{\mathrm{m}}\varepsilon_{ij} &= \frac{1}{2}\left(\frac{\partial {}^{\mathrm{m}}u_i}{\partial X_j} + \frac{\partial {}^{\mathrm{m}}u_j}{\partial X_i}\right) = {}^{\mathrm{m}}u_{(i,j)} \ , \\
{}^{\mathrm{M}}\varepsilon_{ij} &= \frac{1}{2}\left(\frac{\partial {}^{\mathrm{M}}u_i}{\partial X_j} + \frac{\partial {}^{\mathrm{M}}u_j}{\partial X_i}\right) = {}^{\mathrm{M}}u_{(i,j)} \ ,
\end{aligned}
\tag{1.6}
$$

with the usual comma notation for partial derivatives in space and round brackets for symmetrizing over these indices. At both scales, the displacement field as well as the strain field are different. The energy descriptions are also different. But the total amount of the energy used for the deformation is equivalent.

1.3 Determination of Parameters

For the given macroscopic energy in Eq. (1.3), we observe the following symmetry relations based on the quadratic description

$$
{}^{\mathrm{M}}C_{ijkl} = {}^{\mathrm{M}}C_{klij} \ , \quad D_{ijklmn} = D_{lmnijk} \ ,
\tag{1.7}
$$

additionally the following relations based on symmetric strain

$$^{\mathrm{M}}C_{ijkl} = {}^{\mathrm{M}}C_{jikl} \, , \; D_{ijklmn} = D_{jiklmn} \, , \; G_{ijklm} = G_{jiklm} \, , \; G_{ijklm} = G_{ijlkm} \, .$$
(1.8)

Specifically, we restrict to a specific case and assume isotropic and centro-symmetric properties at the macroscale for the linear strain gradient elastic material modeled as in Eq. (1.4) leading to

$$
\begin{aligned}
^{\mathrm{M}}C_{ijkl} =& c_1 \delta_{ij}\delta_{kl} + c_2(\delta_{ik}\delta_{jl} + \delta_{il}\delta_{jk}) \, , \\
D_{ijklmn} =& c_3(\delta_{ij}\delta_{kl}\delta_{mn} + \delta_{in}\delta_{jk}\delta_{lm} + \delta_{ij}\delta_{km}\delta_{ln} + \delta_{ik}\delta_{jn}\delta_{lm}) + \\
& + c_4\delta_{ij}\delta_{kn}\delta_{ml} + \\
& + c_5(\delta_{ik}\delta_{jl}\delta_{mn} + \delta_{im}\delta_{jk}\delta_{ln} + \delta_{ik}\delta_{jm}\delta_{ln} + \delta_{il}\delta_{jk}\delta_{mn}) + \quad (1.9)\\
& + c_6(\delta_{il}\delta_{jm}\delta_{kn} + \delta_{im}\delta_{jl}\delta_{kn}) + \\
& + c_7(\delta_{il}\delta_{jn}\delta_{mk} + \delta_{im}\delta_{jn}\delta_{lk} + \delta_{in}\delta_{jl}\delta_{km} + \delta_{in}\delta_{jm}\delta_{kl}) \, , \\
G_{ijklm} =& 0 \, ,
\end{aligned}
$$

we refer to Appendix of Abali et al (2015) for their derivation. The parameters $c = \{c_1, c_2, c_3, c_4, c_5, c_6, c_7\}$ are unknowns to be determined. For 7 unknowns we need 7 independent deformation cases. For one specific case, by inserting the latter into Eq. (1.4), we obtain

$$\sum_{\alpha=1}^{7} A_{1\alpha} c_\alpha = R_1 \, ,$$
(1.10)

with the coefficient matrix A and right hand side R as follows:

$$A_{11} = \delta_{ij}\delta_{kl} \int_{\mathcal{B}} {}^{\mathrm{M}}\varepsilon_{ij} {}^{\mathrm{M}}\varepsilon_{kl} \, \mathrm{dv} \, ,$$

$$A_{12} = (\delta_{ik}\delta_{jl} + \delta_{il}\delta_{jk}) \int_{\mathcal{B}} {}^{\mathrm{M}}\varepsilon_{ij} {}^{\mathrm{M}}\varepsilon_{kl} \, \mathrm{dv} \, ,$$

$$A_{13} = (\delta_{ij}\delta_{kl}\delta_{mn} + \delta_{in}\delta_{jk}\delta_{lm} + \delta_{ij}\delta_{km}\delta_{ln} + \delta_{ik}\delta_{jn}\delta_{lm}) \int_{\mathcal{B}} {}^{\mathrm{M}}\varepsilon_{ij,k} {}^{\mathrm{M}}\varepsilon_{lm,n} \, \mathrm{dv} \, ,$$

$$A_{14} = \delta_{ij}\delta_{kn}\delta_{ml} \int_{\mathcal{B}} {}^{\mathrm{M}}\varepsilon_{ij,k} {}^{\mathrm{M}}\varepsilon_{lm,n} \, \mathrm{dv} \, ,$$

$$A_{15} = (\delta_{ik}\delta_{jl}\delta_{mn} + \delta_{im}\delta_{jk}\delta_{ln} + \delta_{ik}\delta_{jm}\delta_{ln} + \delta_{il}\delta_{jk}\delta_{mn}) \int_{\mathcal{B}} {}^{\mathrm{M}}\varepsilon_{ij,k} {}^{\mathrm{M}}\varepsilon_{lm,n} \, \mathrm{dv} \, ,$$

$$A_{16} = (\delta_{il}\delta_{jm}\delta_{kn} + \delta_{im}\delta_{jl}\delta_{kn}) \int_{\mathcal{B}} {}^{\mathrm{M}}\varepsilon_{ij,k} {}^{\mathrm{M}}\varepsilon_{lm,n} \, \mathrm{dv} \, ,$$

$$A_{17} = (\delta_{il}\delta_{jn}\delta_{mk} + \delta_{im}\delta_{jn}\delta_{lk} + \delta_{in}\delta_{jl}\delta_{km} + \delta_{in}\delta_{jm}\delta_{kl}) \int_{\mathcal{B}} {}^{\mathrm{M}}\varepsilon_{ij,k} {}^{\mathrm{M}}\varepsilon_{lm,n} \, \mathrm{dv} \, ,$$

$$R_1 = \int_{\mathcal{B}} {}^{\mathrm{m}}\varepsilon_{ij} {}^{\mathrm{m}}C_{ijkl} {}^{\mathrm{m}}\varepsilon_{kl} \, \mathrm{dv} \, .$$
(1.11)

By defining 7 cases, we acquire $\boldsymbol{Ac} = \boldsymbol{R}$ with $\boldsymbol{A}$ of rank 7 leading to unique determination of unknowns by $\boldsymbol{c} = \boldsymbol{A}^{-1}\boldsymbol{R}$. First, we explain one specific case. Second, we propose 7 cases. Third, we compute some example structures for demonstrating the proposed method in action.

1.4 Computation of One Specific Case

For the homogenized structure, we give the desired displacement field as a smooth function, $^{\text{M}}\boldsymbol{u} = {^{\text{M}}}\boldsymbol{u}(\boldsymbol{X})$, and assume that the material model allows to capture this displacement field accurately. In other words, the chosen substructure has to be in such a way that the isotropic and centro-symmetric class is sufficient for the macroscale. As we define the displacement by giving a smooth function, the compatibility is fulfilled such that the second derivative can be interchanged, $^{\text{M}}u_{i,jk} = {^{\text{M}}}u_{i,kj}$. By knowing the macroscale displacement field, we propose to have a microscale displacement field with the same displacement distribution on the boundaries as in the macroscale. This choice is the key difference to the conventional homogenization procedures with an RVE, where the boundary condition is circumvented by applying periodic boundaries. Herein we intend to suggest a scheme without an RVE as we fail to argue that the periodic boundary condition is adequate in the case of higher gradients. Hence, we simply state that both scales:

- provide the same boundary values,
- deform with the same energy.

The latter statement has been used for constructing the inverse analysis as in Eq. (1.10). The former will be used to generate a computational scheme for obtaining $^{\text{m}}\boldsymbol{u}$ in the case of a given substructure.

Consider a continuum body, $\mathcal{B}$, at the microscale; for example Fig. 1.1 includes such a geometry with one inclusion. We search for the displacement field, $^{\text{m}}\boldsymbol{u}$, which lets the action invariant leading to the variational formulation:

$$\delta\mathcal{A} = 0 \Rightarrow \int_{\mathcal{B}} \frac{\partial^{\text{m}}w}{\partial^{\text{m}}u_{i,j}} \delta u_{i,j}\, \mathrm{d}V = 0 \ , \tag{1.12}$$

with the integral measure $\mathrm{d}V = \mathrm{d}X_1\,\mathrm{d}X_2\,\mathrm{d}X_3$ in Cartesian coordinates. After inserting Eqs. (1.2), (1.6), we obtain the integral form:

$$\text{Form} = \int_{\mathcal{B}} {^{\text{m}}}C_{ijkl}\,{^{\text{m}}}u_{(k,l)}\delta u_{i,j}\, \mathrm{d}V \ . \tag{1.13}$$

This so-called weak form contains the unknowns, $^{\text{m}}\boldsymbol{u} = \{{^{\text{m}}}u_1, {^{\text{m}}}u_2, {^{\text{m}}}u_3\}$, and their corresponding test functions, $\delta\boldsymbol{u} = \{\delta u_1, \delta u_2, \delta u_3\}$, both are approximated with the same Hilbertian Sobolev space

$$\hat{\mathcal{V}} = \left\{ {^{\text{m}}}\boldsymbol{u}, \delta\boldsymbol{u} \in [\mathcal{H}^n(\Omega)]^3 : {^{\text{m}}}\boldsymbol{u}, \delta\boldsymbol{u} = \text{given } \forall \boldsymbol{x} \in \partial\mathcal{B} \right\} , \tag{1.14}$$

which is known as the GALERKIN method. On all boundaries, $\partial\mathcal{B}$, the displacement is given by so-called DIRICHLET type boundary conditions being equal to displacement at the macroscale, $^{\mathrm{M}}\boldsymbol{u}$, as follows:

$$^{\mathrm{m}}u_i = {}^{\mathrm{M}}u_i \,, \quad \delta u_i = 0 \; \forall \boldsymbol{x} \in \partial\mathcal{B} \,. \tag{1.15}$$

The computation delivers the displacement at the microscale to be used in Eq. (1.11) for constructing R_1.

1.5 Algorithm for All Deformation Cases

The main objective is to construct the matrices, $\boldsymbol{A}$, $\boldsymbol{R}$, in Eq. (1.11) in order to determine the unknown material parameters, $\boldsymbol{c}$. The following Algorithm 1 demonstrates the necessary repeated steps for assembling the matrices and solving for the parameters. Seven cases are necessary to obtain seven parameters, c_α.

Algorithm 1: Determination of c_α parameters

input : Substructure and material properties at the microscale, homogenized displacement field at the macroscale, $^{\mathrm{M}}\boldsymbol{u}$, for $\alpha = 1 \ldots 7$ cases

output: Calculating the material parameters c_α

begin

    set   : geometry at the microscale

    set   : material data at the microscale

    case is set by $\alpha = 0$

    while case $\alpha \neq 7$ **do**

        $\alpha := \alpha + 1$

        choose: smooth function $^{\mathrm{M}}\boldsymbol{u}$

        set   : boundary conditions with $^{\mathrm{M}}\boldsymbol{u}$

        solve : $^{\mathrm{m}}\boldsymbol{u}$ by using Eq. (1.13)

        solve : $^{\mathrm{m}}\varepsilon_{ij} = {}^{\mathrm{m}}u_{(i,j)}$ and $^{\mathrm{M}}\varepsilon_{ij} = {}^{\mathrm{M}}u_{(i,j)}$

        solve : coefficients in $\boldsymbol{A}$ and $\boldsymbol{R}$ by using Eq. (1.11)

    end

    Determine the coefficients by $\boldsymbol{c} = \boldsymbol{A}^{-1}\boldsymbol{R}$

end

The necessary seven cases are compiled as follows:

$$\text{case1}: {}^{\mathrm{M}}\!\boldsymbol{u} = \left(\frac{y}{2}, \frac{x}{2}, 0\right),$$

$$\text{case2}: {}^{\mathrm{M}}\!\boldsymbol{u} = \left(x, 0, 0\right),$$

$$\text{case3}: {}^{\mathrm{M}}\!\boldsymbol{u} = \left(-xz, 0, xy\right),$$

$$\text{case4}: {}^{\mathrm{M}}\!\boldsymbol{u} = \left(xz, 0, -\frac{x^2}{2}\right), \tag{1.16}$$

$$\text{case5}: {}^{\mathrm{M}}\!\boldsymbol{u} = \left(-yz, 0, xy\right),$$

$$\text{case6}: {}^{\mathrm{M}}\!\boldsymbol{u} = \left(0, -y, \frac{y^2}{2}\right),$$

$$\text{case7}: {}^{\mathrm{M}}\!\boldsymbol{u} = \left(0, \frac{y^2}{2}, 0\right),$$

where the choice of up to quadratic functions is justified by acquiring constant strain gradients. We emphasize that the chosen cases are indeed generating independent deformation responses such that 7 cases let us uniquely determine 7 coefficients. Another set of 7 cases might be found as well; we assume to have several admissible sets to be used. Herein we are interested to demonstrate briefly, how the algorithm works. As long as the computation allows one to find a numerical solution for the suggested cases; there are no limitations to suggest another set than proposed herein. We have constructed the aforementioned cases in such a way that the strain, ${}^{\mathrm{M}}\varepsilon_{ij}$, and the strain gradient tensor, ${}^{\mathrm{M}}\varepsilon_{ij,k}$, possess as less components as possible. The proposed displacement field at the macroscale, ${}^{\mathrm{M}}\!\boldsymbol{u}$, is easy to consider in a computational setting; but it is nearly impossible to construct an experimental setup to obtain the same deformation profile. Hence, we understand that the proposed algorithm is a useful tool for determining structure related coefficients and further experiments are necessary to examine the accuracy.

1.6 Examples

For demonstrating the algorithm in action, we have exploited open-source packages developed under the FEniCS project (Alnaes et al, 2009; Logg et al, 2012) and used a Python code, which can be found in the web site in Abali (2017) to be used under the GNU Public license as in GNU Public (2007). The algorithm suggests to solve 7 cases on an arbitrary geometry representing the inner substructure. In order to examine the methodology as well as comprehend the physical meaning of the length scale used in this context, we performed several studies. We use a standard case where ball shaped inclusions are embedded in a matrix material as shown in Fig. 1.2. This cell, $1\,\mu\mathrm{m} \times 1\,\mu\mathrm{m} \times 1\,\mu\mathrm{m}$, can be understood as a representative volume element such that repeated positioning of the same cell builds up the structure. Technically, we concentrate on the case where different materials are used for the matrix, Ω^{Ma}, and inclusion, Ω^{In}. These domains comprise the whole domain, $\Omega = \Omega^{\mathrm{Ma}} \cup \Omega^{\mathrm{In}}$.

Fig. 1.2 Inner substructure for performing computations, inclusions are positioned along the diagonals within the cell, of diameter $d = 0.5\,\mu\mathrm{m}$, with a ratio of inclusions of approximately 12% regarding the whole domain of size $1\,\mu\mathrm{m}$.

By selecting the same material properties for both domains, i.e., the ball shaped inclusions and embedding matrix, we expect to have a homogenized microstructure leading to zero—the implementation gives out 10^{-5} or smaller—parameters, c_3, c_4, c_5, c_6, c_7. By choosing different materials, these parameters emerge and vary depending on the number of repeated cells measured by the length scale. We simply use that one cell as the lower threshold of the length scale, in order words, we assume that in this length scale and smaller, the microscale approach has to be used by modeling both materials differently. A homogenization approach is not adequate for this structure with the length scale $1\,\mu\mathrm{m}$ or less. This result is obvious; ratio of the geometric length scale per the substructure length scale is approaching unity. The substructure length scale is measured as the diameter of ball shaped inclusions, d. We construct different geometric length scales $1\,\mu\mathrm{m}$, $2\,\mu\mathrm{m}$, $3\,\mu\mathrm{m}$, $4\,\mu\mathrm{m}$, $5\,\mu\mathrm{m}$ by repeating the cells, as presented in Fig. 1.3. A particular example of such a structure relies on additive manufacturing resulting in a porous bulk material out of ABS or PLA, where the bulk material creates Ω^{Ma} and the voids determine Ω^{In}. In this case, we may approximate the structure by using following material parameters, YOUNG's modulus and POISSON's ratio, respectively,

Fig. 1.3 All used domains with the same inner substructure, representing voids or inclusions in a porous material (dark blue areas with a red line along the edges for the sake of visualization), the inclusions are positioned diagonally providing a nearly isotropic distribution.

$$E = 1\,\text{GPa}\,,\ \nu = 0.3\ \forall x \in \Omega^{\text{Ma}}\,,$$
$$E = 1\,\text{MPa}\,,\ \nu = 0.3\ \forall x \in \Omega^{\text{In}}\,,$$
(1.17)

at the microscale; leading to the usual LAME parameters:

$$\lambda = \frac{E\nu}{(1+\nu)(1-2\nu)}\,,\ \mu = \frac{E}{2(1+\nu)}$$
(1.18)

with the isotropic linear HOOKE's law:

$${}^{\text{m}}C_{ijkl} = \lambda \delta_{ij}\delta_{kl} + \mu \delta_{ik}\delta_{jl} + \mu \delta_{il}\delta_{jk}\,.$$
(1.19)

By using the proposed algorithm, we obtain the homogenized parameters for the diameter $d = 0.5\,\mu\text{m}$ as plotted in Fig. 1.4. We observe that the parameters monotonously converge to different than zero values until the length scale of $4\,\mu\text{m}$ such that we understand $4\,\mu\text{m}$ as the smallest length scale for a homogenization with the strain gradient formulation—this observation is valid for $d = 0.5\,\mu\text{m}$. In the case of increasing inclusion diameters to $d = 0.7\,\mu\text{m}$ for the identical configuration, Fig. 1.5 shows that the threshold is decreased approximately to $3\,\mu\text{m}$. This observation tells us that the strain gradient theory is capable of predicting the mechanical response in a length scale greater than the threshold. Of course, this threshold is difficult to determine accurately and it serves to a better understanding of the limitations of the homogenization method. For much larger length scales, the effect of strain gradient terms diminish as discussed in Abali et al (2017).

A similar result is obtained by studying a porous metal structure with the same configuration but higher YOUNG's modulus for the bulk, as follows:

$$E = 100\,\text{GPa}\,,\ \nu = 0.3\ \forall x \in \Omega^{\text{Ma}}\,,$$
$$E = 1\,\text{MPa}\,,\ \nu = 0.3\ \forall x \in \Omega^{\text{In}}\,.$$
(1.20)

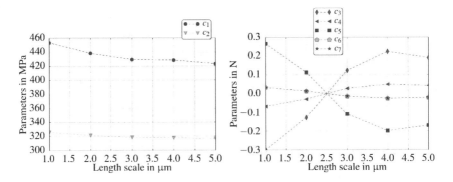

Fig. 1.4 Results for a plastic matrix, parameters are obtained in geometries in different length scales but the same inner substructure with $d = 0.5\,\mu\text{m}$.

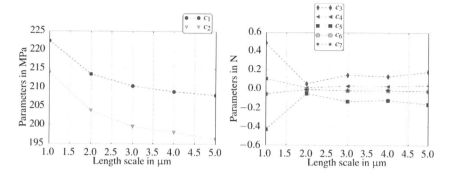

Fig. 1.5 Results for a plastic matrix, parameters are obtained in geometries in different length scales but the same inner substructure with $d = 0.7\,\mu m$.

Practically, parameters are scaled in such a way that we observe alike results for different inclusion diameters as demonstrated in Figs. 1.6, 1.7. We observed no significant effect on the numerical values of parameters in the case of lowering the stiffness by choosing $E = 0.1\,MPa$ or changing $\nu = 0.45$ of voids. The algorithm is robust and as far as the chosen substructure suggests, it is reliable.

1.7 Conclusion

A novel methodology has been developed in order to determine the parameters in higher gradient theories. The proposed approach is yet similar but powerful as it is working for arbitrary geometries with a given substructure. The homogenization procedure is based on FEM computations at the microscale by using 7 distinct pre-

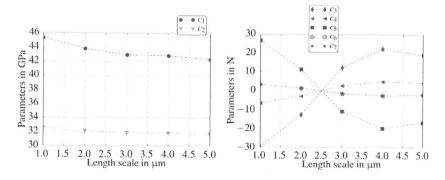

Fig. 1.6 Results for a metal matrix, parameters are obtained in geometries in different length scales but the same inner substructure with $d = 0.5\,\mu m$.

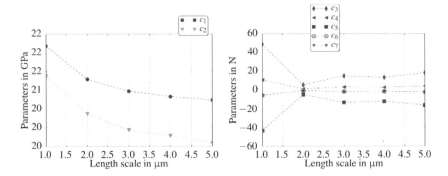

Fig. 1.7 Results for a plastic matrix, parameters are obtained in geometries in different length scales but the same inner substructure with $d = 0.7$ μm.

defined displacement scenarios. These cases seem to be not feasible to examine experimentally. Therefore, we propose to determine the parameters by the computational approach shown herein and then eventually validate this method's results by simulating experiments. Such experiments for different kinds of metamaterials subject to complex loading setups are investigated in the literature, among others we refer to Turco et al (2017); Barchiesi et al (2018); Ganzosch et al (2018); Yang et al (2018). For simulation of higher gradient theories, ample studies have been performed, see for example, Jeong et al (2009); Solyaev et al (2019); Khakalo and Niiranen (2017); Reiher et al (2016); Placidi et al (2016); Abali et al (2017).

References

Abali BE (2017) Technical University of Berlin, Institute of Mechanics, Chair of Continuum Mechanics and Material Theory, Computational Reality. http://www.lkm.tu-berlin.de/ComputationalReality/

Abali BE (2019) Revealing the physical insight of a length-scale parameter in metamaterials by exploiting the variational formulation. Continuum Mechanics and Thermodynamics 31(4):885–894

Abali BE, Müller WH (2016) Numerical solution of generalized mechanics based on a variational formulation. Oberwolfach reports - Mechanics of Materials: Mechanics of Interfaces and Evolving Microstructure 17(1):9–12

Abali BE, Müller WH, Eremeyev VA (2015) Strain gradient elasticity with geometric nonlinearities and its computational evaluation. Mechanics of Advanced Materials and Modern Processes 1(1):1–11

Abali BE, Müller WH, dell'Isola F (2017) Theory and computation of higher gradient elasticity theories based on action principles. Archive of Applied Mechanics 87(9):1495–1510

Abdoul-Anziz H, Seppecher P (2018) Strain gradient and generalized continua obtained by homogenizing frame lattices. Mathematics and Mechanics of Complex Systems 6(3):213–250

Alnaes MS, Logg A, Mardal KA, Skavhaug O, Langtangen HP (2009) Unified framework for finite element assembly. International Journal of Computational Science and Engineering 4(4):231–244

Barboura S, Li J (2018) Establishment of strain gradient constitutive relations by using asymptotic analysis and the finite element method for complex periodic microstructures. International Journal of Solids and Structures 136:60–76

Barchiesi E, Ganzosch G, Liebold C, Placidi L, Grygoruk R, Müller WH (2018) Out-of-plane buckling of pantographic fabrics in displacement-controlled shear tests: experimental results and model validation. Continuum Mechanics and Thermodynamics 31(1):33–45

Bensoussan A, Lions JL, Papanicolaou G (1978) Asymptotic Analysis for Periodic Structures. North-Holland, Amsterdam

Berryman JG (2005) Bounds and self-consistent estimates for elastic constants of random polycrystals with hexagonal, trigonal, and tetragonal symmetries. Journal of the Mechanics and Physics of Solids 53(10):2141–2173

Bigoni D, Drugan W (2007) Analytical derivation of cosserat moduli via homogenization of heterogeneous elastic materials. Journal of Applied Mechanics 74(4):741–753

Castañeda PP, Tiberio E (2000) A second-order homogenization method in finite elasticity and applications to black-filled elastomers. Journal of the Mechanics and Physics of Solids 48(6-7):1389–1411

Chen C, Pei Y, De Hosson JTM (2010) Effects of size on the mechanical response of metallic glasses investigated through in situ tem bending and compression experiments. Acta Materialia 58(1):189–200

Chung PW, Tamma KK, Namburu RR (2001) Asymptotic expansion homogenization for heterogeneous media: computational issues and applications. Composites Part A: Applied Science and Manufacturing 32(9):1291–1301

Dormieux L, Kondo D, Ulm FJ (2006) Microporomechanics. John Wiley & Sons

Eremeyev VA (2016) On effective properties of materials at the nano-and microscales considering surface effects. Acta Mechanica 227(1):29–42

Eremeyev VA, Lebedev LP, Altenbach H (2012) Foundations of Micropolar Mechanics. Springer Science & Business Media

Eringen A (1968) Mechanics of micromorphic continua. In: Kröner E (ed) Mechanics of Generalized Continua, Springer-Verlag, Berlin, pp 18–35

Eringen A, Suhubi E (1964) Nonlinear theory of simple micro-elastic solids. International Journal of Engineering Science 2:189–203

Eshelby JD (1957) The determination of the elastic field of an ellipsoidal inclusion, and related problems. Proceedings of the Royal Society of London Series A, Mathematical and Physical Sciences 241(1226):376–396

Forest S, Dendievel R, Canova GR (1999) Estimating the overall properties of heterogeneous cosserat materials. Modelling and Simulation in Materials Science and Engineering 7(5):829

Forest S, Pradel F, Sab K (2001) Asymptotic analysis of heterogeneous cosserat media. International Journal of Solids and Structures 38(26-27):4585–4608

Ganzosch G, Hoschke K, Lekszycki T, Giorgio I, Turco E, Müller WH (2018) 3D-measurements of 3D-deformations of pantographic structures. Technische Mechanik 38(3):233–245

GNU Public (2007) Gnu general public license. http://www.gnu.org/copyleft/gpl.html

Hashin Z (1991) The spherical inclusion with imperfect interface. Journal of Applied Mechanics 58(2):444–449

Hashin Z, Shtrikman S (1962) On some variational principles in anisotropic and nonhomogeneous elasticity. Journal of the Mechanics and Physics of Solids 10(4):335–342

Hill R (1972) On constitutive macro-variables for heterogeneous solids at finite strain. Proceedings of the Royal Society of London A Mathematical and Physical Sciences 326(1565):131–147

Hollister SJ, Kikuchi N (1992) A comparison of homogenization and standard mechanics analyses for periodic porous composites. Computational Mechanics 10(2):73–95

Irving J, Kirkwood J (1950) The statistical mechanical theory of transport processes. IV. The equations of hydrodynamics. The Journal of Chemical Physics 18:817–829

Ivanova EA, Vilchevskaya EN (2016) Micropolar continuum in spatial description. Continuum Mechanics and Thermodynamics 28(6):1759–1780

Jeong J, Ramézani H, Münch I, Neff P (2009) A numerical study for linear isotropic cosserat elasticity with conformally invariant curvature. ZAMM-Journal of Applied Mathematics and Mechanics/Zeitschrift für Angewandte Mathematik und Mechanik: Applied Mathematics and Mechanics 89(7):552–569

Kachanov M, Sevostianov I (2013) Effective properties of heterogeneous materials, Solid Mechanics and Its Applications, vol 193. Springer

Kanaun S, Kudryavtseva L (1986) Spherically layered inclusions in a homogeneous elastic medium. Journal of Applied Mathematics and Mechanics 50(4):483–491

Khakalo S, Niiranen J (2017) Isogeometric analysis of higher-order gradient elasticity by user elements of a commercial finite element software. Computer-Aided Design 82:154–169

Kouznetsova V, Brekelmans W, Baaijens F (2001) An approach to micro-macro modeling of heterogeneous materials. Computational Mechanics 27:37–48

Kushnevsky V, Morachkovsky O, Altenbach H (1998) Identification of effective properties of particle reinforced composite materials. Computational Mechanics 22(4):317–325

Ladeveze P, Neron D, Passieux J (2010) On multiscale computational mechanics with time-space homogenization. In: Fish J (ed) Multiscale Methods: Bridging the Scales in Science and Engineering, Oxford University Press, New York, pp 247–284

Lam DC, Yang F, Chong A, Wang J, Tong P (2003) Experiments and theory in strain gradient elasticity. Journal of the Mechanics and Physics of Solids 51(8):1477–1508

Lebensohn R, Liu Y, Castaneda PP (2004) On the accuracy of the self-consistent approximation for polycrystals: comparison with full-field numerical simulations. Acta Materialia 52(18):5347–5361

Levin V (1976) Determination of composite material elastic and thermoelastic constants. Mechanics of Solids 11(6):119–126

Li J (2011) Establishment of strain gradient constitutive relations by homogenization. Comptes Rendus Mécanique 339(4):235–244

Logg A, Mardal KA, Wells G (2012) Automated Solution of Differential Equations by the Finite Element Method: The FEniCS Book, vol 84. Springer Science & Business Media

Mandadapu KK, Abali BE, Papadopoulos P (2018) On the polar nature and invariance properties of a thermomechanical theory for continuum-on-continuum homogenization. arXiv preprint arXiv:180802540

Mercer B, Mandadapu K, Papadopoulos P (2015) Novel formulations of microscopic boundary-value problems in continuous multiscale finite element methods. Computer Methods in Applied Mechanics and Engineering 286:268–292

Miehe C, Schotte J, Lambrecht M (2002) Homogenization of inelastic solid materials at finite strains based on incremental minimization principles. Application to the texture analysis of polycrystals. J Mech Phys Solids 50:2123–2167

Mindlin R (1964) Micro-structure in linear elasticity. Archive for Rational Mechanics and Analysis 16(1):51–78

Mori T, Tanaka K (1973) Average stress in matrix and average elastic energy of materials with misfitting inclusions. Acta Metallurgica 21(5):571–574

Namazu T, Isono Y, Tanaka T (2000) Evaluation of size effect on mechanical properties of single crystal silicon by nanoscale bending test using AFM. Journal of Microelectromechanical Systems 9(4):450–459

Nazarenko L (1996) Elastic properties of materials with ellipsoidal pores. International Applied Mechanics 32(1):46–52

Nazarenko L, Khoroshun L, Müller WH, Wille R (2009) Effective thermoelastic properties of discrete-fiber reinforced materials with transversally-isotropic components. Continuum Mechanics and Thermodynamics 20(7):429–458

Nazarenko L, Bargmann S, Stolarski H (2016) Lurie solution for spherical particle and spring layer model of interphases: Its application in analysis of effective properties of composites. Mechanics of Materials 96:39–52

Nazarenko L, Stolarski H, Khoroshun L, Altenbach H (2018) Effective thermo-elastic properties of random composites with orthotropic components and aligned ellipsoidal inhomogeneities. International Journal of Solids and Structures 136:220–240

Nemat-Nasser S, Hori M (2013) Micromechanics: Overall Properties of Heterogeneous Materials, vol 37. Elsevier

Nemat-Nasser S, Srivastava A (2011) Overall dynamic constitutive relations of layered elastic composites. Journal of the Mechanics and Physics of Solids 59:1953–1965

Pideri C, Seppecher P (1997) A second gradient material resulting from the homogenization of an heterogeneous linear elastic medium. Continuum Mechanics and Thermodynamics 9(5):241–257

Pietraszkiewicz W, Eremeyev V (2009) On natural strain measures of the non-linear micropolar continuum. International Journal of Solids and Structures 46(3):774–787

Placidi L, Greco L, Bucci S, Turco E, Rizzi NL (2016) A second gradient formulation for a 2d fabric sheet with inextensible fibres. Zeitschrift für angewandte Mathematik und Physik 67(5):114

Polizzotto C (2013a) A second strain gradient elasticity theory with second velocity gradient inertia–Part I: Constitutive equations and quasi-static behavior. International Journal of Solids and Structures 50(24):3749–3765

Polizzotto C (2013b) A second strain gradient elasticity theory with second velocity gradient inertia–Part II: Dynamic behavior. International Journal of Solids and Structures 50(24):3766–3777

Rahali Y, Giorgio I, Ganghoffer J, dell'Isola F (2015) Homogenization à la Piola produces second gradient continuum models for linear pantographic lattices. International Journal of Engineering Science 97:148–172

Reiher JC, Giorgio I, Bertram A (2016) Finite-element analysis of polyhedra under point and line forces in second-strain gradient elasticity. Journal of Engineering Mechanics 143(2):04016,112

Reuss A (1929) Berechnung der Fließgrenze von Mischkristallen auf Grund der Plastizitätsbedingung für Einkristalle. ZAMM-Journal of Applied Mathematics and Mechanics/Zeitschrift für Angewandte Mathematik und Mechanik 9(1):49–58

Sburlati R, Cianci R, Kashtalyan M (2018) Hashin's bounds for elastic properties of particle-reinforced composites with graded interphase. International Journal of Solids and Structures 138:224–235

Seppecher P, Alibert JJ, dell'Isola F (2011) Linear elastic trusses leading to continua with exotic mechanical interactions. Journal of Physics: Conference Series 319(1):012,018

Sevostianov I, Kachanov M (2006) Homogenization of a nanoparticle with graded interface. International Journal of Fracture 139(1):121–127

Sevostianov I, Kachanov M (2014) On some controversial issues in effective field approaches to the problem of the overall elastic properties. Mechanics of Materials 69(1):93–105

Shafiro B, Kachanov M (2000) Anisotropic effective conductivity of materials with nonrandomly oriented inclusions of diverse ellipsoidal shapes. Journal of Applied Physics 87(12):8561–8569

Solyaev Y, Lurie S, Ustenko A (2019) Numerical modeling of a composite auxetic metamaterials using micro-dilatation theory. Continuum Mechanics and Thermodynamics 31(4):1099–1107

Steinmann P (1994) A micropolar theory of finite deformation and finite rotation multiplicative elastoplasticity. International Journal of Solids and Structures 31(8):1063–1084

Temizer I (2012) On the asymptotic expansion treatment of two-scale finite thermoelasticity. International Journal of Engineering Science 53:74–84

Turco E, Golaszewski M, Giorgio I, D'Annibale F (2017) Pantographic lattices with non-orthogonal fibres: experiments and their numerical simulations. Composites Part B: Engineering 118:1–14

Voigt W (1889) Über die Beziehung zwischen den beiden Elasticitätsconstanten isotroper Körper. Annalen der Physik 274(12):573–587

Willis J (1977) Bounds and self-consistent estimates for the overall properties of anisotropic composites. Journal of the Mechanics and Physics of Solids 25(3):185–202

Yang H, Ganzosch G, Giorgio I, Abali BE (2018) Material characterization and computations of a polymeric metamaterial with a pantographic substructure. Zeitschrift für angewandte Mathematik und Physik 69(4):105

Zheng QS, Du DX (2001) An explicit and universally applicable estimate for the effective properties of multiphase composites which accounts for inclusion distribution. Journal of the Mechanics and Physics of Solids 49(11):2765–2788

Zohdi TI (2004) Homogenization methods and multiscale modeling. In: Stein E, de Borst R, Hughes TJR (eds) Encyclopedia of Computational Mechanics, Wiley Online Library, vol 2: Solids and Structures, pp 407–430

Zohdi TI, Wriggers P (2008) An Introduction to Computational Micromechanics. Springer Science & Business Media

Chapter 2
Extensible Beam Models in Large Deformation Under Distributed Loading: a Numerical Study on Multiplicity of Solutions

Francesco dell'Isola, Alessandro Della Corte, Antonio Battista, and Emilio Barchiesi

Abstract In this paper we present numerical solutions to a geometrically nonlinear version of the extensible Timoshenko beam model under distributed load. The particular cases in which: i) extensional stiffness is infinite (inextensible Timoshenko model), ii) shear stiffness is infinite (extensible Euler model) and iii) extensional and shear stiffnesses are infinite (inextensible Euler model) will be numerically explored. Parametric studies on the axial stiffness in both the Euler and Timoshenko cases will also be shown and discussed.

F. dell'Isola
International Research Center for the Mathematics and Mechanics of Complex Systems - M&MoCS, Università dell'Aquila, L'Aquila
Department of Structural and Geotechnical Engineering, Università di Roma La Sapienza, 18 Via Eudossiana, Rome, Italy
Research Institute for Mechanics, Nizhny Novgorod Lobachevsky State University, 23, Gagarin av. 603950 Nizhny Novgorod, Russia
e-mail: francesco.dellisola.me@gmail.com

A. Della Corte
International Research Center for the Mathematics and Mechanics of Complex Systems - M&MoCS, Università dell'Aquila, L'Aquila
Department of Structural and Geotechnical Engineering, Università di Roma La Sapienza, 18 Via Eudossiana, Rome, Italy
e-mail: alessandro.dellacorte.memocs@gmail.com

A. Battista
International Research Center for the Mathematics and Mechanics of Complex Systems - M&MoCS, Università dell'Aquila, L'Aquila, Italy
LaSIE, Université de La Rochelle, La Rochelle, France
e-mail: antonio.battista@uni-lr.fr

E. Barchiesi
International Research Center for the Mathematics and Mechanics of Complex Systems - M&MoCS, Università dell'Aquila, L'Aquila
Department of Structural and Geotechnical Engineering, Università di Roma La Sapienza, 18 Via Eudossiana, Rome, Italy
e-mail: barchiesiemilio@gmail.com

© Springer Nature Switzerland AG 2019
H. Altenbach et al. (eds.), *Higher Gradient Materials and Related Generalized Continua*, Advanced Structured Materials 120,
https://doi.org/10.1007/978-3-030-30406-5_2

Keywords: Timoshenko beam · Large deformation of beams · Extensional beam model · Shooting technique

2.1 Introduction

In the recent literature the behaviour of a clamped-free nolinear inextensible Euler *Elastica* introduced in Euler (1952); Bernoulli (1843, 1691); see Luongo and Zulli (2013); Eugster (2015); Steigmann and Faulkner (1993) for general reference works, has been mathematically investigated under distributed load (Della Corte et al, 2016). In particular, the set of stable equilibrium configurations has been completely characterized in Della Corte et al (2019). On the other hand, extending such rigorous results to the extensible Euler beam model, or to the Timoshenko beam model is not straightforward. Indeed, the presence of the additional kinematical descriptor accounting for extensibility gives rise to various new mathematical difficulties. The most basic one is that it changes the functional set in which the problem is naturally collocated and in particular prevents it from being a vector space, since a strictly positive local axial deformation has to be prescribed. Of course one can obtain this making suitable assumptions on the energy, but in any case the non-autonomous variational problem that arises in the case of a distributed load will present new difficulties with respect to the inextensible case (Battista et al, 2018).

The problem of existence and stability of equilibrium configurations for extensible Euler and Timoshenko beams in large deformations under distributed load is therefore an open one. Because of this reason, it is interesting to study the behavior of solutions by means of a systematic collection of parametric studies starting from the inextensible *Elastica* and approaching more general beam models. The present work is aimed at performing such kind of investigation. Our main effort will be to show the variety of different (and at times rather exotic) equilibrium configurations that can arise when the value of the load is large enough. Moreover, we want to numerically investigate how fast the number of possible equilibrium solutions increases with the load and how this particular feature is affected by allowing shear deformation.

The study of these exotic configurations is particularly important, nowadays, due to the enhancement of computational methods that make nonlinearity more practically relevant for structural members (Fertis, 2006; Ladevèze, 2012; Antman and Renardy, 1995; Rezaei et al, 2012; Eugster et al, 2014; Steigmann, 2017) as well as in machine mechanics applications (Pepe et al, 2016; Giorgio and Del Vescovo, 2018). Moreover, they are becoming fashionable as an elementary constituent of microstructured objects manufactured with computer-aided techniques (Atai and Steigmann, 1997; dell'Isola et al, 2016b; Ravari and Kadkhodaei, 2015; Ravari et al, 2014; dell'Isola et al, 2016a; Turco et al, 2016, 2017; Milton et al, 2017; Spagnuolo et al, 2017). These objects are potentially advantageous for their mechanical characteristics, as shown in recent literature (Boutin et al, 2017; Giorgio et al, 2017; Scerrato et al, 2016; Eremeyev, 2017; Golaszewski et al, 2019; Turco et al, 2019;

Franciosi et al, 2019) and their theoretical study requires tools from homogenization theory (Boubaker et al, 2007; Ravari et al, 2016; Dos Reis and Ganghoffer,
2012; Reda et al, 2016), discrete mechanics (Turco, 2018; Jawed et al, 2018; Andreaus et al, 2018), theory of generalized continua (Misra et al, 2016; Altenbach
et al, 2013; Altenbach and Eremeyev, 2013; Placidi et al, 2017, 2014, 2016) as well
as developments in nonlinear elasticity of beams. In this last regard, certainly a
full understanding of the onset and the characteristics of multiple solutions for the
static problem under distributed load would be an important step forward, and possibly more general beam models will also have to be considered (Diyaroglu et al,
2015; Challamel, 2013; Challamel et al, 2013) because of the exotic properties of
microstructured continua (Misra et al, 2018; dell'Isola et al, 2018; Diyaroglu et al,
2017).

The paper is organized as follows: we introduce the general model for a nonlinear version of the Timoshenko beam; we deduce the equilibrium equations by
means of the Lagrange multipliers method. Then, imposing stationarity to the energy functional, we deduce an expression for the total energy in stationary points
which depends only on the angles formed by the tangent to the deformed shape.
Then we show and discuss numerical results on the multiplicity of solutions with
large value of the load and on the effect of releasing extensional stiffness in both the
Euler and the Timoshenko case. Finally we propose some future research directions.

2.2 The Model

2.2.1 Kinematics and Deformation Energy

Let $\{D_1, D_2\}$ be an orthogonal reference system in which the beam lies in the
unstressed configuration along D_1. We will denote by s the abscissa along the beam,
by the apex $'$ the differentiation with respect to the reference abscissa[1] and by $\chi(s)$
the placement function. The tangent vector to the current configuration of the beam
is then:

$$\chi' = \alpha(s)\left[\cos(\theta(s))D_1 + \sin(\theta(s))D_2\right] := \alpha(s)e(\theta(s)) \qquad (2.1)$$

where $e(\theta(s))$ represents the unit vector parallel to χ'. Therefore $\alpha(s)$ describes the
local elongation of the beam:

$$\|\chi'(s)\| = \alpha(s) \qquad (2.2)$$

while $\theta(s)$ is the angle between $\chi'(s)$ and D_1.

We will assume the following energy functional:

---

[1] Notice that this means that, denoting by θ the angle formed by the tangent to the deformed shape
and a reference axis, θ' does not coincide with the geometrical curvature but with the so-called
Chebyshev curvature (see Chebyshev, 1878).

$$\mathcal{E}^{\text{def}}(\boldsymbol{\chi}, \varphi) := \int_0^L \left\{ \underbrace{\frac{k_e}{2} \left(\|\boldsymbol{\chi}'(s)\| - 1 \right)^2}_{\text{extensional energy}} + \underbrace{\frac{k_b}{2} (\varphi'(s))^2}_{\text{flexural energy}} + \underbrace{\frac{k_t}{2} (\varphi(s) - \theta(s))^2}_{\text{shear energy}} \right\} ds$$

(2.3)

Here φ is the angle between sections of the beam (supposed rigid) and the normal to the neutral axes, while k_b, k_e and k_t are respectively the bending, extensional and shear stiffness. The energy can be rewritten as:

$$\mathcal{E}^{\text{def}} = \int_0^L \left\{ \frac{k_e}{2} (\alpha - 1)^2 + \frac{k_b}{2} (\varphi'(s))^2 + \frac{k_t}{2} (\varphi(s) - \theta(s))^2 \right\} ds. \qquad (2.4)$$

We introduce now a uniformly distributed load $\boldsymbol{b}(s)$ and a concentrated load and couple in the endpoint of the beam $s = L$, denoted respectively by $\boldsymbol{R}(L)$ and $\boldsymbol{M}(L)$. The total energy of the system is then:

$$\mathcal{E}^{\text{tot}} = \int_0^L \left\{ \frac{k_e}{2} (\alpha - 1)^2 + \frac{k_b}{2} (\varphi'(s))^2 + \frac{k_t}{2} (\varphi(s) - \theta(s))^2 - \boldsymbol{b} \cdot \boldsymbol{\chi} \right\} ds$$
$$- \boldsymbol{R} \cdot \boldsymbol{\chi}(L) - \boldsymbol{M} \varphi(L)$$

(2.5)

The beam model described by the functional (2.5) is a geometrically nonlinear version of the Timoshenko beam model (introduced in Timoshenko, 1921, 1922), which is a particular case of Cosserat continuum introduced in Cosserat and Cosserat (1909); for general references and interesting results see e.g. Altenbach et al (2010); Birsan et al (2012); Forest (2005); Eremeyev and Pietraszkiewicz (2016)). Interesting generalizations of the Timoshenko beam model have been proposed (see e.g. Romano et al, 1992; Serpieri and Rosati, 2014), while a periodic mechanical system whose homogenized limit is the model (2.5) (in the particular case $\alpha \equiv 1$) is shown in Battista et al (2018). It is in fact a microstructured 1D system whose unit cell is an articulated parallelogram and equipped with suitably placed rotational springs, and it can be easily obtained by means of 3D printing. Of course also other (possibly more complex) microstructured systems can have a similar homogenized version (on microstructured continua see e.g. Barchiesi et al, 2019; Engelbrecht and Berezovski, 2015; Engelbrecht et al, 2005; Barchiesi et al, 2018).

2.2.2 Lagrange Multipliers Method

Given the total energy of the system, equilibrium configurations are found as stationary points of the energy functional (2.5). A synthetic formulation of the problem, taking into account together the constrain of Eq. (2.1) and the total energy, is obtained with the introduction of a Lagrange multiplier $\Lambda(s)$. We get the following functional formulation:

$$\mathcal{E}^{\text{tot}} = \int_0^L \left\{ \frac{k_e}{2}(\alpha - 1)^2 + \frac{k_b}{2}(\varphi'(s))^2 + \frac{k_t}{2}(\varphi(s) - \theta(s))^2 - \boldsymbol{b} \cdot \boldsymbol{\chi} + \right.$$
$$\left. + \boldsymbol{\Lambda} \cdot (\boldsymbol{\chi}' - \alpha \left[\cos(\theta(s))\boldsymbol{D_1} + \sin(\theta(s))\boldsymbol{D_2} \right]) \right\} ds - \boldsymbol{R} \cdot \boldsymbol{\chi}(L) - \boldsymbol{M}\varphi(L)$$

$$(2.6)$$

i.e., the energy is a function of the fields $\alpha(s), \varphi(s), \theta(s), \boldsymbol{\chi}(s), \boldsymbol{\Lambda}(s)$. The first variation of the energy with respect to these fields (considered independent) gives the two boundary value problems (BVPs):

$$\begin{cases} -k_b\varphi'' + k_t(\varphi - \theta) = 0 \\ k_b\varphi'(L)\delta\varphi(L) = \boldsymbol{M}\delta\varphi(L) \\ k_b\varphi'(0)\delta\varphi(0) = 0 \end{cases} \qquad (2.7)$$

and

$$\begin{cases} -\boldsymbol{\Lambda}' - \boldsymbol{b} = 0 \\ \boldsymbol{\Lambda}(L) \cdot \delta\boldsymbol{\chi}(L) = \boldsymbol{R} \cdot \delta\boldsymbol{\chi}(L) \\ \boldsymbol{\Lambda}(0) \cdot \delta\boldsymbol{\chi}(0) = 0, \end{cases} \qquad (2.8)$$

as well as the algebraic relations:

$$k_t(\varphi - \theta) + \alpha\boldsymbol{\Lambda} \cdot \boldsymbol{e}_\perp(\theta) = 0 \qquad (2.9)$$

$$k_e(\alpha - 1) - \boldsymbol{\Lambda} \cdot \boldsymbol{e}(\theta) = 0 \qquad (2.10)$$

and the kinematic constraint Eq. (2.1).

In Eq. (2.9) we introduced:

$$\boldsymbol{e}_\perp(\theta(s)) = -\sin(\theta(s))\boldsymbol{D_1} + \cos(\theta(s))\boldsymbol{D_2}$$

and the two boundary conditions in 0 are imposed considering a cantilever beam (lying along $\boldsymbol{D_1}$ in the reference configuration and clamped in the extreme $s = 0$).

2.3 Numerical Simulations

2.3.1 Numerical Methods

An increasingly popular approach for the numerical study of nonlinear beams is isogeometric analysis (see e.g. Balobanov and Niiranen, 2018; Niiranen et al, 2017; Cazzani et al, 2016; Greco et al, 2017; Dortdivanlioglu et al, 2017), which is a suitable variant of the finite element method. This method is very powerful and relatively light from a computational point of view, but just like every energy-related method it is not very suitable to study the multiplicity of arising solutions. For this reason, the numerical technique used here is the same as in Battista et al (2018). Indeed, the boundary value problem for the clamped-free Euler and Timoshenko beams has been solved by means of a shooting technique. We introduce a family of Cauchy problems [2]:

$$\mathcal{P}_k = \begin{cases} \theta'' = -b(1 - s) \cos \theta \\ \theta(0) = 0 \\ \theta'(0) = k \end{cases} \tag{2.11}$$

depending on the parameter k. Then we selected the solutions of (2.11) which satisfy (with prescribed accuracy) $\theta'(1) = 0$, so as to obtain a numerical solution of the equilibrium condition $\delta \mathcal{E}^{tot} = 0$. Clearly the solution for the Cauchy problem exists and is unique for every initial datum k.

In Fig. 2.1, we show the plot of $\theta'(1)$ as a function of $k := \theta'(0)$ for an inextensible Euler beam model for $b = 250$. The graph intersects in five different points the horizontal axis, which means that in this case we have five different solutions of the boundary value problem with $\theta(0) = 0$ and $\theta'(1) = 0$. In the next section, the solutions will be shown in the same order as they appear as intersections between $\theta'(1)$ and the horizontal axis. Therefore the absolute minimum of the total energy will be always the rightmost configuration and, of course, the only solution for b small enough.

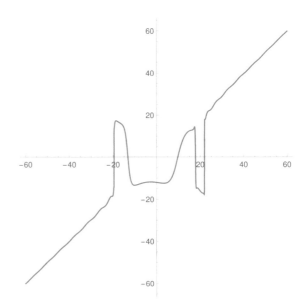

Fig. 2.1 $\theta'(1)$ as a function of $k := \theta'(0)$ for an inextensible Euler beam with a transverse applied load $b = 250$. Solutions to the boundary value problem (2.11) correspond to the intersections of the curve with the horizontal axis.

---

[2] We consider here the inextensible Euler beam for simplicity, but everything is analogous for the general case.

Remark 2.1. Of course this numerical technique can only capture *regular* equilibrium configurations. It is well-known that the minima (and maxima) of nonautonomous functionals may be not regular enough to satisfy the Euler-Lagrange equations (even in the one-dimensional case as in Ball and Mizel, 1987). This, together with the assessment of the stability of the solutions, is the main reason for which a rigorous study of the problem will be crucial.

2.3.2 The Number of Equilibrium Configurations when the Load Increases

It is generally very difficult to address theoretically the problem of evaluating how fast the number of solutions of a nonlinear parametric dynamical system increases with the parameter (Guckenheimer and Holmes, 1983), which in our case is the external load. It is therefore interesting, as a preliminary step, to address the problem numerically. In Fig. 2.2 the number of solutions for an inextensible (left) and extensible (right) Euler beem model is plotted as a function of the non-dimensional external load. As expected, the behavior is that of a step-function, as it is clear that new branches of solutions arise only when the external load overcomes specific thresholds. The number of equilibrium configurations increases significantly with the load, and again as expected it reaches slightly larger values in case an additional kinematic degree of freedom (i.e. α) is included.

Figure 2.3 is consistent with this. Indeed, in this case inextensible (left) and extensible (right) Timoshenko beam models are considered, which produces an even more rapid increase of the possible equilibrium configurations. It has to be remarked, however, that a straight comparison with the Euler case is difficult because adimensionalizing the load is a different procedure in the two cases.

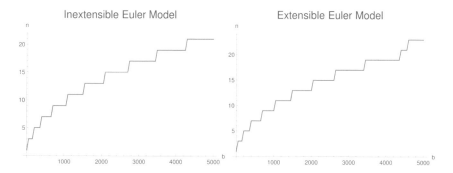

Fig. 2.2 Number of solutions, n, as a function of the transversal applied load, b. Left: inextensible Euler model. Right: extensible Euler model ($k_e = 1.3 \times 10^3$).

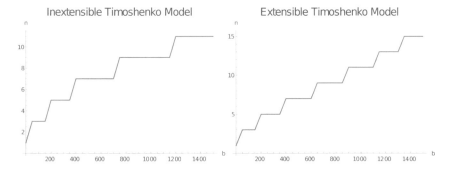

Fig. 2.3 Number of solutions, n, as a function of the transversal applied load, b. Left: inextensible Timoshenko model ($k_t = 1.8 \times 10^4$). Right: extensible Timoshenko model ($k_t = 1.8 \times 10^4$, $k_e = 2 \times 10^3$).

2.3.3 Equilibrium Configurations

In the first gallery of equilibrium solutions we will show the full set of equilibrium configurations when the nondimensional load is $b = 60$, $b = 250$ and $b = 500$. We considered inextensible Euler (Figs. 2.4, 2.8, 2.12), extensible Euler (Figs. 2.5, 2.9, 2.13), inextensible Timoshenko (Figs. 2.6, 2.10, 2.14) and extensible Timoshenko (Figs. 2.7, 2.11, 2.15) beam models.

For the chosen values of the other parameters, we have the same number of solutions (when applying the same load) for the four beam models. It has to be remarked that no branch of solutions appear to bifurcate. Instead, when the load increases, new branches appear at some thresholds. It can be seen that $\theta(s)$ is a positive monotonic function for the rightmost solution in all the cases. For $b = 60$ the central equilib-

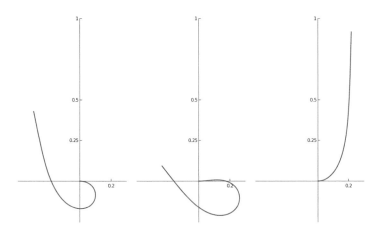

Fig. 2.4 Clamped inextensible Euler beam with a transversal applied load $b = 60$

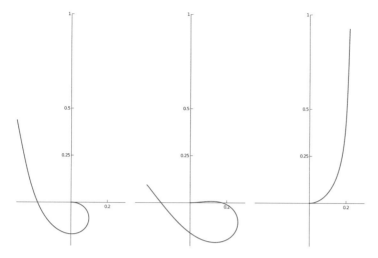

Fig. 2.5 Clamped extensible Euler beam with a transversal applied load $b = 60$ ($k_e = 3000$)

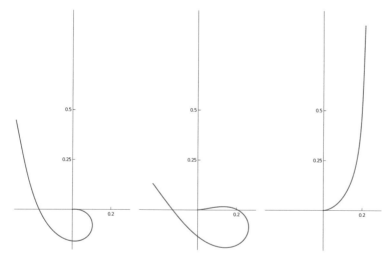

Fig. 2.6 Clamped inextensible Timoshenko beam with a transversal applied load $b = 60$ ($k_t = 7000$)

rium shape is non monotonic, while the leftmost is a negative monotonic function. This is a general trend as the load increases. Indeed, branches that arise with increasingly large values of the load will be made of progressively more numerous monotonic pieces. Of course in the Timoshenko case in general it is $\theta(0) \neq 0$. As the boundary datum for θ in 0 can have more than one solution, we have always chosen the smallest one in absolute value (see also the Appendix). The axial elongation relative to Figs. 2.9 and 2.11 are shown respectively in Figs. 2.16 and 2.18. The shear

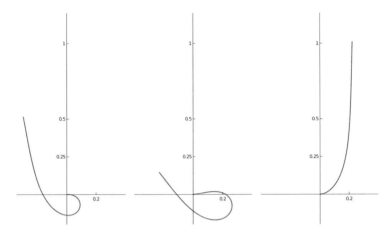

Fig. 2.7 Clamped extensible Timoshenko beam with a transverse applied load $b = 60$ ($k_t = 7000$, $k_e = 3000$)

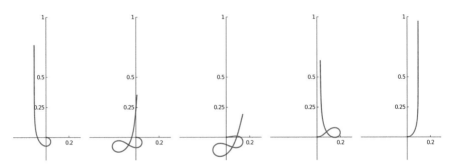

Fig. 2.8 Clamped inextensible Euler beam with a transversal applied load $b = 250$

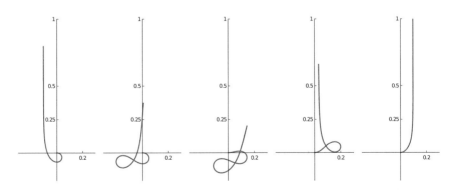

Fig. 2.9 Clamped extensible Euler beam with a transversal applied load $b = 250$ ($k_e = 3000$)

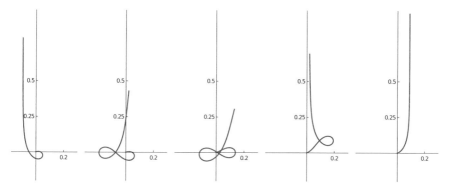

Fig. 2.10 Clamped inextensible Timoshenko beam with a transversal applied load $b = 250$ ($k_t = 7000$)

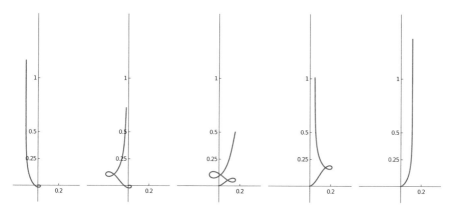

Fig. 2.11 Clamped extensible Timoshenko beam with a transverse applied load $b = 250$ ($k_t = 7000$, $k_e = 3000$)

deformation relative to Figs. 2.10 and 2.11 are shown respectively in Figs. 2.17 and 2.19.

It is not clear whether the solutions shown herein for the extensible Euler and Timoshenko models can be stable—while in Della Corte et al (2019) it has been proved that for the inextensible Euler case only the left and right configurations of Figs. 2.4 and 2.8 can be stable.

2.3.4 Parametric Study on the Extensional Stiffness

When the parameter k_e diverges, the beam model tends to inextensibility. The effect of decreasing k_e is evaluated for the Euler and Timoshenko models respectively in Figs. 2.24 and 2.26 for $b = 120$. In Figs. 2.25 and 2.27 the local elongation $\alpha(s)$ is

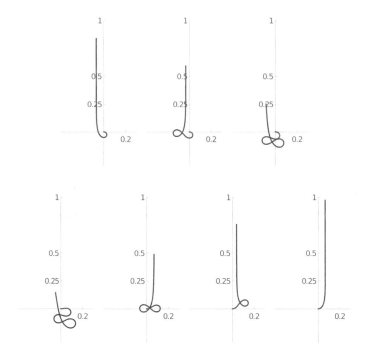

Fig. 2.12 Clamped inextensible Euler beam with a transverse applied load $b = 500$.

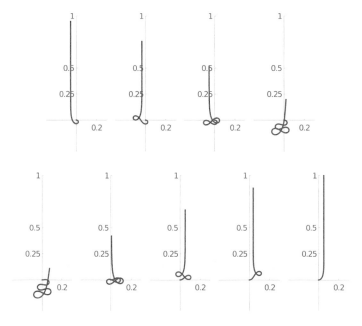

Fig. 2.13 Clamped extensible Euler beam with a transverse applied load $b = 500$ and $k_e = 5000$.

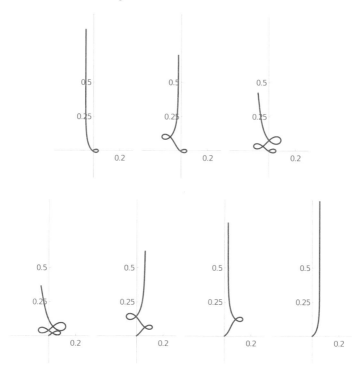

Fig. 2.14 Clamped inextensible Timoshenko beam with a transverse applied load $b = 500$ and $k_t = 7000$.

shown for the two previous cases respectively. The elongation reaches its maximum (minimum) value where the beam lies parallel to the load with the same (opposite) direction. Instead, it is close to 1 where the beam lies orthogonal to the load. While in the Euler case the change in k_e causes a minimal change in the deformed shape, in the Timoshenko case there is a much more relevant influence of k_e on the configuration. In particular, decreasing k_e allows a much larger maximum value of the local geometrical curvature $\gamma := \theta'/\alpha$ of the beam. This maximum is attained when $\theta(s) = -\pi/2$; we will define s_0 the point at which this occurs. It has to be noted, however, that the Chebyshev curvature $\theta'(s_0)$ takes similar values in the two cases. For instance, in the rightmost simulation of Fig. 2.24 we have $\theta'(s_0) \approx -16.4$ and $\gamma(s_0) \approx -17.9$, while in the rightmost simulation of Fig. 2.26 we have $\theta'(s_0) \approx -16.3$ and $\gamma(s_0) \approx -111.9$ (we point out that $s_0 \approx 0.303$ in the right panel of Fig. 2.24 and $s_0 \approx 0.288$ in the right panel of Fig. 2.26).

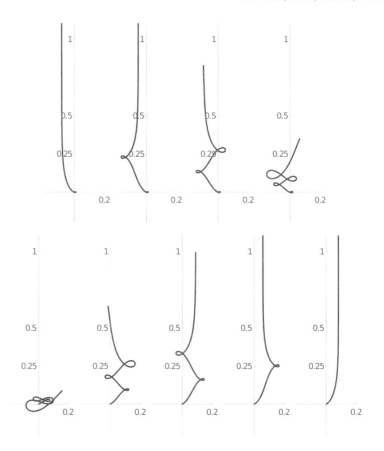

Fig. 2.15 Clamped extensible Timoshenko beam with a transverse applied load $b = 500$ and $k_t = 7000$ and $k_e = 5000$.

Fig. 2.16 Plot of α for the five configurations in Fig. 2.9.

2.4 Conclusions

In this paper we numerically studied clamped-free Euler and Timoshenko beams in large deformation under distributed load. Extensibility has been taken into account and results on the static behavior of the beam under different values of the load and of the axial stiffness has been shown. The main interest of the results consists in the multiplicity of solutions that arise as the load increases, not as a bifurcation of ex-

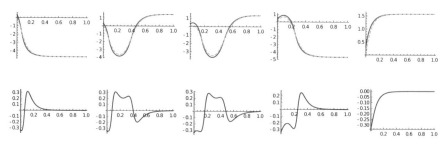

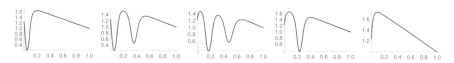

Fig. 2.17 Above: plot of ϕ (dotted) and θ for the five configurations shown in Fig. 2.10. Below: the corresponding plot of $\phi - \theta$.

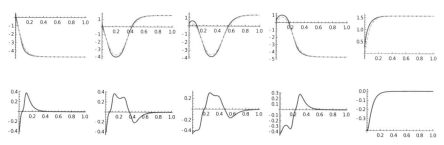

Fig. 2.18 Plot of α for the five configurations in Fig. 2.11.

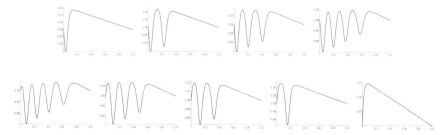

Fig. 2.19 Above: plot of ϕ (dotted) and θ for the five configurations shown in Fig. 2.11. Below: the corresponding plot of $\phi - \theta$.

Fig. 2.20 Plot of α for the configurations in Fig. 2.13.

isting branches of solutions but as new branches that arise when the load overcomes a series of progressively larger threshold-values. Future investigations are required to establish whether these multiple solutions can be stable. In this regard, an analysis of the small oscillations of the beam around candidate stable equilibria would be useful to assess numerically the question. Moreover, addressing theoretically the

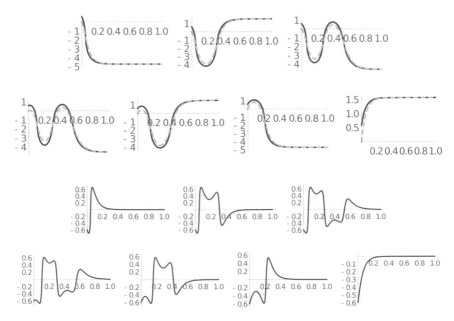

Fig. 2.21 Above: plot of ϕ (dotted) and θ for the five configurations shown in Fig. 2.14. Below: the corresponding plot of $\phi - \theta$.

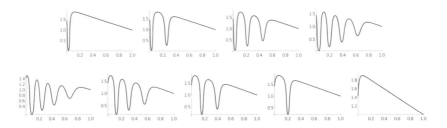

Fig. 2.22 Plot of α for the configurations in Fig. 2.15.

dynamical behavior of the nonlinear version of the Timoshenko beam model proposed here is a challenging task. The results developed in Berezovski et al (2018); Luongo and D'Annibale (2013); Piccardo et al (2015b); Taig et al (2015); Piccardo et al (2015a); Chróścielewski et al (2019) may prove useful in this direction.

Apart from the models introduced here, an independent and novel approach to understand the large deformation of beams to be investigated in the future is to regard a beam as the boundary curve of a two-dimensional manifold in a three-dimensional space. In doing so, not only is the curve endowed with its own energy similar to that in the context of lower-dimensional energetics (Javili et al, 2013b) but also in a geometrically nonlinear framework (Javili et al, 2014) and in accordance with higher gradient elasticity accounting for boundary energetics elaborated in Javili et al (2013a). The advantage of this approach, particularly from a computational

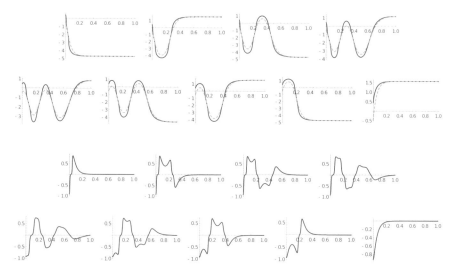

Fig. 2.23 Above: plot of ϕ (dotted) and θ for the five configurations shown in Fig. 2.15. Below: the corresponding plot of $\phi - \theta$.

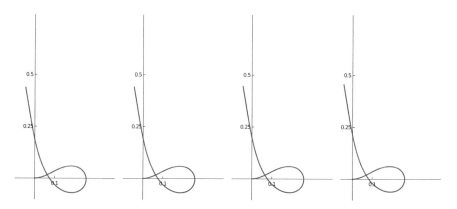

Fig. 2.24 Parametric study on a clamped extensible Euler beam with a transversal applied load $b = 120$ with $k_e = 2500$, $k_e = 2000$, $k_e = 1500$, $k_e = 1000$.

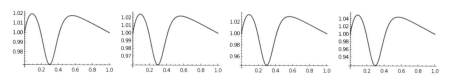

Fig. 2.25 The local elongation $\alpha(s)$ relative to the equilibrium shapes shown in Fig. 2.24.

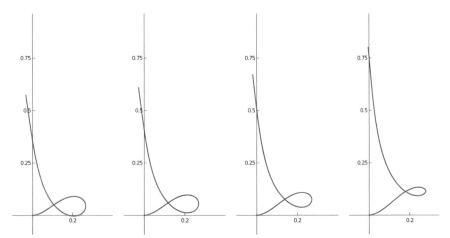

Fig. 2.26 Parametric study on a clamped extensible Timoshenko beam with a transversal applied load $b = 120$ with $k_e = 2500$, $k_e = 2000$, $k_e = 1500$, $k_e = 1000$ ($k_t = 18000$).

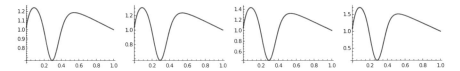

Fig. 2.27 The local elongation $\alpha(s)$ relative to the equilibrium shapes shown in Fig. 2.26.

viewpoint, is that the bulk material acts to regularize the behavior of the beam especially important to analyze the instabilities associated with thin beams similar to the instabilities of thin films on an elastic foundation (Javili et al, 2015). Obviously, in the limiting case of the vanishing bulk one would recover exactly the beam theory.

Appendix

We show here the Euler–Lagrange boundary value problem associated with the functional (2.5). We recall that the solutions $\theta(s)$ of this BVP are the scalar fields of angles formed by the tangent to the deformed configuration and a reference axis at the equilibrium for a clamped-free, extensible Timoshenko beam, in large deformation regime, under distributed load.

$$\theta'' = \left[k_b\left(bk_e(1-s)\sin(\theta) - (b(1-s))^2(\cos(2\theta)) + k_e k_t\right)\right]^{-1}$$
$$\left[bk_b(1-s)\theta^2\cos(\theta)(4b(1-s)\sin(\theta) + k_e) - 2bk_b\theta(2b(1-s)\cos(2\theta)+\right.$$
$$\left.- k_e\sin(\theta)) + \cos(\theta)\left(2b^2 k_b\sin(\theta) - bk_t(1-s)(b(1-s)\sin(\theta) + k_e))\right)\right]$$

$$(2.12)$$

with boundary conditions:

$$\left(\theta - \frac{b^2(1-s)^2\sin(\theta)\cos(\theta)}{k_e k_e} + \frac{b(1-s)\cos(\theta)}{k_e}\right)\Big|_{s=0} = 0$$
$$\left(\theta - \frac{b^2(1-s)^2\sin(\theta)\cos(\theta)}{k_e k_e} + \frac{b(1-s)\cos(\theta)}{k_e}\right)'\Big|_{s=1} = 0$$

$$(2.13)$$

Remark 2.2. The boundary conditions in the previous problem have in general more than one solution. In the numerical simulations, we always considered the value $\theta(0)$ which was smaller in absolute value.

References

Altenbach H, Eremeyev VA (2013) Cosserat-type shells. In: Altenbach H, Eremeyev VA (eds) Generalized Continua from the Theory to Engineering Applications, CISM International Centre for Mechanical Sciences, vol 541, Springer, Vienna, pp 131–178

Altenbach H, Birsan M, Eremeyev VA (2013) Cosserat-type rods. In: Altenbach H, Eremeyev VA (eds) Generalized Continua from the Theory to Engineering Applications, CISM International Centre for Mechanical Sciences, vol 541, Springer, Vienna, pp 179–248

Altenbach J, Altenbach H, Eremeyev VA (2010) On generalized Cosserat-type theories of plates and shells: a short review and bibliography. Archive of Applied Mechanics 80(1):73–92

Andreaus U, Spagnuolo M, Lekszycki T, Eugster SR (2018) A Ritz approach for the static analysis of planar pantographic structures modeled with nonlinear Euler–Bernoulli beams. Continuum Mechanics and Thermodynamics 30(5):1103–1123

Antman SS, Renardy M (1995) Nonlinear problems of elasticity. SIAM Review 37(4):637

Atai AA, Steigmann DJ (1997) On the nonlinear mechanics of discrete networks. Archive of Applied Mechanics 67(5):303–319

Ball JM, Mizel VJ (1987) One-dimensional variational problems whose minimizers do not satisfy the Euler–Lagrange equation. In: Analysis and Thermomechanics, Springer, pp 285–348

Balobanov V, Niiranen J (2018) Locking-free variational formulations and isogeometric analysis for the timoshenko beam models of strain gradient and classical elasticity. Computer Methods in Applied Mechanics and Engineering 339:137–159

Barchiesi E, dell'Isola F, Laudato M, Placidi L, Seppecher P (2018) A 1D continuum model for beams with pantographic microstructure: Asymptotic micro-macro identification and numerical results. In: dell'Isola F, Eremeyev V, Porubov A (eds) Advances in Mechanics of Microstructured Media and Structures, Advanced Structured Materials, vol 87, Springer, Cham, pp 43–74

Barchiesi E, Spagnuolo M, Placidi L (2019) Mechanical metamaterials: a state of the art. Mathematics and Mechanics of Solids 24(1):212–234

Battista A, Della Corte A, dell'Isola F, Seppecher P (2018) Large deformations of 1D microstructured systems modeled as generalized Timoshenko beams. Zeitschrift für angewandte Mathematik und Physik 69(3):52

Berezovski A, Yildizdag M, Scerrato D (2018) On the wave dispersion in microstructured solids. Continuum Mechanics and Thermodynamics pp 1–20, DOI 10.1007/s00161-018-0683-1

Bernoulli D (1843) The 26th letter to Euler. Correspondence Mathématique et Physique 2

Bernoulli J (1691) Quadratura curvae, e cujus evolutione describitur inflexae laminae curvatura. Die Werke von Jakob Bernoulli pp 223–227

Birsan M, Altenbach H, Sadowski T, Eremeyev V, Pietras D (2012) Deformation analysis of functionally graded beams by the direct approach. Composites Part B: Engineering 43(3):1315–1328

Boubaker BB, Haussy B, Ganghoffer J (2007) Discrete models of woven structures. macroscopic approach. Composites Part B: Engineering 38(4):498–505

Boutin C, Giorgio I, Placidi L, et al (2017) Linear pantographic sheets: Asymptotic micro-macro models identification. Mathematics and Mechanics of Complex Systems 5(2):127–162

Cazzani A, Malagù M, Turco E (2016) Isogeometric analysis of plane-curved beams. Mathematics and Mechanics of Solids 21(5):562–577

Challamel N (2013) Variational formulation of gradient or/and nonlocal higher-order shear elasticity beams. Composite Structures 105:351–368

Challamel N, Zhang Z, Wang C (2013) Nonlocal equivalent continua for buckling and vibration analyses of microstructured beams. Journal of Nanomechanics and Micromechanics 5(1):A4014,004

Chebyshev P (1878) Sur la coupe des vetements. Complete works by PL Chebyshev 5:165–170

Chróścielewski J, Schmidt R, Eremeyev VA (2019) Nonlinear finite element modeling of vibration control of plane rod-type structural members with integrated piezoelectric patches. Continuum Mechanics and Thermodynamics 31(1):147–188

Cosserat E, Cosserat F (1909) Théorie des corps déformables. A Hermann et fils

Della Corte A, dell'Isola F, Esposito R, Pulvirenti M (2016) Equilibria of a clamped Euler beam (*Elastica*) with distributed load: Large deformations. Mathematical Models and Methods in Applied Sciences pp 1–31

Della Corte A, Battista A, dell'Isola F, Seppecher P (2019) Large deformations of Timoshenko and Euler beams under distributed load. Zeitschrift für angewandte Mathematik und Physik 70(52), DOI 10.1007/s00033-019-1098-y

dell'Isola F, Giorgio I, Pawlikowski M, Rizzi N (2016a) Large deformations of planar extensible beams and pantographic lattices: heuristic homogenization, experimental and numerical examples of equilibrium. Proceedings of the Royal Society A: Mathematical, Physical and Engineering Sciences 472(2185):20150,790

dell'Isola F, Steigmann D, Della Corte A (2016b) Synthesis of fibrous complex structures: designing microstructure to deliver targeted macroscale response. Applied Mechanics Reviews 67(6):060,804–060,804–21

dell'Isola F, Seppecher P, Alibert JJ, Lekszycki T, Grygoruk R, Pawlikowski M, Steigmann D, Giorgio I, Andreaus U, Turco E, Golaszewski M, Rizzi N, Boutin C, Eremeyev VA, Misra A, Placidi L, Barchiesi E, Greco L, Cuomo M, Cazzani A, Corte AD, Battista A, Scerrato D, Eremeeva IZ, Rahali Y, Ganghoffer JF, Müller W, Ganzosch G, Spagnuolo M, Pfaff A, Barcz K, Hoschke K, Neggers J, Hild F (2018) Pantographic metamaterials: an example of mathematically driven design and of its technological challenges. Continuum Mechanics and Thermodynamics 31(4):851–884

Diyaroglu C, Oterkus E, Oterkus S, Madenci E (2015) Peridynamics for bending of beams and plates with transverse shear deformation. International Journal of Solids and Structures 69:152–168

Diyaroglu C, Oterkus E, Oterkus S (2017) An Euler–Bernoulli beam formulation in an ordinary state-based peridynamic framework. Mathematics and Mechanics of Solids 24(2):361–376

Dortdivanlioglu B, Javili A, Linder C (2017) Computational aspects of morphological instabilities using isogeometric analysis. Computer Methods in Applied Mechanics and Engineering 316:261–279

Dos Reis F, Ganghoffer J (2012) Construction of micropolar continua from the asymptotic homogenization of beam lattices. Computers & Structures 112:354–363

Engelbrecht J, Berezovski A (2015) Reflections on mathematical models of deformation waves in elastic microstructured solids. Mathematics and Mechanics of Complex Systems 3(1):43–82

Engelbrecht J, Berezovski A, Pastrone F, Braun M (2005) Waves in microstructured materials and dispersion. Philosophical Magazine 85(33-35):4127–4141

Eremeyev VA (2017) On characterization of an elastic network within the six-parameter shell theory. In: Pietraszkiewicz W, Witkowski W (eds) Shell Structures: Theory and Applications Volume 4: Proceedings of the 11th International Conference in Shell Structures: Theory and Applications, SSTA 2017, CRC Press, pp 81–84

Eremeyev VA, Pietraszkiewicz W (2016) Material symmetry group and constitutive equations of micropolar anisotropic elastic solids. Mathematics and Mechanics of Solids 21(2):210–221

Eugster S, Hesch C, Betsch P, Glocker C (2014) Director-based beam finite elements relying on the geometrically exact beam theory formulated in skew coordinates. International Journal for Numerical Methods in Engineering 97(2):111–129

Eugster SR (2015) Geometric Continuum Mechanics and Induced Beam Theories, Lecture Notes in Applied and Computational Mechanics, vol 75. Springer

Euler L (1952) Methodus inveniendi lineas curvas maximi minimive proprietate gaudentes sive solutio problematis isoperimetrici latissimo sensu accepti (ed. by C. Carathéodory), Opera mathematica, vol 1. Birkhäuser, Basel

Fertis DG (2006) Nonlinear Structural Engineering. Springer

Forest S (2005) Mechanics of Cosserat media - an introduction. Ecole des Mines de Paris, Paris pp 1–20

Franciosi P, Spagnuolo M, Salman OU (2019) Mean Green operators of deformable fiber networks embedded in a compliant matrix and property estimates. Continuum Mechanics and Thermodynamics 31(1):101–132

Giorgio I, Del Vescovo D (2018) Non-linear lumped-parameter modeling of planar multi-link manipulators with highly flexible arms. Robotics 7(4):60

Giorgio I, Rizzi N, Turco E (2017) Continuum modelling of pantographic sheets for out-of-plane bifurcation and vibrational analysis. Proceedings of the Royal Society A: Mathematical, Physical and Engineering Sciences 473(2207):20170,636

Golaszewski M, Grygoruk R, Giorgio I, Laudato M, Di Cosmo F (2019) Metamaterials with relative displacements in their microstructure: technological challenges in 3D printing, experiments and numerical predictions. Continuum Mechanics and Thermodynamics 31(4):1015–1034

Greco L, Cuomo M, Contrafatto L, Gazzo S (2017) An efficient blended mixed b-spline formulation for removing membrane locking in plane curved Kirchhoff rods. Computer Methods in Applied Mechanics and Engineering 324:476–511

Guckenheimer J, Holmes P (1983) Nonlinear Oscillations, Dynamical Systems, and Bifurcations of Vector Fields, Applied Mathematical Sciences, vol 42. Springer

Javili A, dell'Isola F, Steinmann P (2013a) Geometrically nonlinear higher-gradient elasticity with energetic boundaries. Journal of the Mechanics and Physics of Solids 61(12):2381–2401

Javili A, McBride A, Steinmann P (2013b) Thermomechanics of solids with lower-dimensional energetics: on the importance of surface, interface, and curve structures at the nanoscale. a unifying review. Applied Mechanics Reviews 65(1):010,802

Javili A, McBride A, Steinmann P, Reddy B (2014) A unified computational framework for bulk and surface elasticity theory: a curvilinear-coordinate-based finite element methodology. Computational Mechanics 54(3):745–762

Javili A, Dortdivanlioglu B, Kuhl E, Linder C (2015) Computational aspects of growth-induced instabilities through eigenvalue analysis. Computational Mechanics 56(3):405–420

Jawed MK, Novelia A, O'Reilly OM (2018) A Primer on the Kinematics of Discrete Elastic Rods. Springer

Ladevèze P (2012) Nonlinear Computational Structural Mechanics: New Approaches and Nonincremental Methods of Calculation. Springer Science & Business Media

Luongo A, D'Annibale F (2013) Double zero bifurcation of non-linear viscoelastic beams under conservative and non-conservative loads. International Journal of Non-Linear Mechanics 55:128–139

Luongo A, Zulli D (2013) Mathematical Models of Beams and Cables. John Wiley & Sons

Milton G, Briane M, Harutyunyan D (2017) On the possible effective elasticity tensors of 2-dimensional and 3-dimensional printed materials. Mathematics and Mechanics of Complex Systems 5(1):41–94

Misra A, Placidi L, Scerrato D (2016) A review of presentations and discussions of the workshop computational mechanics of generalized continua and applications to materials with microstructure that was held in Catania 29–31 October 2015. Mathematics and Mechanics of Solids 22(9):1891–1904

Misra A, Lekszycki T, Giorgio I, Ganzosch G, Müller WH, dell'Isola F (2018) Pantographic metamaterials show atypical poynting effect reversal. Mechanics Research Communications 89:6–10

Niiranen J, Balobanov V, Kiendl J, Hosseini S (2017) Variational formulations, model comparisons and numerical methods for Euler–Bernoulli micro-and nano-beam models. Mathematics and Mechanics of Solids 24(1):312–335

Pepe G, Carcaterra A, Giorgio I, Del Vescovo D (2016) Variational feedback control for a nonlinear beam under an earthquake excitation. Mathematics and Mechanics of Solids 21(10):1234–1246

Piccardo G, Pagnini LC, Tubino F (2015a) Some research perspectives in galloping phenomena: critical conditions and post-critical behavior. Continuum Mechanics and Thermodynamics 27(1-2):261–285

Piccardo G, Tubino F, Luongo A (2015b) A shear–shear torsional beam model for nonlinear aeroelastic analysis of tower buildings. Zeitschrift für angewandte Mathematik und Physik 66(4):1895–1913

Placidi L, Rosi G, Giorgio I, Madeo A (2014) Reflection and transmission of plane waves at surfaces carrying material properties and embedded in second-gradient materials. Mathematics and Mechanics of Solids 19(5):555–578

Placidi L, Greco L, Bucci S, Turco E, Rizzi NL (2016) A second gradient formulation for a 2D fabric sheet with inextensible fibres. Zeitschrift für angewandte Mathematik und Physik 67(5):114

Placidi L, Andreaus U, Giorgio I (2017) Identification of two-dimensional pantographic structure via a linear d4 orthotropic second gradient elastic model. Journal of Engineering Mathematics 103(1):1–21

Ravari MRK, Kadkhodaei M (2015) A computationally efficient modeling approach for predicting mechanical behavior of cellular lattice structures. Journal of Materials Engineering and Performance 24(1):245–252

Ravari MRK, Kadkhodaei M, Badrossamay M, Rezaei R (2014) Numerical investigation on mechanical properties of cellular lattice structures fabricated by fused deposition modeling. International Journal of Mechanical Sciences 88:154–161

Ravari MRK, Kadkhodaei M, Ghaei A (2016) Effects of asymmetric material response on the mechanical behavior of porous shape memory alloys. Journal of Intelligent Material Systems and Structures 27(12):1687–1701

Reda H, Rahali Y, Ganghoffer JF, Lakiss H (2016) Wave propagation in 3D viscoelastic auxetic and textile materials by homogenized continuum micropolar models. Composite Structures 141:328–345

Rezaei DAH, Kadkhodaei M, Nahvi H (2012) Analysis of nonlinear free vibration and damping of a clamped–clamped beam with embedded prestrained shape memory alloy wires. Journal of Intelligent Material Systems and Structures 23(10):1107–1117

Romano G, Rosati L, Ferro G (1992) Shear deformability of thin-walled beams with arbitrary cross sections. International Journal for Numerical Methods in Engineering 35(2):283–306

Scerrato D, Giorgio I, Rizzi NL (2016) Three-dimensional instabilities of pantographic sheets with parabolic lattices: numerical investigations. Zeitschrift für angewandte Mathematik und Physik 67(3):53

Serpieri R, Rosati L (2014) A frame-independent solution to Saint-Venant's flexure problem. Journal of Elasticity 116(2):161–187

Spagnuolo M, Barcz K, Pfaff A, dell'Isola F, Franciosi P (2017) Qualitative pivot damage analysis in aluminum printed pantographic sheets: numerics and experiments. Mechanics Research Communications 83:47–52

Steigmann D, Faulkner M (1993) Variational theory for spatial rods. Journal of Elasticity 33(1):1–26

Steigmann DJ (2017) Finite Elasticity Theory. Oxford University Press

Taig G, Ranzi G, D'annibale F (2015) An unconstrained dynamic approach for the generalised beam theory. Continuum Mechanics and Thermodynamics 27(4-5):879–904

Timoshenko SP (1921) Lxvi. on the correction for shear of the differential equation for transverse vibrations of prismatic bars. The London, Edinburgh, and Dublin Philosophical Magazine and Journal of Science 41(245):744–746

Timoshenko SP (1922) X. on the transverse vibrations of bars of uniform cross-section. The London, Edinburgh, and Dublin Philosophical Magazine and Journal of Science 43(253):125–131

Turco E (2018) Discrete is it enough? the revival of Piola–Hencky keynotes to analyze three-dimensional Elastica. Continuum Mechanics and Thermodynamics 30(5):1039–1057

Turco E, Golaszewski M, Cazzani A, Rizzi NL (2016) Large deformations induced in planar pantographic sheets by loads applied on fibers: experimental validation of a discrete lagrangian model. Mechanics Research Communications 76:51–56

Turco E, Golaszewski M, Giorgio I, D'Annibale F (2017) Pantographic lattices with non-orthogonal fibres: Experiments and their numerical simulations. Composites Part B: Engineering 118:1–14

Turco E, Misra A, Sarikaya R, Lekszycki T (2019) Quantitative analysis of deformation mechanisms in pantographic substructures: experiments and modeling. Continuum Mechanics and Thermodynamics 31(1):209–223

Chapter 3
On the Characterization of the Nonlinear Reduced Micromorphic Continuum with the Local Material Symmetry Group

Victor A. Eremeyev

Abstract Following the definition of the local material symmetry group for micromorphic media given in Eremeyev (2018), we discuss here the constitutive equations of the reduced micromorphic continuum introduced in Neff et al (2014). With this definition we demonstrate that the reduced micromorphic model can be characterized as a micromorphic subfluid that is an intermediate class between micromorphic solids and fluids.

3.1 Introduction

Recently the interest is growing to generalized models of continua with respect to advances in manufacturing of complex microstructured materials as well as the necessity to describe various physical phenomena in details, see Maugin (2011, 2013, 2017); dell'Isola et al (2017); Auffray et al (2015a); dell'Isola and Eremeyev (2018) for history of developments in the field. Among generalized continua models the micromorphic medium introduced by Mindlin (1964) and Eringen and Suhubi (1964) found many applications. In particular, the micromorphic model found applications to modeling acoustic metamaterials with local resonant properties, see, e.g., Sridhar et al (2016, 2017); Madeo et al (2016); Misra and Poorsolhjouy (2016) and the reference therein. The kinematics of the micromorphic medium is described by the placement vector $\mathbf{x}$ and the micro-distortion second-order tensor $\mathbf{P}$ and their gradients, see Eringen (1999); Forest (2013); Maugin (2013, 2017); Eremeyev et al (2018a). The latter responds to microdeformations of the medium. As a result, for

V. A. Eremeyev
Faculty of Civil and Environmental Engineering, Gdańsk University of Technology, ul. Gabriela Narutowicza 11/12, 80-233 Gdańsk, Poland
Research Institute for Mechanics, Nizhny Novgorod Lobachevsky State University, 23, Gagarin av. 603950 Nizhny Novgorod, Russia
e-mail: eremeyev.victor@gmail.com

H. Altenbach et al. (eds.), *Higher Gradient Materials and Related Generalized Continua*, Advanced Structured Materials 120,
https://doi.org/10.1007/978-3-030-30406-5_3

43

a hyperelastic micromorphic medium the strain energy function depends on two second-order tensors and one third-order tensor. This leads to quite complex representation of this function. According to Truesdell and Noll (2004) the determination of effective constitutive equations for specific materials or classes of materials constitutes a general problem of continuum mechanics. For simple materials the notion of material symmetry may lead to significant reduction of the form of the constitutive equations. The intuitive notion of the material symmetry was formalized through the local material symmetry group definition, see Truesdell and Noll (2004); Lurie (1990); Eremeyev et al (2018a). For hyperelastic materials, this group consists of such transformations of a reference placement which do not change the strain energy density. Knowing a priori a material symmetry, one can significantly reduce the number of arguments in a strain energy function. For example, for an isotropic material the strain energy density depends on three principal invariants of the strain tensor.

For generalized continua the notion of the local material symmetry group is even more important as for these media the strain energy density depends on many arguments, in general. For micropolar solids the local material group definition was introduced by Eringen and Kafadar (1976) and was further extended by Eremeyev and Pietraszkiewicz (2012, 2013, 2016). For shells and material surfaces it was analyzed by Murdoch and Cohen (1979); Eremeyev and Pietraszkiewicz (2006), whereas for strain-gradient continua the group was discussed by Elzanowski and Epstein (1992); Bertram (2017); Reiher and Bertram (2018), see also reference therein. The local material symmetry group for micromorphic media was introduced by Eremeyev (2018) where its application to granular materials was discussed. Here we discuss the characterization of the reduced micromorphic model introduced by Neff et al (2014) within the material symmetry group as a micromorphic subfluid.

The paper is organized as follows. In Sect. 3.2 we briefly recall the basic constitutive equations of micromorphic media under finite deformations. In Sect. 3.3 we formulate the definitions of the local material symmetry group and the micromorphic subfluid. Finally, we derive the group which leads to the constitutive equations of the reduced micromorphic material.

In what follows we use the direct (coordinate-free) tensor calculus as in Eringen (1980); Lurie (1990); Simmonds (1994); Lebedev et al (2010); Eremeyev et al (2018a).

3.2 Micromorphic Continua

Following Eringen (1999); Eremeyev et al (2018a,b) we introduce the hyperelastic micromorphic continuum trough the strain energy density

$$\mathcal{W} = \mathcal{W}(\mathbf{E}, \mathbf{C}, \mathbf{K}), \tag{3.1}$$

where

$$\mathbf{E} = \mathbf{F} \cdot \mathbf{P}^{-1}, \quad \mathbf{C} = \mathbf{P} \cdot \mathbf{P}^T, \quad \mathbf{K} = \nabla \mathbf{P} \cdot \mathbf{P}^{-1} \tag{3.2}$$

are the strain measures. Here

$$\mathbf{F} = \nabla \mathbf{x}$$

is the deformation gradient, $\mathbf{P}$ is the micro-distortion second-order tensor, ∇ is the 3D nabla-operator, $\mathbf{x} = \mathbf{x}(\mathbf{X})$ and $\mathbf{X}$ are position vectors in an current and reference placements, respectively. In addition, the dot stands for the scalar product, $\mathbf{P}^T$ and $\mathbf{P}^{-1}$ denote the transpose and inverse of $\mathbf{P}$, respectively. It is worth to underline that $\mathbf{E}$ and $\mathbf{C}$ are second-order tensors whereas $\mathbf{K}$ is a third-order tensor. The general representation of (3.1) is quite complex even in the case small deformations. As an example, let us mention here similar constitutive equations with third-order tensors for strain gradient elasticity, see, e.g. dell'Isola et al (2009); Bertram (2017); Auffray et al (2013, 2015b, 2017).

In order to reduce this complexity as well for modelling some acoustic meta-materials (Neff et al, 2014) proposed new micromorphic model called the relaxed micromorphic continuum. Its generalization for the case of finite deformations is given by Eremeyev et al (2018b)

$$\mathcal{W}_R = \mathcal{W}_R(\mathbf{E}, \mathbf{C}, \mathbf{B}), \quad \mathbf{B} = (\nabla \times \mathbf{P}) \cdot \mathbf{P}^{-1}, \tag{3.3}$$

where $\times$ stands for the cross product. Note that unlike $\mathbf{K}$, here $\mathbf{B}$ is a second-order tensor. As $\mathbf{B}$ is introduced through the cross product it is a pseudotensor, see Eremeyev et al (2018a). As a result, Eq. (3.3) constitute a dependence of a scalar function of three second-order tensors. For such functions some representation theorems can be found in Spencer (1971); Boehler (1987); Zheng (1994). These theorem were used by Eremeyev and Pietraszkiewicz (2012, 2016) for the reduction of constitutive equations of micropolar materials.

With the identity

$$\nabla \times \mathbf{P} = -(\mathbf{I} \times \mathbf{I}) : \nabla \mathbf{P}$$

we find the relation between $\mathbf{K}$ and $\mathbf{B}$

$$\mathbf{B} = -(\mathbf{I} \times \mathbf{I}) : \mathbf{K}. \tag{3.4}$$

Hereinafter $\mathbf{I}$ is the unit tensor, ":" denotes the double contraction operation, for example, we have

$$(\mathbf{a} \otimes \mathbf{b}) : (\mathbf{c} \otimes \mathbf{d}) = (\mathbf{a} \cdot \mathbf{c})(\mathbf{b} \cdot \mathbf{d}),$$

$$(\mathbf{a} \otimes \mathbf{b} \otimes \mathbf{c}) : (\mathbf{d} \otimes \mathbf{e}) = (\mathbf{b} \cdot \mathbf{d})(\mathbf{c} \cdot \mathbf{e})\mathbf{a},$$

for any vectors $\mathbf{a}$, $\mathbf{b}$, $\mathbf{c}$, $\mathbf{d}$, and $\mathbf{e}$, and we apply the cross product between tensors following Gibbs' definition for dyads

$$(\mathbf{a} \otimes \mathbf{b}) \times (\mathbf{c} \otimes \mathbf{d}) = \mathbf{a} \otimes (\mathbf{b} \times \mathbf{c}) \otimes \mathbf{d},$$

where "·" and "⊗" denote the dot and tensor products, respectively, see (Wilson, 1901, p. 281). Obviously, these formulae for dyads can be easily extended for tensors of any order.

3.3 Local Material Symmetry Group

As in the case of simple (Truesdell and Noll, 2004; Lurie, 1990; Eremeyev et al, 2018a) or non-simple (Murdoch and Cohen, 1979; Eremeyev and Pietraszkiewicz, 2012, 2016; Reiher and Bertram, 2018) materials, for micromorphic media we introduce the local material symmetry group as an ordered set of tensors which make the strain energy density invariant under corresponding changes of a reference placement. In what follows we use the following tensor groups, see Table 3.1.

Following Eremeyev (2018) we introduce

Definition 3.1. *A set of ordered triples of two second-order unimodular tensors* $\mathbf{S}$ *and* $\mathbf{R}$ *and third-order tensor* $\mathbf{L}$

$$\mathcal{G}_{\varkappa}(\mathbf{X}) = \{\mathbb{X} = (\mathbf{S}, \mathbf{R}, \mathbf{L})\}$$

is called the local material symmetry group if the following relation is valid

$$\mathcal{W}(\mathbf{E}, \mathbf{C}, \mathbf{K}) = \mathcal{W}(\mathbf{S} \cdot \mathbf{E} \cdot \mathbf{R}^{-1}, \mathbf{R} \cdot \mathbf{C} \cdot \mathbf{R}^T, \mathbf{S} \cdot [\mathbf{R} * \mathbf{K}] + \mathbf{L})$$

for a given point $\mathbf{X}$ *in the reference placement* $\varkappa$ *and for all admissible tensors* $\mathbf{E}$, $\mathbf{C}$, $\mathbf{K}$ *from the domain of the definition of* W, *where operation* $*$ *between second-order* $\mathbf{R}$ *and third-order* $\mathbf{K}$ *tensors is defined as follows*

$$\mathbf{R} * \mathbf{K} = \mathbf{R} * (K_{mnk}\mathbf{i}_m \otimes \mathbf{i}_n \otimes \mathbf{i}_k) = K_{mnk}\mathbf{i}_m \otimes (\mathbf{R} \cdot \mathbf{i}_n) \otimes (\mathbf{i}_k \cdot \mathbf{R}^{-1}),$$

The set $\mathcal{G}_{\varkappa}(\mathbf{X})$ *is a group relative to the operation defined as follows*

$$\mathbb{X}_1 \circ \mathbb{X}_2 \equiv (\mathbf{S}_1, \mathbf{R}_1, \mathbf{L}_1) \circ (\mathbf{S}_2, \mathbf{R}_2, \mathbf{L}_2) = (\mathbf{S}_1 \cdot \mathbf{S}_2, \mathbf{R}_1 \cdot \mathbf{R}_2, \mathbf{L}_1 + \mathbf{S}_1 \cdot [\mathbf{R}_1 * \mathbf{L}_2]).$$

Table 3.1 Tensor groups

Notation	Title	Description
$Unim$	Unimodular group	$Unim = \{\mathbf{S}: \quad \det \mathbf{S} = \pm 1\}$
$Orth$	Orthogonal group	$Orth = \{\mathbf{Q}: \quad \mathbf{Q} \cdot \mathbf{Q}^T = \mathbf{I}\}$
$Orth_+$	Proper orthogonal group	$Orth_+ = \{\mathbf{Q}: \quad \mathbf{Q} \cdot \mathbf{Q}^T = \mathbf{I}, \quad \det \mathbf{Q} = +1\}$
Lin_3	Linear group of third-order tensors	$Lin_3 = \{\mathbf{L} = L_{mnk}\mathbf{i}_m \otimes \mathbf{i}_n \otimes \mathbf{i}_k\}$
Lin_3^S	Linear group of third-order tensors with symmetry of two indices	$Lin_3^S = \{\mathbf{L} = L_{mnk}\mathbf{i}_m \otimes \mathbf{i}_n \otimes \mathbf{i}_k : L_{mnk} = L_{nmk}\}$

So the local material symmetry group for a micromorphic medium consists of ordered triples of tensors. Nevertheless, one can easily establish some analogies between classic tensor groups summarized in Table 3.1 and symmetry groups for micromorphic material. For example, "orthogonal" and "unimodular" groups are given by

$$\mathcal{O} = \{\mathbb{X} = (\mathbf{Q}, \mathbf{Q}, \mathbf{0}), \mathbf{Q} \in Orth\},$$

$$\mathcal{U} = \{\mathbb{X} = (\mathbf{S}, \mathbf{R}, \mathbf{L}), \mathbf{S} \in Unim, \mathbf{R} \in Unim, \mathbf{L} \in Lin_3\}.$$

Hereinafter $\mathbf{0}$ stands for zero vectors and tensors of any order. In particular, in $\mathcal{O}$ tensor $\mathbf{0}$ is the zero third-order tensor. These groups are used for definition of isotropic and fluid micromorphic materials, see Eremeyev (2018) for details.

Let us note that the given definition constitutes quite reach structure of the group. In particular, there is an intermediate class of micromorphic materials called micromorphic subfluids or micromorphic liquid crystals defined as follows

Definition 3.2. *A micromorphic elastic continuum is called a micromorphic elastic subfluid or micromorphic elastic liquid crystal if its material symmetry group $\mathcal{G}_\varkappa$, related to a reference placement $\varkappa$ called undistorted, contains elements which are not members of the orthogonal group $\mathcal{O}$ and $\mathcal{G}_\varkappa$ does not coincide with the maximal group $\mathcal{U}$:*

$$\mathcal{G}_\varkappa \neq \mathcal{U}, \quad \exists \mathbb{X} \in \mathcal{G}_\varkappa : \quad \mathbb{X} \notin \mathcal{O}.$$

Originally the notion of subfluid was proposed by Wang (1965) for simple materials, see also Truesdell and Noll (2004). As here the structure of the material symmetry group is more complex, the number of possible micromorphic subfluids is greater than the number of simple subfluids. In order to describe granular materials behaviour, Eremeyev (2018) discussed some specific examples of micromorphic subfluids.

3.4 Relaxed Micromorphic Medium as a Micromorphic Subfluid

In order to characterize the possible invariance properties of the strain energy density leading to (3.3), let us consider some preliminary results on representation of scalar functions of tensorial arguments. First we consider the invariance property for functions depending on a second-order tensor. We prove the following lemma.

Lemma 3.1. *Let f be a scalar function of a second-order tensor with the following invariance property*

$$f = f(\mathbf{A}) = f(\mathbf{A} + \mathbf{S}) \quad \forall \mathbf{S} : \quad \mathbf{S} = \mathbf{S}^T. \tag{3.5}$$

Then f depends only on the skew-symmetric part of $\mathbf{A}$:

$$f = f(\mathbf{A}_A), \quad \mathbf{A}_A = \frac{1}{2}(\mathbf{A} - \mathbf{A}^T). \tag{3.6}$$

Proof. Indeed, let us represent $\mathbf{A}$ as a sum of its symmetric and skew-symmetric parts

$$\mathbf{A} = \mathbf{A}_A + \mathbf{A}_S, \quad \mathbf{A}_S = \frac{1}{2}(\mathbf{A} + \mathbf{A}^T). \tag{3.7}$$

So f can be considered as a function of $\mathbf{A}_S$ and $\mathbf{A}_A$

$$f = \tilde{f}(\mathbf{A}_S, \mathbf{A}_A).$$

Invariance (3.5) transforms into

$$\tilde{f}(\mathbf{A}_S, \mathbf{A}_A) = \tilde{f}(\mathbf{A}_S + \mathbf{S}, \mathbf{A}_A) \quad \forall \mathbf{S}: \quad \mathbf{S} = \mathbf{S}^T. \tag{3.8}$$

This results in

$$\frac{\partial f}{\partial \mathbf{A}_S} = \mathbf{0},$$

that is to independence on $\mathbf{A}_S$. Indeed, as $\mathbf{S}$ is an arbitrary tensor we transform (3.8) into

$$\tilde{f}(\mathbf{A}_S, \mathbf{A}_A) = \tilde{f}(\mathbf{A}_S + t\mathbf{S}, \mathbf{A}_A) \quad \forall t, \quad \forall \mathbf{S}: \quad \mathbf{S} = \mathbf{S}^T. \tag{3.9}$$

Differentiating (3.9) with respect to t we get that

$$\frac{\partial f}{\partial \mathbf{A}_S} : \mathbf{S} = 0, \quad \forall \mathbf{S}: \quad \mathbf{S} = \mathbf{S}^T, \tag{3.10}$$

where ":" stands for the double contraction of two second-order tensors which plays a role of a scalar product in the space of second-order tensors. Note that by definition the derivative of a scalar function with respect to a symmetric tensor is also a symmetric tensor. Left part of Eq. (3.9) is the scalar product in the vector space of symmetric second-order tensors, so $\partial f / \partial \mathbf{A}_S$ must be orthogonal to any symmetric tensor. Thus, it is the zero tensor and we get (3.6). $\qquad\square$

Note that any skew-symmetric tensor can be represented through its axial vector $\mathbf{A}_A = \mathbf{a} \times \mathbf{I}$, where $\mathbf{I}$ is the unit tensor and $\mathbf{a}$ is the axial vector of $\mathbf{A}_A$. It can be expressed through $\mathbf{A}_A$ with the use of the vectorial invariant

$$\mathbf{a} = -\frac{1}{2}(\mathbf{A}_A)_\times.$$

The vectorial invariant for a diad is defined as follows

$$(\mathbf{a} \otimes \mathbf{b})_\times = \mathbf{a} \times \mathbf{b}.$$

Obviously, this definition can be easily extended for any second-order tensor, see for details original lectures by Gibbs in (Wilson, 1901, p. 275), and more recent contributions by Lebedev et al (2010); Eremeyev et al (2018a) and the appendices by Naumenko and Altenbach (2007). In particular, the vectorial invariant of $\mathbf{A}$ can be also introduced by the formula

$$\mathbf{A}_\times = -(\mathbf{I} \times \mathbf{I}) : \mathbf{A}. \tag{3.11}$$

So instead of (3.6) one can use $f = f(\mathbf{a} \times \mathbf{I})$ or even more simple dependence $f = f(\mathbf{a})$.

Now let us consider a function of a third-order tensor which has similar invariance property. For any third-order tensor $\mathbf{T}$ we introduce the following permutation operation

$$\mathbf{T}^{T(1,2)} \equiv [T_{mnk}\mathbf{i}_m \otimes \mathbf{i}_n \otimes \mathbf{i}_k]^{T(1,2)} = T_{mnk}\mathbf{i}_n \otimes \mathbf{i}_m \otimes \mathbf{i}_k. \tag{3.12}$$

Here for simplicity we use the representation of $\mathbf{T}$ in Cartesian base vector bases $\mathbf{i}_j$, $j = 1, 2, 3$. For example, we have

$$(\mathbf{a} \otimes \mathbf{b} \otimes \mathbf{c})^{T(1,2)} = \mathbf{b} \otimes \mathbf{a} \otimes \mathbf{c}, \quad (\mathbf{S} \otimes \mathbf{a})^{T(1,2)} = \mathbf{S}^T \otimes \mathbf{a}.$$

With this operation we can introduce the partially symmetric and partially skew-symmetric third-order tensors by formulas

$$\mathbf{T}^{T(1,2)} = \mathbf{T}, \quad \mathbf{T}^{T(1,2)} = -\mathbf{T},$$

respectively. Obviously, these classes constitute orthogonal subspaces in the whole vector space of third-order tensors. Moreover, as for second-order tensors we have the following representation

$$\mathbf{T} = \mathbf{T}_S + \mathbf{T}_A, \quad \mathbf{T}_S \,\vdots\, \mathbf{T}_A = 0, \tag{3.13}$$

where

$$\mathbf{T}_S = \mathbf{T}_S^{T(1,2)} = \frac{1}{2}(\mathbf{T}_S + \mathbf{T}_S^{T(1,2)}), \quad \mathbf{T}_A = -\mathbf{T}_A^{T(1,2)} = \frac{1}{2}(\mathbf{T}_S - \mathbf{T}_S^{T(1,2)}),$$

and "$\cdot$" is the triple contraction operation. Eq. (3.13) is a full analogy to (3.7). We denote a set of tensors such as $\mathbf{T}^{T(1,2)} = \mathbf{T}$ as a Lin_3^S which is also a group with respect to addition.

We prove next lemma.

Lemma 3.2. *Let f be a scalar-valued function of a third-order tensor with the following invariance property*

$$f = f(\mathbf{T}) = f(\mathbf{T} + \mathbf{L}) \quad \forall \mathbf{L} \in Lin_3^S \quad (\mathbf{L}^{T(1,2)} = \mathbf{L}). \tag{3.14}$$

Then

$$f = f(\mathbf{T}_A), \quad \mathbf{T}_A = \frac{1}{2}(\mathbf{T}_S - \mathbf{T}_S^{T(1,2)}). \tag{3.15}$$

Proof. The proof mimics the proof of Lemma 3.1. Using (3.13) we transform f into the dependence

$$f = \tilde{f}(\mathbf{T}_S, \mathbf{T}_A).$$

So (3.14) results in

$$\tilde{f}(\mathbf{T}_S, \mathbf{T}_A) = \tilde{f}(\mathbf{T}_S + t\mathbf{L}, \mathbf{T}_A) \quad \forall t, \quad \forall \mathbf{L}: \quad \mathbf{L}^{T(1,2)} = \mathbf{L}. \tag{3.16}$$

Differentiating (3.16) with respect to t we get

$$\frac{\partial f}{\partial \mathbf{T}_S} \because \mathbf{L} = 0, \quad \forall \mathbf{L}: \quad \mathbf{L}^{T(1,2)} = \mathbf{L} \tag{3.17}$$

As $\partial f / \partial \mathbf{T}_S$ has the same symmetry as $\mathbf{L}$ that is

$$\left(\frac{\partial f}{\partial \mathbf{T}_S}\right)^{T(1,2)} = \frac{\partial f}{\partial \mathbf{T}_S},$$

from (3.17) it follows that

$$\frac{\partial f}{\partial \mathbf{T}_S} = \mathbf{0}. \tag{3.18}$$

Thus, $\tilde{f}$ depends on $\mathbf{T}_A$ only. $\qquad\qquad\qquad\qquad\qquad\qquad\qquad \square$

Now we are almost ready to characterize the constitutive equations of the reduced micromorphic continuum with the corresponding local material symmetry group. Let the material symmetry group $\mathcal{G}_{\varkappa}$ contains the following subgroup

$$\mathcal{R}_{\varkappa} = \{\mathbb{X} = (\mathbf{I}, \mathbf{I}, \mathbf{L}): \quad \mathbf{L} \in Lin_3^S\}.$$

So the following relation is satisfied

$$\mathcal{W}(\mathbf{E}, \mathbf{C}, \mathbf{K}) = \mathcal{W}(\mathbf{E}, \mathbf{C}, \mathbf{K} + \mathbf{L}), \quad \forall \mathbf{L} \in Lin_3^S.$$

With Lemma 3.2 this means that $\mathcal{W}$ has the form

$$\mathcal{W} = \mathcal{W}(\mathbf{E}, \mathbf{C}, \mathbf{K}_A), \quad \mathbf{K}_A = \frac{1}{2}(\mathbf{K} - \mathbf{K}^{T(1,2)}). \tag{3.19}$$

Let us analyze the form of $\mathbf{K}_A$. Any third-order tensor $\mathbf{K}_A$ can be represented as a sum

$$\mathbf{K}_A = \mathbf{A}_k \otimes \mathbf{i}_k, \tag{3.20}$$

where

$$\mathbf{A}_k \equiv \mathbf{K}_A \cdot \mathbf{i}_k, \quad k = 1, 2, 3, \tag{3.21}$$

are second-order tensors. The relation

$$\mathbf{K}_A^{T(1,2)} = -\mathbf{K}_A \tag{3.22}$$

means that tensors $\mathbf{A}_k$ are skew-symmetric:

$$\mathbf{A}_k^T = -\mathbf{A}_k. \tag{3.23}$$

So they can be represented through their axial vectors

$$\mathbf{A}_k = \mathbf{a}_k \times \mathbf{I} = \mathbf{I} \times \mathbf{a}_k.$$

Thus, (3.20) transforms into

$$\mathbf{K}_A = \mathbf{I} \times \mathbf{A}, \quad \mathbf{A} = \mathbf{a}_k \otimes \mathbf{i}_k.$$

Tensor $\mathbf{A}$ can be determined through $\mathbf{K}_A$ as follows

$$\mathbf{A} = -\frac{1}{2}(\mathbf{I} \times \mathbf{I}) : \mathbf{K}_A. \tag{3.24}$$

As $(\mathbf{I} \times \mathbf{I}) : \mathbf{K}_S = \mathbf{0}$ we get

$$\mathbf{A} = -\frac{1}{2}(\mathbf{I} \times \mathbf{I}) : \mathbf{K}. \tag{3.25}$$

Comparing (3.25) with (3.4), we conclude that

$$\mathbf{A} = \frac{1}{2}\mathbf{B} \tag{3.26}$$

and instead of

$$\mathbf{K}_A = \frac{1}{2}\mathbf{I} \times \mathbf{B} \tag{3.27}$$

we can use $\mathbf{B}$ as the strain measure.

So we prove the main theorem.

Theorem 3.1. *The reduction of the full micromorphic continuum model with the strain energy*

$$\mathcal{W} = \mathcal{W}(\mathbf{E}, \mathbf{C}, \mathbf{K}), \quad \mathbf{E} = \mathbf{F} \cdot \mathbf{P}^{-1}, \quad \mathbf{C} = \mathbf{P} \cdot \mathbf{P}^T, \quad \mathbf{K} = \nabla \mathbf{P} \cdot \mathbf{P}^{-1}$$

to the reduced micromorphic continuum model with the constitutive relation

$$\mathcal{W}_{\mathrm{R}} = \mathcal{W}_{\mathrm{R}}(\mathbf{E}, \mathbf{C}, \mathbf{B}), \quad \mathbf{B} = (\nabla \times \mathbf{P}) \cdot \mathbf{P}^{-1},$$

corresponds to inclusion into the material symmetry group $\mathcal{G}_{\varkappa}$ *the following group*

$$\mathcal{R}_{\varkappa} = \{\mathbb{X} = (\mathbf{I}, \mathbf{I}, \mathbf{L}) : \mathbf{L} \in Lin_3^S\}$$

as a subgroup:

$$\mathcal{R}_{\varkappa} \subset \mathcal{G}_{\varkappa}.$$

As any element $\mathbb{X} \in \mathcal{R}_{\varkappa}$ *does not belongs to the orthogonal group* $\mathcal{O}$ *this means that the reduced micromorphic continuum is an micromorphic subfluid.*

Remark 3.1. Let us note that the definition of the local material symmetry group including the group operation depends essentially on the choice of strain measures in the constitutive equations. This means that choosing another set of strain measures as arguments of the strain energy density one comes to another group definition. In particular, one can directly analyze invariance properties of (3.3). This may lead to another definition of the symmetry group which will be valid for reduced micromorphic media. Such analysis will be performed in the forthcoming papers.

3.5 Conclusions

Here we considered the model of reduced micromorphic continuum and shown that it can be characterized as a micromorphic subfluid. Micromorphic subfluids constitute a wide class of materials which local material symmetry group contains non-orthogonal elements and does not coincide with the unimodular group $\mathcal{U}$. In other words such material may present both fluid-like and solid-like properties. Note that such peculiar behavior of the materials described within the reduced micromorphic model is similar to the analysis of the acceleration waves given by Eremeyev et al (2018b), where it was shown that the number of propagating waves is lesser than for the full micromorphic model. Unlike simple material where subfluids constitute quite particular case of materials, for such generalized media as micropolar and micromorphic ones these subfluids may find more application for modelling of microstructured materials. In particular, additional symmetries which possesses a subfluid may lead to additional non-trivial conservation laws for such models.

Acknowledgements Author acknowledges the support of the Government of the Russian Federation (contract No. 14.Y26.31.0031).

References

Auffray N, Le Quang H, He QC (2013) Matrix representations for 3D strain-gradient elasticity. Journal of the Mechanics and Physics of Solids 61(5):1202–1223

Auffray N, dell'Isola F, Eremeyev VA, Madeo A, Rosi G (2015a) Analytical continuum mechanics à la Hamilton–Piola least action principle for second gradient continua and capillary fluids. Mathematics and Mechanics of Solids 20(4):375–417

Auffray N, Dirrenberger J, Rosi G (2015b) A complete description of bi-dimensional anisotropic strain-gradient elasticity. International Journal of Solids and Structures 69:195–206

Auffray N, Kolev B, Olive M (2017) Handbook of bi-dimensional tensors: Part I: Harmonic decomposition and symmetry classes. Mathematics and Mechanics of Solids 22(9):1847–1865

Bertram A (2017) Compendium on Gradient Materials. TU Berlin, www.redaktion.tu-berlin.de /fileadmin/fg49/publikationen/bertram/Compendium_on_Gradient_Materials_Dec_2017.pdf

Boehler JP (ed) (1987) Applications of Tensor Functions in Solid Mechanics, CISM International Centre for Mechanical Sciences, vol 292. Springer, Wien

dell'Isola F, Eremeyev VA (2018) Some introductory and historical remarks on mechanics of microstructured materials. In: dell'Isola F, Eremeyev VA, Porubov A (eds) Advances in Mechanics of Microstructured Media and Structures, Advanced Structured Materials, vol 87. Springer, Cham, pp 1–20

dell'Isola F, Sciarra G, Vidoli S (2009) Generalized Hooke's law for isotropic second gradient materials. Proceedings of the Royal Society of London Series A 495:2177 – 2196

dell'Isola F, Corte AD, Giorgio I (2017) Higher-gradient continua: The legacy of Piola, Mindlin, Sedov and Toupin and some future research perspectives. Mathematics and Mechanics of Solids 22(4):852–872

Elzanowski M, Epstein M (1992) The symmetry group of second-grade materials. International Journal of Non-linear Mechanics 27(4):635–638

Eremeyev VA (2018) On the material symmetry group for micromorphic media with applications to granular materials. Mechanics Research Communications 94:8–12

Eremeyev VA, Pietraszkiewicz W (2006) Local symmetry group in the general theory of elastic shells. Journal of Elasticity 85(2):125–152

Eremeyev VA, Pietraszkiewicz W (2012) Material symmetry group of the non-linear polar-elastic continuum. International Journal of Solids and Structures 49(14):1993–2005

Eremeyev VA, Pietraszkiewicz W (2013) Material symmetry group and consistently reduced constitutive equations of the elastic Cosserat continuum. In: Altenbach H, Forest S, Krivtsov A (eds) Generalized Continua as Models for Materials: with Multi-scale Effects or Under Multi-field Actions, Advanced Structured Materials, vol 22. Springer, Berlin, pp 77–90

Eremeyev VA, Pietraszkiewicz W (2016) Material symmetry group and constitutive equations of micropolar anisotropic elastic solids. Mathematics and Mechanics of Solids 21(2):210–221

Eremeyev VA, Cloud MJ, Lebedev LP (2018a) Applications of Tensor Analysis in Continuum Mechanics. World Scientific, New Jersey

Eremeyev VA, Lebedev LP, Cloud MJ (2018b) Acceleration waves in the nonlinear micromorphic continuum. Mechanics Research Communications 93:70–74

Eringen AC (1980) Mechanics of Continua. Robert E. Krieger Publishing Co., Huntington, NY

Eringen AC (1999) Microcontinuum Field Theory. I. Foundations and Solids. Springer, New York

Eringen AC, Kafadar CB (1976) Polar field heories. In: Eringen AC (ed) Continuum Physics, vol IV, Academic Press, New York, pp 1–75

Eringen AC, Suhubi ES (1964) Nonlinear theory of simple micro-elastic solids–I. International Journal of Engineering Science 2(2):189–203

Forest S (2013) Micromorphic media. In: Altenbach H, Eremeyev VA (eds) Generalized Continua from the Theory to Engineering Applications, CISM International Centre for Mechanical Sciences, vol 541, Springer Vienna, pp 249–300

Lebedev LP, Cloud MJ, Eremeyev VA (2010) Tensor Analysis with Applications in Mechanics. World Scientific, New Jersey

Lurie AI (1990) Nonlinear Theory of Elasticity. North-Holland, Amsterdam

Madeo A, Neff P, Ghiba ID, Rosi G (2016) Reflection and transmission of elastic waves in nonlocal band-gap metamaterials: a comprehensive study via the relaxed micromorphic model. Journal of the Mechanics and Physics of Solids 95:441–479

Maugin GA (2011) A historical perspective of generalized continuum mechanics. In: Altenbach H, Erofeev VI, Maugin GA (eds) Mechanics of Generalized Continua. From the Micromechanical Basics to Engineering Applications, Advanced Structured Materials, vol 7, Springer, Berlin, pp 3–19

Maugin GA (2013) Generalized Continuum Mechanics: Various Paths, Solid Mechanics and Its Applications, vol. 196. Springer Netherlands, Dordrecht, pp 223–241

Maugin GA (2017) Non-Classical Continuum Mechanics: A Dictionary. Springer, Singapore

Mindlin RD (1964) Micro-structure in linear elasticity. Archive for Rational Mechanics and Analysis 16(1):51–78

Misra A, Poorsolhjouy P (2016) Granular micromechanics based micromorphic model predicts frequency band gaps. Continuum Mechanics and Thermodynamics 28(1-2):215

Murdoch AI, Cohen H (1979) Symmetry considerations for material surfaces. Archive for Rational Mechanics and Analysis 72(1):61–98

Naumenko K, Altenbach H (2007) Modeling of Creep for Structural Analysis. Springer, Berlin

Neff P, Ghiba ID, Madeo A, Placidi L, Rosi G (2014) A unifying perspective: the relaxed linear micromorphic continuum. Continuum Mechanics and Thermodynamics 26(5):639–681

Reiher JC, Bertram A (2018) Finite third-order gradient elasticity and thermoelasticity. Journal of Elasticity 133(2):223–252

Simmonds JG (1994) A Brief on Tensor Analysis, 2nd edn. Springer, New Yourk

Spencer AJM (1971) Theory of invariants. In: Eringen AC (ed) Continuum Physics, Vol. 1, Academic Press, New-York, pp 239–353

Sridhar A, Kouznetsova VG, Geers MGD (2016) Homogenization of locally resonant acoustic metamaterials towards an emergent enriched continuum. Computational mechanics 57(3):423–435

Sridhar A, Kouznetsova VG, Geers MGD (2017) A semi-analytical approach towards plane wave analysis of local resonance metamaterials using a multiscale enriched continuum description. International Journal of Mechanical Sciences 133:188–198

Truesdell C, Noll W (2004) The Non-linear Field Theories of Mechanics, 3rd edn. Springer, Berlin

Wang CC (1965) A general theory of subfluids. Archive for Rational Mechanics and Analysis 20(1):1–40

Wilson EB (1901) Vector Analysis, Founded upon the Lectures of G. W. Gibbs. Yale University Press, New Haven

Zheng QS (1994) Theory of representations for tensor functions – a unified invariant approach to constitutive equations. Applied Mechanics Reviews 47(11):545–587

Chapter 4
Structural Modeling of Nonlinear Localized Strain Waves in Generalized Continua

Vladimir I. Erofeev, Anna V. Leontyeva, Alexey O. Malkhanov, and Igor S. Pavlov

Abstract The basic principles of structural modeling for the construction of mathematical models of microstructured media (generalized continua) are given. Here, microstructure means not the smallness of absolute values, but the smallness of some medium scale with respect to other scales, and the particles are considered to be non-deformable and homogeneous, without their own internal structure, presenting realistic materials. A nonlinear dynamically consistent model of a gradient-elastic medium has been elaborated by the method of structural modeling and using the continualization method involving nonlocality of coupling between the displacements of the lattice sites and the obtained continuum. The formation of spatially localized nonlinear strain waves in such media has been investigated.

Keywords: Structural modeling · Gradient theory · Nonlinear strain waves

4.1 Introduction

The mechanics of microstructured media and the theory of generalized continua, which are interrelated, remained unclaimed for many years due to their complexity and lack of practical applications. Nowadays, there is yet another stage of its development as a consequence of works of Cosserat and Cosserat (1909) that is

V. I. Erofeev, A. V. Leontyeva, I. S. Pavlov
Mechanical Engineering Research Institute of Russian Academy of Sciences
Research Institute for Mechanics, Nizhny Novgorod Lobachevsky State University, 23, Gagarin av. 603950 Nizhny Novgorod, Russia
e-mail: erof.vi@yandex.ru; aleonav@mail.ru; ispavlov@mail.ru

A. O. Malkhanov
Mechanical Engineering Research Institute of Russian Academy of Sciences, Nizhny Novgorod, Russia
e-mail: alexey.malkhanov@gmail.com

© Springer Nature Switzerland AG 2019   55
H. Altenbach et al. (eds.), *Higher Gradient Materials and Related Generalized Continua*, Advanced Structured Materials 120,
https://doi.org/10.1007/978-3-030-30406-5_4

discovered in 1960s, when this theory was expected to achieve great success in the field of the theory of dislocations and the mechanics of composites. Interest to this theory began to grow again from the middle of 1990s with applications to fracture mechanics (Krivtsov, 2007; de Borst and van der Giessen, 1998), geomechanics (Nikolaevskiy, 1996; Vardoulakis and Sulem, 1995), mechanics of granular materials (Chang and Ma, 1994; Christoffersen et al, 1981), gradient plasticity theory (Lippman, 1995). This interest continued at the beginning of the 21st century, primarily due to development of nanotechnologies (Cleland, 2003; Ghoniem et al, 2003; Li and Wang, 2018) and construction of metamaterials (Shining and Xiang, 2018; Kolken and Zadpoor, 2017; Bobrovnitsky and Tomilina, 2018; Gulyaev et al, 2008)—a new class of substances with a complexly organized internal structure. At present, generalized continua, such as micropolar or oriented media, high-gradient materials, micromorphic media, composites, solids with weak or strong non-local interactions, are intensively studied by both theoretical and experimental researchers specializing in various branches of mechanics and physics (Altenbach et al, 2011). The wave dynamics of microstructured media is developed, which allows, in particular, to propose new methods of nondestructive testing of the stress-strain state, structure and properties of materials (Erofeyev, 2003; Vakhnenko, 1996; Potapov et al, 1999, 2009; Pavlov, 2010). However, an adequate description of intensive two-dimensional and three-dimensional processes in structured materials by means of nonlinear wave dynamics necessitates developing of new mathematical models. Such models can be elaborated using the structural modeling method (Berglund, 1982; Pavlov et al, 2006; Pavlov and Potapov, 2008; Erofeev and Pavlov, 2018; Vasiliev and Pavlov, 2018), since structural models contain parameters characterizing geometry of a material (a lattice period, a particle size, and a shape) and therefore they are the most suitable models for studying influence of size effects on macroproperties of a material.

4.2 Principles of Structural Modeling

During the structural modeling, penetration deep into the material is multilevel (multi-scale). However, frequently, it is very difficult to attribute a theory (a model) to a certain scale expressed in units of distance. Nevertheless, such classifications exist and are useful for estimation of the applicability areas of various theories (Ghoniem et al, 2003).

The structural theory of the continuum mechanics is based on the molecular, atomistic, or subatomistic structure of a solid. Most of the early structural theories, which are used in the solid mechanics, considered particles of a medium as centers of forces possessing mass. These elements of a solid act on each other using the central forces. It is assumed that the forces of interaction between the structural elements of a solid rapidly decrease with distance and they can be neglected if the distance between the elements exceeds "the radius of molecular action." Applying of the method of central forces to crystalline media, depending on their symmetry, leads to certain relations between the second-order elasticity constants. These relations are called Cauchy relations. Within the scope of structural models, the Cauchy relations between the elasticity constants can be not valid.

It should be noted that as early as in 1842, Poisson made the assumption that molecules of a crystal can not be simply points, but small solid bodies possessing both translational and rotational degrees of freedom (Love, 1920). In 1887, this idea was developed by Voigt (1887). In 1890, W. Thomson (Lord Kelvin) remarked that the Cauchy relations could be eliminated if crystals are represented as two homogeneous point formations (two sublattices) penetrating into each other. In 1915, Max Born proposed more general structural schemes of crystalline materials, where each crystal element—the unit cell—consists of attracting and repelling particles (Born and Huang, 1954). Inside each cell, the particles are uniformly located with respect to each other. Crystal structures are classified according to the type of crystal lattice and the nature of the interparticle (interatomic) connections. The classification by crystalline systems gives an idea of the geometric characteristics of crystals, but does not consider the question about the nature of the forces holding particles (atoms, ions, molecules, or nanoclusters) in the lattice sites. The classification based on the types of connection forces enables one performing some generalizations of the properties and behavior of crystals that was impossible by considering only the geometry of the lattices.

Elaboration of a structural model starts with the selection of a certain minimum volume—a structural cell (that is analog of the periodicity cell in the crystalline material) in the bulk of a material. Such a cell is capable of reflecting the main features of the macroscopic behavior of this material. As a rule, a structural cell represents a particle, which behavior is characterized by interaction with its environment and is described by kinematic variables (Berglund, 1982; Born and Huang, 1954; Broberg, 1997). In contrast to the standard theory of crystal lattices (Born and Huang, 1954), the structural modeling involves presence in the lattice sites of finite-size bodies having internal degrees of freedom instead of material points. For instance, domains, granules, fullerenes, nanotubes, or clusters of nanoparticles can play the role of such bodies. If to consider their micromotions, then new types of motion arise in a microstructured medium. For example, account of microrotations with respect to the mass centers of particles leads to the appearance of microrotations in granular media (Pavlov et al, 2006).

The force interaction between elements is described by means of model potentials used in the molecular mechanics and the solid state physics. Due to the presence of finite-size bodies in lattice sites it is possible to introduce into consideration the central and moment interactions between particles. The method of structural modeling takes into account the parameters characterizing the lattice period, the size and shape of particles. Therefore, this is the most suitable method for studying the influence of size effects on material properties.

Let us formulate basic principles of the structural modeling:

1. *Minimality of generalizations.* Such generalizations are necessary, which would lead to qualitatively new results. At the same time, the number of new parameters containing in a model should be as small as possible.

2. *Variability of the model.* The possibility of a fairly wide variation of linear and nonlinear parameters of the model due to the choice of kinematic and force schemes of the interaction of structural elements.
3. *Identification and verification of the model.* Verification of the adequacy of the constructed model to real systems and determination of relationships between the model parameters and the physical constants of a material (density, porosity, macro- and microelasticity modules, etc.). For this purpose, the relationships between the parameters of the micromodel and the main physical and mechanical characteristics of a medium should be found.
4. *The principle of compliance.* In the limiting cases, a new model, as a rule, should be degenerated into the known (classical) theories of a deformable solid.

The advantages of the structural modeling are the clear coupling between a structure and macroparameters of a medium and the possibility of purposeful design of materials with specified properties. Shortcomings of the structural modeling are absence of universality of modeling procedure and complexity of the accounting of nonlinear and nonlocal effects of interparticle interactions.

The structural modeling consists of the following stages:

- **1st stage.** *The geometric description of a structure.* In a regular lattice consisting of particles of a given shape, a periodicity cell is determined, its characteristic sizes and kinematic variables describing the current state of the cell are introduced. A kinetic energy is calculated.
- **2nd stage.** *The simulation of force interactions.* Since small deviations of the particles from the equilibrium states are considered, the force and moment interactions of the particles can be described by a power-law potential. In the harmonic approximation, the interaction potential is a quadratic form of the variables of a state of the system. The potential energy per lattice cell is equal to the potential energy of the particle interacting with its neighbors. Particularly, for a system with two spatial variables it can be represented by the following expression:

$$
U\left(\Delta_{nr}q^k,\ \varphi,\ \Delta_{nr}\varphi\right) = \sum_{k,s=1}^{2} \sum_{n,r,l,m} \frac{\partial^2 U}{\partial\left(\Delta_{nr}q^k\right)\partial\left(\Delta_{lm}q^s\right)}\Delta_{nr}q^k\Delta_{lm}q^s +
$$

$$
+ \sum_{n,r,l,m} \frac{\partial^2 U}{\partial\left(\Delta_{nr}\varphi\right)\partial\left(\Delta_{lm}\varphi\right)}\Delta_{nr}\varphi\,\Delta_{lm}\varphi +
$$

$$
+ \sum_{k=1}^{2} \sum_{n,r,l,m} \frac{\partial^2 U}{\partial\left(\Delta_{nr}q^k\right)\partial\left(\Delta_{lm}\varphi\right)}\Delta_{nr}q^k\Delta_{lm}\varphi +
$$

$$
+ \sum_{k=1}^{2} \sum_{n,r} \frac{\partial^2 U}{\partial\left(\Delta_{nr}q^k\right)\partial\varphi}\Delta_{nr}q^k\varphi + \frac{\partial^2 U}{(\partial\varphi)^2}\varphi^2.
$$

$$(4.1)$$

Here $q^k = \left\{q^1,\ q^2\right\} = \{u_{i,j},\ w_{i,j}\}$ are the components of the vector of the mass centre displacements of a particle located at a site with indices (i,

j), $\Delta_{nr}q^k = \left(q^k_{i+n,\,j+r} - q^k_{i,j}\right)/a$ are quantities for the relative variation of the distances between the interacting particles, $\Delta_{nr}\varphi = \left(\varphi_{i+n,\,j+r} - \varphi_{i,j}\right)/a$ are quantities for the relative variation of the orientation angles of the particles, and coefficients n and r, as well as l and m determine the spatial positions of neighbouring particles. Summarizing is performed over all quasi-elastic connections in the cell. The second-order derivatives of the potential energy are the constants of quasi-elastic interactions of the particles and represent elements of force matrices of the crystalline structure (Fedorov, 1968).

In the phenomenological theories, the force constants should be found experimentally. Their connection with the geometrical structure and with the scheme of force interactions in a concrete crystalline lattice is not clear. From general energy reasoning and the requirements of symmetry of the lattice, it is possible to receive only some restrictions on the values of the force constants. Usage of the structural approach enables one to find an explicit dependence between the elements of the force matrices and the parameters characterising the inner structure of the lattice, i.e. its period, sizes, and shape of its particles.

For structural modelling of solids, an equivalent force scheme is introduced as a system of rods or springs that incorporates the transmission of forces and moments between the structural elements instead of a field description of the interaction of the particles (Berglund, 1982; Ostoja-Starzewski et al, 1996; Suiker et al, 2001). The mechanical characteristics of the connecting rods and springs should be generally determined from the requirement of equality of the strain energy in the investigated object and in its model. A spring model is used in this paper.

- **3rd stage.** *Derivation of dynamical equations for a discrete system.* Due to expressions for the kinetic and potential energy that were obtained at the previous stages, it is possible to make up differential-difference equations describing the lattice dynamics. Such equations represent Lagrange equations of the second kind

$$\frac{d}{dt}\left(\frac{\partial L}{\partial \dot{q}^{(l)}_{i,j}}\right) - \frac{\partial L}{\partial q^{(l)}_{i,j}} = 0, \tag{4.2}$$

where $L = T_{i,j} - U_{i,j}$ is the Lagrange function, which is equal to the difference between the kinetic and potential energy of a cell, $q^{(l)}_{i,j}$ are the generalized coordinates,

$$q^{(1)}_{i,j} = u_{i,j}, \quad q^{(2)}_{i,j} = w_{i,j}, \quad q^{(3)}_{i,j} = \varphi_{i,j}, \tag{4.3}$$

where $\dot{q}^{(l)}_{i,j}$ are the generalized velocities.

- **4th stage.** *The continuum approximation.* A transition from discrete models to continual ones is performed by extrapolating the functions specified at discrete points by continuous fields of displacements and micro-rotations. For long-wavelength perturbations, when $\lambda \gg a$ (where λ is a characteristic spatial scale of deformation), discrete labels i and j can be changed by means of a continuous spatial variables $x=ia$ and $y=ja$. In this case, the functions specified at discrete points are interpolated by the continuous functions and their partial derivatives in accordance with the standard Taylor formula. Depending on the number of interpolation terms, one can consider various approximations of a discrete model of a microstructured medium and elaborate a hierarchy of quasi-continuum models.

- **5th stage.** *Identification of a model.* The goal of identification is the construction of the best (optimal) model on the basis of experimental observations. Identification is divided by *structural* and *parametric. The structural identification is a choice of the optimal form of equations for a mathematical model. The parametric identification* is a determination by the experimental data of the values of the parameters of the mathematical model ensuring the agreement of the model values with the experimental data, provided that the model and the object are subjected to similar influences. Such identification also includes a numerical simulation of experiment and a choice of the model parameters from the condition of the best coincidence of calculation and experimental results.

The main problem of parametric identification is the choice of variables possessing information about the medium. Such variables should be measured experimentally. For example, in acoustic spectroscopy, the measured values are the acoustic characteristics of a microstructured medium, using which the parameters of its micromodel can be determined. In the linear approximation, the acoustic characteristics of the medium can be determined due to dispersion dependences between the frequency and the length of the elastic wave. In order to determine the nonlinear characteristics of the medium, it is necessary to consider the anharmonic interactions of acoustic waves of various types with each other (Vakhnenko, 1996; Potapov et al, 1999), as well as their interaction with the magnetic (Erofeev and Mal'khanov, 2017) and electric fields (Maugin, 1988).

4.3 One-dimensional Model of a Nonlinear Gradient-elastic Continuum.

To obtain a dynamically consistent model of a nonlinear gradient-elastic continuum, we use the structural modeling method and the continualization method proposed by Askes and Metrikine (2002); Metrikine and Askes (2002), which consists in the assumption of a nonlocal connection between the displacements of lattice sites and the resulting continuum.

Consider a long one-dimensional chain of alternating identical masses and springs. We assume that the masses can move only along the chain. This direction we denote as x-axis. Let us also denote: m – mass, k – spring stiffness, l – distance between masses, $u_n = u_n(t)$ – displacement of n-th mass.

The potential energy of the chain consists of potential energies $\phi(\xi_n)$ of all springs, which depends on their displacements $\xi_n = u_n - u_{n-1}$, and have the following form:

$$W = \sum_n \phi(\xi_n) = \sum_n \phi(u_n - u_{n-1}). \tag{4.4}$$

Since the aim of the work is to study nonlinear waves, we will consider cubical item in $\phi(\xi)$:

$$\phi(\xi) = \frac{k\xi^2}{2} + \frac{K\xi^3}{3}. \tag{4.5}$$

The equation of the dynamics of the mass has the following form:

$$m\frac{d^2 u_n}{dt^2} = -\frac{\partial W}{\partial u_n}. \tag{4.6}$$

or after transformations taking into account the previous relations

$$m\frac{d^2 u_n}{dt^2} = k\left(u_{n+1} - 2u_n + u_{n-1}\right) + K\left(u_{n+1} - u_{n-1}\right)\left(u_{n+1} - 2u_n + u_{n-1}\right). \tag{4.7}$$

Let us construct an interpolation polynomial of the second degree with respect to x which coincides with function $u(x, t)$ in points

$$u\left(x_{n-1}, t\right) = u_{n-1}, u\left(x_n, t\right) = u_n, u\left(x_{n+1}, t\right) = u_{n+1}.$$

$$\hat{u}\left(x, t\right) = u_{n-1}\frac{\left(x - x_n\right)\left(x - x_{n+1}\right)}{2l^2} - u_n\frac{\left(x - x_{n-1}\right)\left(x - x_{n+1}\right)}{l^2} + \tag{4.8}$$

$$+ u_{n+1}\frac{\left(x - x_{n-1}\right)\left(x - x_n\right)}{2l^2}.$$

The values of the interpolation polynomial in points $x_n - \theta l$ and $x_n + \theta l$, where $0 < \theta < 1$, can be calculated

$$\hat{u}(x_n + \theta l, t) = u_{n-1}\frac{\theta(\theta - 1)}{2} + u_n(1 - \theta^2) + u_{n+1}\frac{\theta(\theta + 1)}{2},$$

$$\hat{u}(x_n - \theta l, t) = u_{n-1}\frac{\theta(\theta + 1)}{2} + u_n(1 - \theta^2) + u_{n+1}\frac{\theta(\theta - 1)}{2}. \tag{4.9}$$

As a continuous function describing the motion of the continual model of masses and springs, we take the average value $\hat{u}(x_n + \theta l, t)$ and $\hat{u}(x_n - \theta l, t)$. We get the following

$$u(x, t) \approx \frac{\hat{u}(x_n + \theta l, t) + \hat{u}(x_n - \theta l, t)}{2} = u_{n-1}\frac{\theta^2}{2} + u_n(1 - \theta^2) + u_{n+1}\frac{\theta^2}{2}. \tag{4.10}$$

Note that the authors of the alternative continualization (Askes and Metrikine, 2002; Metrikine and Askes, 2002) consider a continuous function as the average between the displacement of three neighboring particles

$$u(x, t) \approx \frac{1}{1 + 2a_1}\left(a_1 u_{n-1} + u_n + a_1 u_{n+1}\right), \tag{4.11}$$

where dimensionless "weight" constant a_1 is in range $0 \leq a_1 < 1$. It is easy to see that relations (4.10) and (4.11) are identical, and the parameters a_1 and θ are connected with the following relations:

$$\theta^2 = \frac{2a_1}{1 + 2a_1}, a_1 = \frac{\theta^2}{2(1 - \theta^2)}. \tag{4.12}$$

When implementing the alternative continualization method, it is also assumed that

$$u_n(t) = u(x, t) + l^2 f_2(x, t) + l^4 f_4(x, t) + O(L^5), \tag{4.13}$$

where

$$f_2(x, t) = -\frac{a_1}{1 + 2a_1} u_{xx}(x, t) = -\frac{\theta^2}{2} u_{xx}(x, t),$$

$$f_4(x, t) = \frac{a_1}{12} \frac{10a_1 - 1}{(1 + 2a_1)^2} u_{xxxx}(x, t) = \frac{\theta^2}{24}(6\theta^2 - 1) u_{xxxx}(x, t). \tag{4.14}$$

Substituting (4.11) into (4.2) and take into account that

$$u_{n+1} = u(x + l, t) + l^2 f_2(x + l, t) + l^4 f_4(x + l, t) + O(L^5),$$

$$u_{n-1} = u(x - l, t) + l^2 f_2(x - l, t) + l^4 f_4(x - l, t) + O(L^5), \tag{4.15}$$

and expressing $u(x \pm l, t), u_{xx}(x \pm l, t), u_{xxxx}(x \pm l, t)$ through function $u(x, t)$ and its derivatives, using a Taylor series decomposition, we get

$$u_{tt} - \frac{l^2 \theta^2}{2} u_{xxtt} + \frac{l^4 \theta^2 (6\theta^2 - 1)}{24} u_{xxxxtt} - \frac{kl^2}{m} u_{xx} - \frac{(1 - 6\theta^2) kl^4}{12 \ m} u_{xxxx} =$$

$$= \frac{2Kl^3}{m} u_x u_{xx} + \frac{1 - 3\theta^2}{3} \frac{Kl^5}{m} u_{xx} u_{xxx} + \frac{1 - 6\theta^2}{6} \frac{Kl^5}{m} u_x u_{xxxx}. \tag{4.16}$$

4.4 Nonlinear Strain Waves

In dimensionless variables, the Eq. (4.16) can be written as follows:

$$\frac{\partial^2 U}{\partial \tau^2} - \frac{\partial^2 U}{\partial z^2} - \frac{\partial^4 U}{\partial z^2 \partial \tau^2} - d_1 \left(\frac{\partial^4 U}{\partial z^4} + \frac{\partial^6 U}{\partial z^4 \partial \tau^2} \right) -$$

$$- d_2 \left(\frac{\partial U}{\partial z} \frac{\partial^2 U}{\partial z^2} + d_1 \frac{\partial U}{\partial z} \frac{\partial^4 U}{\partial z^4} + d_3 \frac{\partial^2 U}{\partial z^2} \frac{\partial^3 U}{\partial z^3} \right) = 0, \tag{4.17}$$

where $U = u/u_0$, $z = x/X$, $\tau = t/T$ – dimensionless values of the displacement, the coordinate and time, respectively. Characteristic values of length and time are equal

$$X^2 = \frac{l^2 \theta^2}{2}, T^2 = \frac{m\theta^2}{2k}, \tag{4.18}$$

dimensionless parameters have the following form:

$$d_1 = \frac{1}{\theta^2}\left(\frac{1}{6} - \theta^2\right), \quad d_2 = \frac{2\sqrt{2}Ku_0}{\theta k}, \quad d_3 = \frac{1}{\theta^2}\left(\frac{1}{3} - \theta^2\right). \tag{4.19}$$

In the variables of a traveling wave, Eq. (4.17) takes the form:

$$W_{\chi\chi\chi\chi\chi} + \frac{(d_1 + v^2)}{d_1 v^2}W_{\chi\chi\chi} + \frac{(1 - v^2)}{d_1 v^2}W_\chi +$$
$$+ \frac{d_2}{d_1 v^2}\left(WW_\chi + d_1 WW_{\chi\chi\chi} + d_3 W_\chi W_{\chi\chi}\right) = 0, \tag{4.20}$$

where $\chi = z - v\tau$ – traveling coordinate, v – velocity of a nonlinear wave. The introduction of a new function $W = dU/d\chi$ allows us to reduce the order of the equation to the fifth.

By the method of the simplest equations (Kudryashov, 2010) we can find the solution of Eq. (4.20):

$$W(\chi) = \frac{20d_1 B_0 v^2}{d_2(2d_1 + d_3)}\left(2 - 3th^2\left(\sqrt{B_0}\chi\right)\right) - \frac{d_1 d_3 - 3d_1 v^2 + d_3 v^2 + 2d_1^2}{d_1 d_2(2d_1 + d_3)}, \tag{4.21}$$

where

$$B_0^2 = -\frac{2d_1^2 + d_1 d_3 - 3d_1 + d_3}{16d_1^2(3d_1 - d_3)}. \tag{4.22}$$

Solution (4.21) has a profile in the form of a symmetric bell with a changing along the vertical axis offset (Fig. 4.1, dashed line). Fixing the sole of the bell at the zero mark (Fig. 4.1, solid line), we find the constraint imposed on the square of the wave

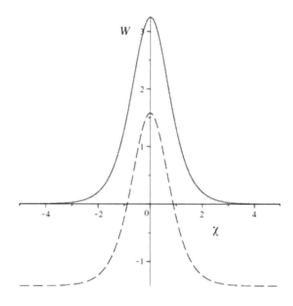

Fig. 4.1 Dependencies $W(\chi)$

velocity:

$$v^2 = -\frac{d_1\left(2d_1 + d_3\right)}{20d_1^2 B_0 - 3d_1 + d_3}. \tag{4.23}$$

At such speeds, the solution (4.21) takes the form:

$$W\left(\chi\right) = -\frac{60d_1^2 B_0}{d_2\left(20d_1^2 B_0 - 3d_1 + d_3\right)\cosh^2\left(\sqrt{B_0}\chi\right)}, \tag{4.24}$$

moreover, this solution has a physical meaning only in the interval

$$\frac{1}{6} < \theta^2 < \frac{2}{9}, \quad \theta > 0. \tag{4.25}$$

The dependences of the amplitude and width of the soliton on its velocity are shown in Fig. 4.2. When increasing the parameter θ within the considered interval, the velocity firstly increases, then decreases, the amplitude and width of the soliton monotonically increase. The curve $v\left(\theta\right)$ has a maximum point, i.e. soliton velocity is bounded from above.

The sign of the dimensionless parameter d_2 influences the polarity of the soliton. For positive values of the parameter (rigid nonlinearity), the soliton has a negative polarity. For negative parameter values (soft nonlinearity), the soliton has a positive polarity. The magnitude of the nonlinearity does not affect the speed of propagation of waves and their width, but affects their amplitude. The smaller the value of nonlinearity, the greater the amplitude of the wave, i.e. in weakly non-linear media propagate waves of greater amplitude.

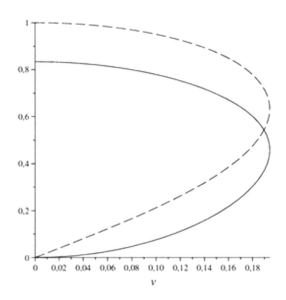

Fig. 4.2 Dependencies $A\left(v\right)$ (solid), $\Delta\left(v\right)$ (dashed)

In the limiting case, when θ^2 is close to its bottom bound, the dimensionless parameters equal

$$d_1 = 0, \quad d_2 = \frac{4\sqrt{3}Ku_0}{k}, \quad d_3 = 1, \tag{4.26}$$

such that we obtain

$$W_{\chi\chi} + b_1 W + b_2 \left(W^2 + W_\chi^2\right) = 0, \tag{4.27}$$

where

$$b_1 = \frac{1 - v^2}{v^2}, \quad b_2 = \frac{d_2}{2v^2} = \frac{2\sqrt{3}d}{v^2}, \quad d = \frac{Ku_0}{k}, \tag{4.28}$$

are introduced to shorten the expressions, d characterizes the elastic nonlinearity.

Equation (4.27) is an equation of the anharmonic oscillator with two quadratic nonlinearities. Phase portrait of the equation when $b_1 > 0$ and $b_2 < 0$ is depicted in the Fig. 4.3. It can be seen that there are two equilibrium states on the phase plane: the "center," which is located at the origin of coordinates, and the "saddle," whose location (to the right or left of the center) depends on the type of nonlinearity. As in the classical case, anharmonic oscillator with quadratic nonlinearity (Erofeev et al, 2002), if the nonlinearity corresponds to "hard" ones, then the saddle is located to the left of the center. When $b_1 < 0$ the phase portrait is shifted along the horizontal axis, so that the saddle is located at the origin of coordinates. Solitary soliton-type waves and periodic waves can propagate in the system as well. The exact analytical solution in the form of solitons for Eq. (4.27) cannot be found, the pole of this equation is zero (Kudryashov, 2010).

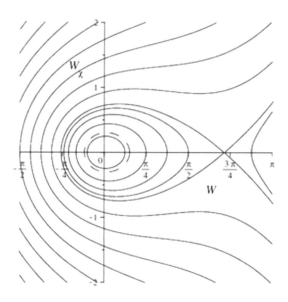

Fig. 4.3 Phase portrait (W, W_χ)

In the case of weak nonlinearity ($d \ll 1$) the solution to (4.27) can be found with the help of asymptotic expansions (Bogoliubov and Mitropolsky, 1961):

$$W(\chi) = A_0 \cos\left(\sqrt{b_1}\chi + \phi_0\right) + \\ + \frac{A_0^2 b_2}{2b_1}\left[\frac{(1-b_1)}{3}\cos 2\left(\sqrt{b_1}\chi + \phi_0\right) - (1+b_1)\right], \tag{4.29}$$

where the amplitude and the phase of oscillations are defined with the help of the following expressions:

$$A_0 = \sqrt{w_1^2 + \frac{w_2^2}{b_1}} + \varepsilon\frac{b_2}{3\sqrt{b_1}}\frac{w_1\left(w_1^2 + 3w_2^2 + 2w_1^2 b_1\right)}{\sqrt{w_1^2 b_1 + w_2^2}},$$

$$\phi_0 = -\arctan\left(\frac{w_2}{\sqrt{b_1}w_1}\right) + \varepsilon\frac{b_2}{3\sqrt{b_1}}\frac{w_2\left(2w_2^2 + 3w_1^2 b_1 + w_2^2 b_1\right)}{b_1\left(w_1^2 b_1 + w_2^2\right)}, \tag{4.30}$$

while w_1 and w_2 are set by initial conditions:

$$W(0) = w_1, W_\chi(0) = w_2. \tag{4.31}$$

A dotted line in Fig. 4.3 marks the approximate solution (4.29) near the stable equilibrium state. The found solution is valid only in the subsonic case and soft nonlinearity. With increasing rigidity, the amplitude of the waves (with fixed initial conditions) decreases, and the period increases. The closer the velocity of a nonlinear wave to the speed of sound, the smaller the amplitude of the periodic waves.

4.5 Conclusions

The equation for a nonlinear quasicontinuum motion, which is obtained using a set of alternating masses and springs and taking into account the non-locality of the relationship between displacements of lattice sites, does not coincide with a similar equation for the nonlinear theory of elasticity and, as distinct from the last one, enables one describing the longitudinal wave dispersion. Thus, two factors influence on the propagation of plane longitudinal waves: dispersion and nonlinearity. The nonlinearity leads to the generation of new harmonics in the wave. The energy is continuously pumped from the main perturbation to these harmonics. This effect contributes to the appearance of sharp differences in the moving profile. The dispersion, on the contrary, smooths the distinguishes due to the difference in the phase velocities of the harmonic components of the wave. The combined effect of these two factors and their "competition" can contribute to the formation of stationary non-sinusoidal waves. Such waves propagate with a constant velocity without changing their shape.

Acknowledgements The research was carried out under the financial support of the Russian Foundation for Basic Research (projects NN 18-29-10073-mk and 19-08-00965-a).

References

Altenbach H, Maugin GA, Erofeev VI (eds) (2011) Mechanics of Generalized Continua, Advanced Structured Materials, vol 7. Springer-Verlag, Berlin, Heidelberg

Askes H, Metrikine A (2002) One-dimensional dynamically consistent gradient elasticity models derived from a discrete microstructure. Part 1: Generic formulation. European Journal of Mechanics A/Solids 21(4):573–588

Berglund K (1982) Structural models of micropolar media. In: Brulin O, Hsieh RKT (eds) Mechanics of Micropolar Media, World Scientific, Singapore, pp 35–86

Bobrovnitsky YI, Tomilina TM (2018) Sound absorption and metamaterials: A review. Acoustical Physics 64(5):519–526

Bogoliubov NN, Mitropolsky YA (1961) Asymptotic Methods in the Theory of Non-Linear Oscillations. Gordon and Breach, New York

Born M, Huang K (1954) Dynamical Theory of Crystal Lattices. Int. Series of Monographs on Physics, Clarendon Press, Oxford

Broberg KB (1997) The cell model of materials. Computational Mechanics 19:447–452

Chang C, Ma L (1994) A micromechanical-based micropolar theory for deformation of granular solids. Intern J Solids and Structures 28(1):67–87

Christoffersen J, Mehrabadi MM, Nemat-Nasser SA (1981) A micromechanical description of granular material behavior. Trans ASME J Appl Mech 48(2):339–344

Cleland AN (2003) Foundations of Nanomechanics. Advanced Texts in Physics, Springer-Verlag, Berlin

Cosserat E, Cosserat F (1909) Theorié des Corps Déformables. Librairie Scientifique A. Hermann et Fils, Paris

de Borst R, van der Giessen E (eds) (1998) Material Instabilities in Solids. J. Wiley and Sons, Chichester-New York-Wienheim-Brisbane-Singapore-Toronto

Erofeev VI, Mal'khanov AO (2017) Localized strain waves in a nonlinearly elastic conducting medium interacting with a magnetic field. Mechanics of Solids 52(2):224–231

Erofeev VI, Pavlov IS (2018) Rotational waves in microstructured materials. In: dell'Isola F, Eremeyev A, Porubov A (eds) Advances in Mechanics of Microstructured Media and Structures, Springer, Cham, Advanced Structured Materials, vol 87, pp 103–124

Erofeev VI, Kazhaev VV, Semerikova NP (2002) Volny v sterzhnyakh. Dispersiya. Dissipatsiya (Nelineitost' (Waves in Pivots. Dispersion. Dissipation. Nonlinearity, in Russ.). Fizmatlit, Moscow

Erofeyev VI (2003) Wave Processes in Solids with Microstructure. World Scientific Publishing, New Jersey-London-Singapore-Hong Kong-Bangalore-Taipei

Fedorov VI (1968) Theory of Elastic Waves in Crystals. Plenum Press, New York

Ghoniem NM, Busso EP, Kioussis N, Huang H (2003) Multiscale modelling of nanomechanics and micromechanics: an overview. Phil Magazine 83(31–34):3475–3528

Gulyaev YV, Lagar'kov AN, Nikitov SA (2008) Metamaterials: basic research and potential applications. Herald of the Russian Academy of Sciences 78(3):268–278

Kolken HMA, Zadpoor AA (2017) Auxetic mechanical metamaterials. RSC Advances 7(9):5111–5129

Krivtsov AM (2007) Deformirovanie i razrushenie tverdykh tels mikrostrukturoi (Deformation and Fracture of Solids with Microstructure, in Russ.). Fizmatlit Publishers, Moscow

Kudryashov NA (2010) Methods of Nonlinear Mathematical Physics (in Russ.). Intellect, Dolgoprudny

Li S, Wang G (2018) Introduction to Micromechanics and Nanomechanics, 2nd edn. World Scientific Co.

Lippman H (1995) Cosserat plasticity and plastic spin. ASME Appl Mech Rev 48(11):753–762

Love AEH (1920) A Treatise on the Mathematical Theory of Elasticity, 3rd edn. The University Press, Cambridge

Maugin GA (1988) Continuum Mechanics of Electromagnetic Solids. Elsevier Science Publisher, Amsterdam

Metrikine A, Askes H (2002) One-dimensional dynamically consistent gradient elasticity models derived from a discrete microstructure. Part 2: Static and dynamic response. European Journal of Mechanics A/Solids 21(4):589–596

Nikolaevskiy VN (1996) Geomechanics and Fluidodynamics, Theory and Applications of Transport in Porous Media, vol 8. Springer, Dordrecht

Ostoja-Starzewski M, Sheng Y, Alzebdeh K (1996) Spring network models in elasticity and fracture of composites and polycrystals. Computational Materials Science 7:82–93

Pavlov IS (2010) Acoustic identification of the anisotropic nanocrystalline medium with non-dense packing of particles. Acoustical Physics 56(6):924–934

Pavlov IS, Potapov AI (2008) Structural models in mechanics of nanocrystalline media. Doklady Physics 53(7):408–412

Pavlov IS, Potapov AI, , Maugin GA (2006) A 2D granular medium with rotating particles. Int J of Solids and Structures 43(20):6194–6207

Potapov AI, Pavlov IS, Maugin GA (1999) Nonlinear wave interactions in 1D crystals with complex lattice. Wave Motion 29(4):297–312

Potapov AI, Pavlov IS, Lisina SA (2009) Acoustic identification of nanocrystalline media. Journal of Sound and Vibration 322(3):564–580

Shining Z, Xiang Z (2018) Metamaterials: artificial materials beyond nature. National Science Review 5(2):131

Suiker ASJ, Metrikine A, de Borst R (2001) Comparison of wave propagation characteristics of the Cosserat continuum model and corresponding discrete lattice models. Int J of Solids and Structures 38:1563–1583

Vakhnenko A (1996) Diagnosis of the properties of a structurized medium by long nonlinear waves. Journal of Applied Mechanics and Technical Physics 37(5):643–649

Vardoulakis I, Sulem J (1995) Bifurcation Analysis in Geomechanics. Blackie Academic and Professional, London

Vasiliev AA, Pavlov IS (2018) Structural and mathematical modeling of Cosserat lattices composed of particles of finite size and with complex connections. IOP Conference Series: Materials Science and Engineering 447:012,079

Voigt W (1887) Theoretische Studien über die Elasticitätsverhältnisse der Kristalle. Abn der Königl Ges Wiss Göttingen 34

Chapter 5
A Diffusion Model for Stimulus Propagation in Remodeling Bone Tissues

Ivan Giorgio, Ugo Andreaus, Faris Alzahrani, Tasawar Hayat, and Tomasz Lekszycki

Abstract The mechanically driven biological stimulus in bone tissues regulates and controls the action of special cells called osteoblasts and osteoclasts. Different models have been proposed to describe the important and not yet completely understood phenomena related to this 'feedback' process. In Lekszycki and dell'Isola (2012) an integro-differential system of equations has been studied to describe the remodeling process in reconstructed bones where the biological stimulus in a given instant t depends on the deformation state of the tissue at the same instant. Instead biological knowledge suggests that the biological stimulus, once produced, is 'diffused' in bone tissue to reach the target cells. In this paper, we propose a model for de-

I. Giorgio
Department of Structural and Geotechnical Engineering, Università di Roma La Sapienza, 18 Via Eudossiana, Rome
International Research Center for the Mathematics and Mechanics of Complex Systems - M&MoCS, Università dell'Aquila, L'Aquila, Italy.
e-mail: ivan.giorgio@uniroma1.it

U. Andreaus
Department of Structural and Geotechnical Engineering,
Università di Roma La Sapienza, 18 Via Eudossiana, Rome, Italy
e-mail: ugo.andreaus@uniroma1.it

F. Alzahrani
NAAM Research Group, Department of Mathematics, King Abdulaziz University, Jeddah 21589, Saudi Arabia
e-mail: faris.kau@hotmail.com

T. Hayat
Department of Mathematics, Quaid-I-Azam University, Islamabad, Pakistan
NAAM Research Group, Department of Mathematics, King Abdulaziz University, Jeddah 21589, Saudi Arabia
e-mail: pensy_t@yahoo.com

T. Lekszycki
Faculty of Production Engineering, Warsaw University of Technology, Narbutta 85, 02-524 Warsaw, Poland
e-mail: t.lekszycki@wip.pw.edu.pl

© Springer Nature Switzerland AG 2019
H. Altenbach et al. (eds.), *Higher Gradient Materials and Related Generalized Continua*, Advanced Structured Materials 120,
https://doi.org/10.1007/978-3-030-30406-5_5

scribing biological stimulus diffusion in remodeling tissues in which diffusive time dependent phenomena are taken into account. Some preliminary numerical simulations are presented which suggest that this model is promising and deserves further investigations.

Keywords: Mechanical–biological coupling · Bone functional adaptation · Growth and resorption processes · Bone remodeling

5.1 Introduction

Modeling growth of soft and hard biological tissues relates with very complex physical, chemical, and mechanical phenomena and represents a challenging task for continuum physics and mechanics (see, *e.g.*, Cowin, 2001; Holzapfel and Ogden, 2006; Fung, 2006; Taber, 2009, 1995; Ganghoffer, 2012; Prakash et al, 2018). There are many different approaches which have been proposed to describe the complex coupled phenomena occurring in growing living tissues. All of them, to our knowledge, assume that the biological stimulus to growth or resorption in a given material particle depends either on the local value of deformation fields or on a space average in its neighborhood. While, we are aware of the fact that stimulus formation is not yet a completely understood phenomenon, it seems evident (see, *e.g.*, Pinson et al, 2000; Gong et al, 2001; Kühl et al, 2000) that the production of the signal involves some biochemical factors which are diffusing inside the tissue which is being remodeled. Therefore, it is natural to imagine a process of stimulus formation which, in a first time, involves the generation of a signal in a given material particle and a given time instant, and which subsequently presupposes a signal diffusion in the remodeling tissue.

Due to the difficulty of describing such signal and stimulus processes and their coupling with mechanical phenomena, different simplified models of growing continua were proposed and studied (see, *e.g.*, Allena and Cluzel, 2018; Cluzel and Allena, 2018; Di Carlo and Quiligotti, 2002; Epstein and Maugin, 2000; Garikipati et al, 2006; Goriely et al, 2008; Imatani and Maugin, 2002; Menzel, 2005; George et al, 2018). Some extended theories were applied involving generalizations of the theory of mixtures (Ambrosi et al, 2010; Ateshian, 2007; Franciosi et al, 2019; Spagnuolo et al, 2017), the micropolar continua (Goda et al, 2014; Park and Lakes, 1986; Yang and Lakes, 1982; Yoo and Jasiuk, 2006; Diebels and Steeb, 2003), see also Eremeyev et al (2013); Altenbach and Eremeyev (2015); Eremeyev and Pietraszkiewicz (2016); Yeremeyev and Zubov (1999); Altenbach and Eremeyev (2014); Eremeyev et al (2015); Altenbach et al (2015), functionally graded material (Altenbach and Eremeyev, 2009) or even the second gradient continua both for the solid constituents (Madeo et al, 2012; Giorgio et al, 2017a) and for the interstitial fluid (Seppecher, 2000, 1993; Camar-Eddine and Seppecher, 2001) as well as higher gradient materials (dell'Isola et al, 2012; Niiranen et al, 2019; Niiranen and Niemi, 2017). In particular, during the last decades the progress in modeling bone mechanics has been very impressive. Nowadays, it is commonly accepted that the mechanical loading plays a key role in bones growth. From a conceptual point of view the analysis is approached by means of a logical distinction of the process of stimulus generation and signal processing inducing stimulus, while mechanical deformation phenomena are modeled with more or less standard continua: the coupling between biological and mechanical phenomena are obtained in a twofold way. The biological stimulus induces a change in the mechanical constitutive equations while the deformation state of the tissue determines its biological activity. Therefore living bone tissue growth is determined by a specific factor which we call the *biological stimulus*: the mechanically driven biological stimulus in bone tissues regulates and controls the action of special cells called osteoblasts and osteoclasts (Turner, 1991). Many investigations have been dedicated to the aim of understanding all phenomena involved in stimulus generation and stimulus activity. Its production, propagation, effects on active or capable of being activated cells which are acting in bone remodeling process (see, *e.g.*, Komori, 2013; Bonewald and Johnson, 2008; Vatsa et al, 2008; van Hove et al, 2009; Dallas and Bonewald, 2010; Lemaire et al, 2010; Sansalone et al, 2013) have been intensively studied.

We believe that it is reasonable, see (Chen et al, 2005; Mlodzik, 2002), to formulate a model for stimulus generation and propagation which is not too explicitly related to one or another aspect of considered biological and mechanical phenomena and is, therefore, flexible enough to catch their overall features. In this spirit, in Lekszycki and dell'Isola (2012); Giorgio et al (2016) integro-differential system of equations has been proposed and studied to describe the remodeling process in reconstructed bones where the biological stimulus in a given instant t depends on the deformation state of the tissue at the same instant, although the deformation state of a material point of a tissue may influence the stimulus generation in its whole neighborhood. Actually this assumption implies an instantaneous transmission of the signal originating from sensor cells to the active cells.

Instead, biological knowledge, and evident logical considerations, suggest that the stimulus, once produced, is "diffused" in bone tissue to reach the target cells (see, *e.g.*, Carpentier et al, 2012; Stern and Nicolella, 2013; Bonucci, 2009; Himeno-Ando et al, 2012; George et al, 2017; Spingarn et al, 2017), with mechanisms involving release of "signaling chemical species," their diffusion troughs the reconstructing tissue and their adsorption in the pro-active cells.

In this paper, by refraining to describe in a detailed way such complex and not yet clear enough mechanisms, we propose a model for describing biological stimulus diffusion in tissue remodeling in which diffusive time delay phenomena are taken into account. Actually, we simply propose to assume that the stimulus propagate in space and time following the Fourier–Fick diffusion balance law in which a source and a sink term appear. These source and sink are given with some constitutive equations respectively given in terms of deformation energy density field and stimulus itself. Indeed, the production of stimulus we heuristically associate to the local value of the deformation energy while the metabolic action of the living tissue, determining the degradation of the stimulus carrier, we assume can be described with a simple decay mechanism driven by the stimulus concentration.

While being aware that in this way a whole biological complexity is completely neglected, we are confident that the macroscopic averaged fields, which are introduced and coupled, are still capable to take into account the overall effects we aim to describe.

Some preliminary numerical simulations, fully described in the sequel, were performed, which suggest that this model is promising and deserves further investigations.

5.2 Accepted Assumptions and Main Variables

In order to proper describe the biological phenomena occurring in a bone tissue, we start from the fundamental observation that, for a living bone tissue porous material, the mass density (and, in a more general context, the internal microstructure) can change in time and in space in response to mechanical external loads which drive the biological activities of appropriate cells in a self-organization process (Roux, 1895). Indeed, the bone remodeling can be interpreted as a feedback phenomenon in the framework of the control theory (see Frost, 1987; Turner, 1991). In this biological process, the material properties of stiffness and strength can be considered as controlled quantities, while the amount and distribution of bone mass can be considered as controlling quantity. In fact, based on the external request for mechanical resistance, some cells, *i.e.*, process actuators, can resorb or synthesize bone tissue. Their action aims to optimize the distribution of bone mass to maintain an appropriate level of deformation in the tissue, and/or rearrange the internal bone microstructure efficiently. In other words, bone structure is the result of a functional adaptation which is realized in an optimum balance between the cost of excessive bone mass and the cost of excessive bone fragility. Another type of cell, *i.e.*, sensor, acquires a feedback signal representative of the current mechanical state. This signal is compared with a reference value, *i.e.*, a set point, and the difference, *i.e.*, error, is used to activate the actuator cells. The whole process can be described using the diagram in Fig. 5.1.

The main hypotheses standing at the basis of the mathematical model of the bone remodeling process can be summarized as follows

- Bone tissue is treated regardless of the particular type of its character, *i.e.*, woven, trabecular or compact bone, *etc*. The remodeling occurs at different rates and proportion for different kind of bones but we assume that the mechanism of feedback control is pervasive in all types of bone.
- A characteristic time of about 120–200 days is assumed for the remodeling process, the longest period being more suited for trabecular bone. This time represents the average period of a single cycle of bone remodeling responsible for bone turnover (Agerbaek et al, 1991; Eriksen, 2010);
- Bone tissue is supposed to be a non-linear elastic or viscoelastic porous material with internal microstructure regulated by the activity of three types of cells, namely *osteoblasts*, *osteoclasts*, and *osteocytes*;

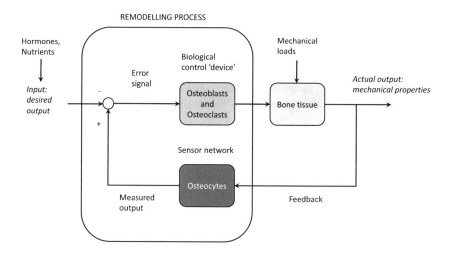

Fig. 5.1 Feedback control representation of the remodeling process.

- *Osteoblast* and *osteoclast* are the cells responsible for the synthesis or resorption of bone tissue, respectively. Their activities are regulated by a stimulus originated from the *osteocytes*, *i.e.* the sensor cells;
- *Osteocytes* are responsible for a feedback signal which we assume mainly originated from the density of strain energy because it can be thought as a good indicator of mechanical state;
- The number of sensor cells present in a 'material particle' of considered porous continuum is proportional to the apparent density of the bone tissue;
- The signal sent from osteocytes to the actuator cells decreases with their distance (Mullender et al, 1994);
- The actuator cells receive and integrate all signals send by surrounding sensor cells which reach them;
- The *reference signal* or set point is assumed to be regulated by hormone activity or by the presence of nutrients (Bednarczyk and Lekszycki, 2016; Lu and Lekszycki, 2018);
- The number of the actuator cells in a given place depends on the porosity of host material. This number approaches zero in the neighborhood of maximum porosity and of porosity absence. Indeed, the activity of these cells is carried out on the inner surface of pores, therefore if such a surface on which the actuator cells can deposit or resorb bone is too small their activity is prevented. As a result, we can assume that there exists an optimal value of porosity associated with the largest number of actuator cells, namely the maximum porous surface available, and in turn, with the highest level of their activities (Martin, 1984; Beaupre et al, 1990a).

All listed assumptions of the considered bio-mechanical system will be incorporated into the proposed model which we present in the following sections and apply
to perform numerical simulations in order to show the potentialities of such a formulation.

Considering a living bone-tissue, the following Eulerian fields, which are all evaluated at the position x in the three-dimensional Euclidean space $\mathcal{E}$ and at the time
instant t (see Lekszycki and dell'Isola, 2012; Andreaus et al, 2014, for details) are
introduced:

1. the macroscopic or "apparent" mass density of bone-tissue $\rho(x, t)$ *i.e.* the mass
   density referred to the whole volume of the body which is inclusive of empty
   and full zones;
2. the porosity of the bone tissue $n(x, t)$, *i.e.* the Eulerian volume fraction which
   is not occupied by bone tissue;
3. the stimulus $S(x, t)$, *i.e.* a scalar quantity which measures the activation signal
   collected at (x, t) by the actor cells, that is osteoblasts, responsible for synthesis
   of bone tissue, and osteoclasts, responsible for resorption of bone tissue;
4. the volume strain-energy density $\mathcal{U}(x, t)$, *i.e.* the energy needed to deform the
   elementary volume of the bone tissue from the free-stress configuration to the
   current one;
5. the density of active sensor cells $d_{oc}(x, t)$, *i.e.* the number of cells per unit
   volume, called osteocytes, which effectively measure the mechanical state and
   send a signal that causes a particular physical response of the actuator cells.

In addition to introduced above four scalar fields $\rho(x, t)$, $S(x, t)$, $\mathcal{U}(x, t)$,
$d_{oc}(x, t)$, we have to present the kinematics of bone tissue which is assumed as
in the finite elasticity of media with microstructure. We introduce the reference $\mathcal{C}^*$
and the actual $\mathcal{C}_t$ configurations which describe the initial and actual state of the
bone tissue in $\mathcal{E}$, respectively. Thus, the deformation of the bone is the mapping χ
given by the differentiable invertible vector-function

$$x = \chi(X, t), \tag{5.1}$$

where x is the position of the material particle of the bone tissue at instant t while
X is the initial position of the same particle. We use standard notations

$$F = \mathrm{Grad}\chi; \qquad J = \det(F), \qquad 2G = F^T F - I$$

where F is called deformation gradient, J its determinant, describing the volume
variation between Lagrangian and Eulerian configurations, $\mathrm{Grad}(\cdot)$ is the gradient
operator in Lagrangian coordinates, G is the Green-Lagrange strain tensor, and I is
the 3D unit tensor, respectively.

In what follows using Eq. (5.1) we will use the Lagrangian description of the
bone tissue. Therefore, we define the following Lagrangian fields

$$\varrho(\boldsymbol{X},t) = J\rho(\chi(\boldsymbol{X},t),t),$$
$$\varphi(\boldsymbol{X},t) = Jn(\chi(\boldsymbol{X},t),t),$$
$$\mathfrak{S}(\boldsymbol{X},t) = S(\chi(\boldsymbol{X},t),t),$$
$$d_{OC}(\boldsymbol{X},t) = Jd_{oc}(\chi(\boldsymbol{X},t),t),$$
$$\mathscr{W}(\boldsymbol{X},t) = J\mathscr{U}(\chi(\boldsymbol{X},t),t).$$

We remark that the transformation from the Eulerian to the Lagrangian framework are mapped with $J = \det(\boldsymbol{F})$ when we consider volume densities, but this not the case for the so-called stimulus $\mathfrak{S}$ because it is an integral quantity defined on all the considered domain. From the principle of material frame-indifference, it follows that $\mathscr{W}$ depends on $\chi(\boldsymbol{X},t)$ via strain tensor $\boldsymbol{G}$ only.

At this level of the bone tissue description, we introduce, following a "homogenization" procedure, a new kinematical descriptor producing, after suitable averages, an expression for deformation energy in terms of 'microscopic' phenomena occurring at a smaller scale. In the framework of poromechanics (see, *e.g.*, Lurie et al, 2018a,b; Khalili and Selvadurai, 2003; Misra et al, 2015, 2013), and according to the Biot model, we consider the field $\zeta(\boldsymbol{X},t)$ defined as the change of the effective volume of the fluid content per unit volume—accounted for by the change of the Lagrangian porosity; thus, it is then given by

$$\zeta(\boldsymbol{X},t) = \varphi(\boldsymbol{X},t) - \varphi^{*}(\boldsymbol{X},t), \qquad (5.2)$$

where the apex * denotes the reference configuration with no stress.

The introduced reference configuration is the domain in which all fields describing the remodeling process under study will be defined. It is most suitable at least from the computational point of view and may be preferred to any kind of Eulerian configuration, as this last is, in general, a time-varying domain.

5.3 Poromechanical Formulation

From a mechanical point of view, we assume that the Lagrangian elastic energy density depends on the Green-Lagrange strain tensor, the change of porosity and eventually on the considered material particle $\boldsymbol{X}$

$$\mathscr{W} = \mathscr{W}(\mathbf{G},\zeta,\boldsymbol{X}). \qquad (5.3)$$

We explicitly note that the assumed deformation energy does not depend explicitly on time and that, differently to what happens in standard continuum mechanics, the Lagrangian mass density is evolving with time. However, these changes are assumed to be slow enough so that possibly inertia effects are negligible so that the initial rate of change of mass densities and the consequent transient phenomena are not accounted for in the considered model.

We focus our attention on a particular case of the constitutive equation (5.3) taking into account the porous nature of the bone. The explicit expression of the strain-energy density can be written as follows

$$\mathscr{W}(\mathbf{G}, \zeta, \boldsymbol{X}) = \mu \operatorname{tr}\left(\mathbf{G}^2\right) + \frac{\lambda}{2}\left(\operatorname{tr}\mathbf{G}\right)^2 + \frac{1}{2}K_c\,\zeta^2 - K_{cp}\,\zeta\operatorname{tr}\mathbf{G} + \frac{1}{2}K_{nl}\left(\nabla\zeta\right)^2,$$
(5.4)

where we assumed the following formulas for the Lamé moduli:

$$\mu = \hat{\mu}(\varphi^*, \boldsymbol{X}); \qquad \lambda = \hat{\lambda}(\varphi^*, \boldsymbol{X}).$$

The last equations are usually made explicit as:

$$\mu = \mu_0\left(1 - \varphi^*\right)^\beta = \mu_0\left(\frac{\varrho^*}{\varrho_{\text{Max}}}\right)^\beta; \qquad \lambda = \lambda_0\left(1 - \varphi^*\right)^\beta = \lambda_0\left(\frac{\varrho^*}{\varrho_{\text{Max}}}\right)^\beta,$$
(5.5)

where μ_0, λ_0 may depend on $\boldsymbol{X}$ for non-homogeneous material, and ϱ_{Max} is the maximum bone density. These expressions for the Lamé moduli correspond to a constant Poisson ratio, ν, and the Young modulus given by

$$Y = Y_0\left(1 - \varphi^*\right)^\beta.$$
(5.6)

Following Giorgio et al (2016) we use the value $\beta = 2$ and Y_0 is related with the Young modulus of compact bone. The power low (5.6) is widely used for modeling of cellular solids, see Gibson and Ashby (1997); Ashby et al (2000). The compressibility stiffness K_c can be evaluated using the drained bulk modulus of the porous matrix, $K_{dr} = Y/(3(1 - 2\nu))$, and the bulk modulus of the fluid inside pores, K_f, by the formula

$$K_c = \left(\frac{\varphi^*}{K_f} + \frac{(\alpha_B - \varphi^*)(1 - \alpha_B)}{K_{dr}}\right)^{-1},$$
(5.7)

where $\alpha_B \in [\varphi^*, 1]$ is the Biot–Willis coefficient. The coupling parameter K_{cp} related to the interaction between the macro- and the micro-deformation is assumed

$$K_{cp} = \sqrt{\hat{g}(\varphi^*)\,\lambda\,K_c},$$
(5.8)

where the function $\hat{g}(\varphi^*)$ is assumed to be

$$\hat{g}(\varphi^*) = \frac{A_{k_3}}{\pi}\left\{\operatorname{atan}\left[s_{k_3}\left(\varphi^* - \frac{1}{2}\right)\right] + \operatorname{atan}\left(\frac{s_{k_3}}{2}\right)\right\},$$
(5.9)

in which $A_{k_3} \in (0, 1]$ and s_{k_3} are suitable material coefficients. The parameter K_{nl} is a modulus related to the non-local interactions between neighborhood pores and is assumed to be constant for the sake of simplicity. It is worth noting that the last term in Eq. (5.4) which takes into account the gradient of the change of porosity allows us to also apply boundary conditions on the porosity which otherwise are not sustainable. Some studies have been proposed to better characterize the complex behavior of systems as bone tissue (see, *e.g.*, Li et al, 2019; Scala et al, 2018; Lek-szycki et al, 2017; Misra and Poorsolhjouy, 2015; Abali et al, 2012; Chatzigeorgiou et al, 2014).

In the literature are known more complex descriptions of stress tensors and strain energy density of growing tissues, see for a comprehensive view Amar and Goriely (2005); Ambrosi and Guillou (2007); Di Carlo and Quiligotti (2002); Epstein and Maugin (2000); Garikipati et al (2006); Goriely et al (2008); Imatani and Maugin (2002); Menzel (2005).

Concerning the mechanical governing equations of the considered body we assume that no inertia effects are of relevance, when considering applied load at the time scale of remodeling process. Consistently, we assume that the loads applied to the system are varying slowly in order to assume that a quasi-static deformation process occurs. Therefore, according to the Principle of Virtual Work we can write the following equation

$$\delta \mathfrak{W}^{int} + \delta \mathfrak{W}^{ext} = 0 \tag{5.10}$$

for any portion $\mathcal{B}$ of the considered body. Specifically, the work done by the inside interactions is

$$\delta \mathfrak{W}^{int} = -\int_{\mathcal{B}} \delta \mathscr{W} d\mathcal{V}, \tag{5.11}$$

while the work done by the external loads is

$$\delta \mathfrak{W}^{ext} = \int_{\partial_\tau \mathcal{B}} \tau_i \delta u_i d\mathcal{S} + \int_{\partial \mathcal{B}} \Xi\, \delta\zeta d\mathcal{S}, \tag{5.12}$$

where τ_i, namely the surface traction on the boundary $\partial_\tau \mathcal{B}$ spends work on the displacement u_i, and Ξ, namely a micro-structural action related to the local dilatant behavior of pores acts on the change of porosity ζ.

5.4 Evolutionary Equations for Bone Remodeling

In this section, in order to describe the main features of the bone adaptation process, the classical approach based on a first order ordinary differential equation for the evolution of the bone mass density is employed (see Beaupre et al, 1990b; Mullender et al, 1994; Lekszycki and dell'Isola, 2012). In this framework, the evolution rule for the apparent mass density in the reference configuration is given by

$$\frac{\partial \varrho^*}{\partial t} = \mathcal{A}(\mathfrak{S}, \varphi) \tag{5.13}$$

where the function $\mathcal{A}$ accounts for phenomena of both biological and mechanical nature. The rate of change of the mass density is supposed to depend on the stimulus, $\mathfrak{S}$, which is originated from the osteocytes on the basis of the current mechanical state and upon the specific surface of bone (Martin, 1984) which is available for the resorption or synthesis and, in turn, on the effective porosity. In this context, the effective porosity is that portion of porosity that actually is involved in the remodeling process. The internal architecture of the bone tissue, characterized by different types of porosity (Cowin, 1999), plays a crucial role in this evolutionary phenomenon. Indeed, the main active cells, namely osteoblasts and osteoclasts, can carry out their activity only on the inner porous surface, specifically the inter-trabecular surface for cancellous bone and Haversian canals as well as the endosteal or periosteal surface for cortical bone. Therefore, the porosity guarantees that the bone could be biologically active that is to allow the diffusion of the above-mentioned bone cells, or their precursors, through the bone-tissue to reach the regions where they are needed, and the nutrition supply made available by the vascular network.

A possible explicit form for the function $\mathcal{A}$ is assumed to be

$$\mathcal{A}(\mathfrak{S}, \varphi) = a(\mathfrak{S})H(\varphi), \tag{5.14}$$

where the function $a(\mathfrak{S})$ gives the rate of the mass density as a piece-wise linear function and the function H is a weight term related to the specific surface which gives a geometric feedback from the internal architecture of the bone and therefore, whose shape is linked for physiologic reasons to the porosity. The function $a(\mathfrak{S})$ can be further specified as

$$a(\mathfrak{S}) = \begin{cases} s_{\mathrm{b}}\left(\mathfrak{S}(\boldsymbol{X},t) - P_{\mathrm{ref}}^{\mathrm{s}}\right) & \text{for } \mathfrak{S}(\boldsymbol{X},t) > P_{\mathrm{ref}}^{\mathrm{s}} \\ 0 & \text{for } P_{\mathrm{ref}}^{\mathrm{r}} \leqslant \mathfrak{S}(\boldsymbol{X},t) \leqslant P_{\mathrm{ref}}^{\mathrm{s}} \\ r_{\mathrm{b}}\left(\mathfrak{S}(\boldsymbol{X},t) - P_{\mathrm{ref}}^{\mathrm{r}}\right) & \text{for } \mathfrak{S}(\boldsymbol{X},t) < P_{\mathrm{ref}}^{\mathrm{r}} \end{cases} \tag{5.15}$$

taking into account the so-called "lazy zone" (see, *e.g.*, Beaupre et al, 1990a; Ruimerman et al, 2005; Giorgio et al, 2017b), namely, the interval between the two thresholds $P_{\mathrm{ref}}^{\mathrm{r}}$ and $P_{\mathrm{ref}}^{\mathrm{s}}$ for resorption and synthesis, respectively, in which an home-ostatic physiologic equilibrium persists. The constants s_{b} and r_{b}, possibly different from each other, are the rates of synthesis and resorption, respectively. As a note, we emphasize that the function $a(\mathfrak{S})$ is similar to the kinetic function describing the velocity of a phase interface in the theory of phase transitions, see Abeyaratne and Knowles (2006); Berezovski et al (2008); Engelbrecht and Berezovski (2015); Eremeyev and Pietraszkiewicz (2009, 2011).

The function H is heuristically evaluated considering a cellular square lattice representing schematically the trabecular pattern for different values of porosity as described in Giorgio et al (2016) (see Fig. 5.2). We note that the function H is evaluated as the specific 'effective' surface of trabeculae which actually takes a role in the remodeling process providing the support surface for actuator cells. Therefore, the curve in Fig. 5.2 refers to the trabecular bone. The zone between null porosity and the beginning of the curve refers to cortical bone and there, as a first approximation, we consider a small threshold different from zero in order to avoid inhibiting the remodeling process once the trabecular bone becomes cortical.

Fig. 5.2 Weighting function
$H(\varphi)$.

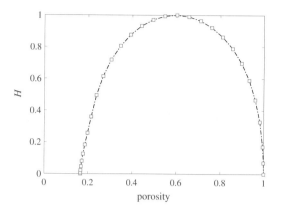

5.5 Stimulus Modeling Without Time Delay and Diffusion Phenomena

The mechanical stimulus is one of the main features characterizing the remodeling process. As it is widely accepted, we assume that all osteocytes located in a given material particle produce a signal which triggers the activity of the osteoblasts and osteoclasts in the neighborhood of the sensor cells. Many possible choices have been proposed for making explicit in a mathematical way the stimulus, some models are based on an energy formulation, others on an effective level of stress or strain, and others again on levels of damage. As proposed in Lekszycki and dell'Isola (2012), we focus our attention on an integral formulation in which the stimulus is represented as a linear functional of the strain energy density $\mathscr{W}$. Indeed, the signal delivered to the actuator cells is the 'sum' of all the contributions stemmed from the surrounding osteocytes. This functional is further characterized by two weight functions, namely the density of osteocytes, d_{OC} (Lekszycki and dell'Isola, 2012), considered, as a first approximation, proportional to their activity, and a spatial influence function, $K(\boldsymbol{X}, \boldsymbol{Y}, t, \tau)$, which takes into account the distances between sensor and actuator cells and therefore the influence zone of the osteocytes, possibly in space and time. This functional can be expressed as follows

$$\mathfrak{S}(\boldsymbol{X}, t) = \int\limits_{0}^{t} \int\limits_{V^*} K(\boldsymbol{X}, \boldsymbol{Y}, t, \tau) \mathscr{W}(\boldsymbol{Y}, \tau) d_{OC}(\boldsymbol{Y}, \tau) \, d\boldsymbol{Y} \, d\tau, \qquad (5.16)$$

where V^* is the volume of the bone tissue in the reference configuration, the integration dummy variable τ stands for time, $\boldsymbol{Y}$ is the generic location of osteocytes and $\boldsymbol{X}$ is the place in which osteoblast or osteoclast act. In the other words, $\mathfrak{S}$ represents a non-local behavior both in time and space, like in the viscoelasticity and the non-local elasticity (see also Carlen et al, 2009). The following constitutive relation for the density of sensor cells can be assumed

$$d_{OC} = \eta\left(1 - \varphi^*\right), \qquad 0 < \eta \le 1, \tag{5.17}$$

where η is a constitutive parameter. This assumption entails that there is a uniform distribution of the sensor cells which is proportional to the volume fraction of the bone.

Under the hypothesis of instantaneous transmission of the signal, *i.e.* the transmission time scale is negligible when compared with the characteristic time of the remodeling phenomenon, the stimulus can be simplified. Its expression becomes

$$\mathfrak{S}(\boldsymbol{X}, t) = \int_{V^*} k(\boldsymbol{X}, \boldsymbol{Y}) \mathscr{W}(\boldsymbol{Y}, t) d_{OC}(\boldsymbol{Y}, t) \, \mathrm{d}\boldsymbol{Y}, \tag{5.18}$$

where the function $k(\boldsymbol{X}, \boldsymbol{Y})$ represents the range of influence of sensor cells with no dependence upon time. This function $k(\boldsymbol{X}, \boldsymbol{Y})$ must be decreasing with the distance between $\boldsymbol{X}$ and $\boldsymbol{Y}$ and in the literature have been proposed (Mullender et al, 1994; Andreaus et al, 2014) both the following functions

$$\begin{aligned} k(\boldsymbol{X}, \boldsymbol{Y}) &= \exp\left\{-D^{-1} \|\boldsymbol{X} - \boldsymbol{Y}\|\right\} \quad \text{or} \\ k(\boldsymbol{X}, \boldsymbol{Y}) &= \exp\left\{-\|\boldsymbol{X} - \boldsymbol{Y}\|^2 / (2D^2)\right\}. \end{aligned} \tag{5.19}$$

Here D is the characteristic length-scale parameter. The reader is invited to remark that the aforementioned influence functions resemble greatly Green function for heat equation in their structure.

5.6 An Improved Version of Stimulus Modeling

Herein, the proposed idea is to replace the integral formulation (5.16) or (5.18) for $\mathfrak{S}$ with a differential equation. This approach has an advantage from the computational point of view because, in commercial FEM programs, various types of partial differential equations are already implemented as well as the convolution integral formulation is avoided. In addition, using such an equation gives the possibility to take into account possible surface effects and consequently various boundary conditions.

Considering the more convenient Lagrangian formulation, a parabolic evolution equation for stimulus $\mathfrak{S}$ is proposed

$$\frac{\partial \mathfrak{S}}{\partial t} = \mathrm{Div}\left(\kappa \nabla \mathfrak{S}\right) + s + r, \tag{5.20}$$

where κ is the permeability to stimulus of considered tissue, which can be a second-order tensor field, in general. The sink s is a density field which describes the resorption of stimulus because of metabolic activity. For instance

$$s = -R\,\mathfrak{S}\,H_v(\mathfrak{S}), \tag{5.21}$$

where R is the constant of stimulus resorption and $H_v(\cdot)$ is the Heaviside function. Clearly, the signal $\mathfrak{S}$ cannot be negative and thus no resorption of it can occurs when it becomes zero. The stimulus driving source r depends on the state of mechanical deformation. One could choose

$$r = \varpi(\varrho^*)\mathscr{W}(\mathbf{G}), \qquad (5.22)$$

where $\varpi(\varrho^*)$ is a suitable weighting function which play the same role of d_{OC}. Specifically, this function can be interpreted from a control point of view as a function of efficiency of the sensor network. In numerical simulations, we have introduced a particular form for this function (see Fig. 5.3) suggested by Giorgio et al (2017c)

$$\varpi(\varrho^*) = \arctan(\xi\varrho^*)\, H_v(\varrho^*), \qquad (5.23)$$

where ξ is a constitutive parameter. The reasoning behind this particular trend is that when a given amount of sensors cells, *i.e.* a suitable amount of bone, is available the sensor network offers the best efficiency, whilst with few cells the acquired signal results degraded. Of course, the absence of cells implies no signal.

We note that in equation (5.22), other possible stimulus sources can be implemented and investigated in future works. For example, a source that takes into account cracks or damage in the bone tissue can be defined, see for more insight on this subject Hambli (2014); Prendergast and Taylor (1994) and also Lu and Lekszycki (2017); Placidi et al (2018b,a); Placidi and Barchiesi (2018); Contrafatto and Cuomo (2006); Cuomo et al (2014). It is also worth noting that taking into account the finite speed of the stimulus propagation, a modification of Eq. (5.20) can be implemented in the following form

$$t_* \frac{\partial^2 \mathfrak{S}}{\partial t^2} + \frac{\partial \mathfrak{S}}{\partial t} = \mathrm{Div}\,(\kappa\nabla\mathfrak{S}) + s + r, \qquad (5.24)$$

where t_* is the characteristic "wave" time.

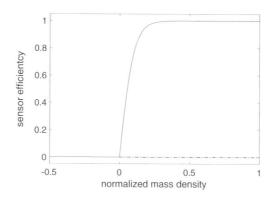

Fig. 5.3 Efficiency function of the sensor network $\varpi(\varrho^*)$.

A possible generalization of the mechanical part of the presented model may include the introduction of internal variable descriptors as those presented in Yildizdag et al (2019); Berezovski et al (2018). The diffusion model which we use in the present paper can be surely improved and better adapted to try to catch the features observed in considered biological systems. Moreover the intrinsic discrete nature of the sensor network system of the osteocytes present in remodeling bone may require a treatment similar to the one presented in Colangeli et al (2016, 2017) in the context of generalizing the diffusive part of presented model Allen–Cahn type equations may be of use: for the may mathematical properties of this equations see De Masi et al (1995, 1986).

5.7 Numerical Simulations

In this section, we show numerical simulations to explore the actual forecasting capability of the proposed model. Particularly, we consider two representative cases in which the mechanical stimulus drives the evolution of the remodeling process in a way that the results are easily comparable with the real behavior of the bone tissue. Firstly, we consider a specimen of bone tissue which undergoes extension tests with different values of the external load exploring the cases in which the deformation is uniform or non-uniform. The test is supposed to explain the physiological functioning of a bone tissue under continuous load over time. Secondly, we consider a sample with a wide area of necrotic tissue and the consequent healing process. Herein, the necrotic tissue is characterized by the absence of osteocytes, and therefore no stimulus is originated from this area.

All the numerical simulations are performed by means of a commercial software, namely COMSOL Multiphysics®, which is based on the Finite Element Method. Although the formulation presented is applicable to the general case of a three-dimensional sample, in this paper we study a two-dimensional problem with a very simple geometry to avoid introducing unnecessary complexities that make the results obtained difficult to interpret and to understand. The numerical methods used in this paper are absolutely standard. We exploited the great potentiality and flexibility of COMSOL Multiphysics®. Of course the code obtained in this way cannot be expected to be an optimized one. However we believed that before developing ad hoc integration schemes and numerical codes it was necessary to get some preliminary results motivating such an effort. These positive preliminary results have been indeed obtained so that we intend to formulate adapted integration schemes and more efficient numerical code, being inspired by the methodologies exploited in Yildizdag et al (2018); Cuomo et al (2014); Cazzani et al (2016b,a); Balobanov et al (2016).

5.7.1 A Physiological Case

A rectangular sample with aspect ratio 2:1 of cancellous bone is tested. The length of the specimen is assumed to be $L = 1$ cm. The initial porosity is 0.5 and uniform over the sample. The material parameters characterizing the sample used in the simulations are: maximum bone Young's modulus $Y_0 = 17$ GPa; Poisson's ratio $\nu = 0.3$; the real density of the bone $\varrho_{Max} = 1800$ kg/m^3; the stiffness of the marrow $K_f = 0.1 Y_0$; the porous non-local stiffness $K_{nl} = 1.7 \times 10^5$ N, the synthesis and resorption rates $s_b = 4.9 \times 10^{-15}$ s/m^2, $r_b = 6.13 \times 10^{-15}$ s/m^2, respectively; the reference thresholds for stimulus $P^s_{ref} = 3.4 \times 10^4$ kN/m^2 and $P^r_{ref} = 3.4 \times 10^3$ kN/m^2.

5.7.1.1 Uniform Tension Test

The first tests are performed with boundary conditions selected to produce a uniform distribution of elastic energy density on the entire specimen, namely the longitudinal displacement of a short edge is fixed as well as the transversal displacement of a point of the same side (to avoid rigid motions) whereas a load uniform in space and constant in time is applied on the opposite side in order to produce an extension (see Fig. 5.4). It is worthy of remark that, limiting our attention to only one relevant time scale, a load constant in time should be considered in the analyzed context as an effective action which, in an average sense, is continuously applied. Five experiments are carried out with as many levels of applied force. The level of the forces are selected in order to obtain a homeostatic equilibrium (in Fig. 5.5 refer to the label '1'), a growth —increasing such a value of 1.3 and 1.5 (labels in Fig. 5.5)—, and a resorption —decreasing the homeostatic force by a multiplicative factor of 0.4 and 0.3 (labels in Fig. 5.5). The amplitude of the external load at the homeostatic state is $3.1 \times 10^{-3} Y_0 L$.

The Fig. 5.5 displays the evolution of the apparent mass density of the bone in a probe point located at the center of the sample. The values of the mass density are normalized by the real mass density of the bone, while the time is normalized by the characteristic time (t_{ref}) of 200 days. Because the distribution of the energy density

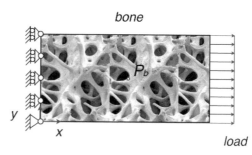

Fig. 5.4 Physiological case under a uniform extension test.

Fig. 5.5 Apparent bone mass
density at the center of the
sample with the proposed
stimulus model.

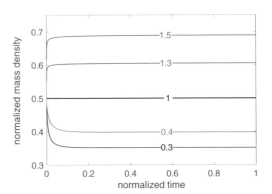

is uniform, we observe that also the stimulus evaluated by the equation (5.20) is
uniform, and therefore all the points of the sample share the same evolution. From
the graphs, we notice that:

1. the higher is the level of force, the greater is the growth of the mass density;
2. for a certain value of the applied force we have an homeostatic equilibrium;
3. whilst for low values of the forces (less than those homeostatic), the smaller is
   the level, the greater is the resorption.

We note also that the transient behavior of the growth and resorption can be identified by the same characteristic time for the different values of forces, being the
process of resorption somewhat slower than the growth.

The same numerical experiments are also conducted with the integral model
of the stimulus (5.18) specifying the influence spatial function k as in the equation $(5.19)_2$ and using an influence range length $D = 0.25L$. Figure 5.6 shows the
associated results in terms of apparent mass density with the same setting of the
previous experiments. Comparing the results of Figs. 5.5 and 5.6, we observe that

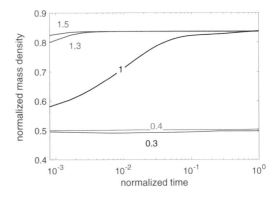

Fig. 5.6 Apparent bone mass
density at the center of the
sample with the integral
model of stimulus. The normalized time is represented in
logarithmic scale.

the evolutions at the end of the process are quite different. Indeed, while in the integral formulation the final result is always a limit case (equilibrium or saturation of the level of porosity), in the new formulation we have different levels of porosity achieved with different values of the load. In the integral formulation the effect of different levels of force turns in a different characteristic time: the larger is the force, the faster is the process. We also note that the stimulus in the integral formulation is not uniform, thus the onset of non-physiological border effects are present (see for more details on this issue Giorgio et al, 2016). Clearly, the results of the new model are more consistent with the real distribution of the porosity in bone tissue. Indeed, according with one of the rules of the bone adaptation (Turner, 1998): "*Bone cells accommodate to a customary mechanical loading environment, making them less responsive to routine loading signals*", we obtain that even if the loading continues to be applied the remodeling process tends to become less responsive. On the contrary, the integral formulation predicts that if the bone is constantly loaded the possible values of the porosity could be associated with a cortical bone, an empty area or the homeostatic state. Possible deviations from this scenario are due to a transient regime in which the level of forces varies. Although such situations are possible, the integral model shows all its limitations.

5.7.1.2 Non-uniform Tension Test

To test the effect of a non-uniform distribution of the deformation energy density, we applied to the same sample a linear distribution of force (see Fig. 5.7). Figure 5.8 shows the mechanical stimulus at the beginning of the process. As it is desirable, the distribution of the stimulus over the sample follows that of the energy density. This behavior of the stimulus is mainly due to the permeability constant κ. Indeed, this parameter drives the rapidity of the diffusion process (if it is a tensor quantity also the direction). In other words, if κ is excessively large the transient due to diffusion is too fast and therefore the non-uniform space distribution of the stimulus becomes very soon negligible. For this reason, the value of the permeability κ should be carefully identified (in the simulations it is $\kappa = 1.0 \times 10^{-4} L^2/t_{\mathrm{ref}}$) because otherwise the distribution of the mass density never will be affected by this non-uniform initial distribution as it is observable in real cases. Figures 5.9 and 5.10 display, respectively, in two probe points the history of mass density, and its distribution in two

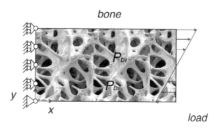

Fig. 5.7 Physiological case under a non-uniform extension test.

Fig. 5.8 Mechanical stimulus at the beginning of the non-uniform tension test.

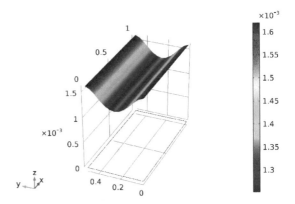

Fig. 5.9 Evolution of the apparent bone mass density in two probe points located at the center P_{b1} and near a long side of the sample P_{b2} for the non-uniform tension test. The normalized time is represented in logarithmic scale.

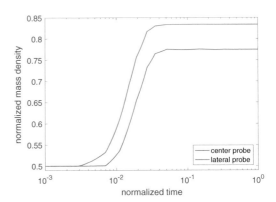

notable time steps, namely an intermediate (2 days) and the final ones (200 days); these results give a clear demonstration of the effect of a non-uniform distribution of the energy density and consequently of the stimulus.

5.7.2 Simulation of a Healing Process

Similarly to a previous example, in what follows we study a rectangular sample of cancellous bone which undergoes a uniform extension test. Also in this case we consider an initial porosity equals to 0.5, but now the specimen presents all the right part characterized by the lack of osteocytes as a result of injury or some type of disease (see Fig. 5.11). Therefore, in this zone we set the initial stimulus, $\mathfrak{S}$, equal to zero. During the evolution in the necrotic area, the function $\varpi(\varrho^*)$ which is related to the activity of the osteocytes, is set to zero and it is activated only when new bone

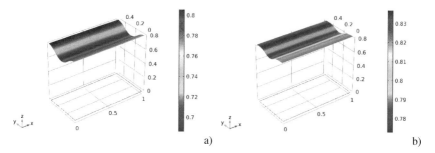

Fig. 5.10 Normalized mass density in non-uniform tension test. a) intermediate stage; b) final stage.

Fig. 5.11 Healing process
under an extension test.

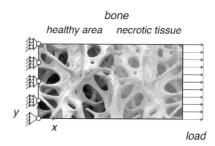

is produced since this entail that new osteocytes have formed. The amplitude of the external load is $5.5 \times 10^{-3} Y_0 L$.

Figure 5.12 shows the distribution of the apparent mass density of the bone in four representative stages of the remodeling process. After the initial condition in which the porosity is uniformly equal to 0.5, we note a first evolution in which the healthy area is subject to growth, then new bone is synthesized in the necrotic area resulting in the birth of new osteocytes that colonize the diseased area. Finally, the entire sample achieve the same uniform porosity and full functionality.

These results show clearly that the proposed model is able to capture the non-local behavior of the stimulus. Indeed, even if at the beginning the necrotic area cannot undergo any evolution, the signal collects by the osteocytes in the healthy area can diffuse in the necrotic area and activate the osteoblasts which, for the sake of simplicity, we suppose already migrated in that area thanks to the porous nature of the tissue.

5.8 Conclusions

In this paper, we have proposed a new model for the generation and propagation of the biological stimulus responsible for the feedback process occurring in the bone

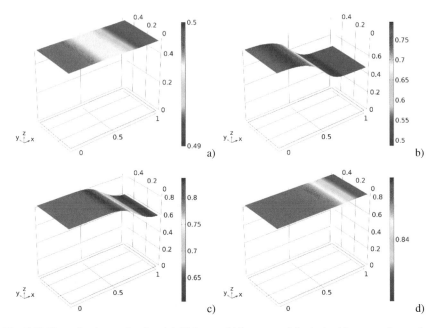

Fig. 5.12 Normalized mass density. a) initial stage; b) bone growth in the healthy zone; c) growth of new bone in the necrotic zone; d) final stage.

remodeling. Specifically, we propose to model the stimulus space/time propagation by using the diffusion equation (Fourier's equation for the heat, Fick's equation for the concentration of chemical species in a fluid) instead of using an integral formulation as done in the previous papers (see Lekszycki and dell'Isola, 2012; Giorgio et al, 2016). Moreover the stimulus generation and resorption are assumed to be driven by deformation energy density and a standard decay rule respectively. This proposal introduces a characteristic time related to the 'diffusion' of the stimulus differently from the previously developed model which neglects any time delay associated to the transient in the evolution of the stimulus.

Herein, we analyze two illustrative cases to characterize the behavior of the stimulus:

1. a *physiological case*, where a specimen of bone tissue undergoes an extension test with different levels of the external load;
2. a *healing process*, where a sample presenting a large area of necrotic tissue is mechanically loaded.

Mathematically speaking, necrosis is described by the absence of generation of stimulus. This means that in the source term for diffusion equation of stimulus the coefficient depending on osteocyte concentration is vanishing. Instead in a healthy bone any material particle is biologically active and therefore generates a source of stimulus proportional to the local deformation energy. In the present paper we assume that the density of osteocytes is proportional to the bone density in physiological case while it is vanishing in necrotic tissues. We will extend this model by introducing an evolution equation for the concentration of osteocytes in future investigations. More specifically, we test the sample in the physiological case with a uniform tension test with different values of the load in order to simulate the different evolutions of the mass density of the bone. The numerical results clearly show that there is a range of values of the external load which corresponds to a physiological homeostatic equilibrium, while for values grater than those related to this equilibrium we have a growth of the bone tissue and for values lesser a resorption occurs, instead. Besides, a test performed with a non-uniform load, *i.e.* with a linear distribution of the external mechanical action, displays a non-uniform evolution of the mass density of the bone tissue which resembles the distribution of the deformation energy density in the subsequent equilibrium configurations attained. Eventually, the healing process, after necrosis is simulated. The numerical results exhibit a reasonable progress of the evolutionary history of the mass density of the bone in which we can distinguish three characteristic stages: in the first place the bone tissue grows in the healthy zone; then new bone tissue is synthesized and goes to occupy the 'sick' area (where the osteocytes are supposed to be died and the inert bone behaves as a scaffold); in the end the whole sample regains the full functionality given that, with the newly reformed bone tissue, also new osteocytes colonize the initially necrotic zone. The qualitative behavior of the proposed model closely matches, from a qualitative point of view, the actual remodeling process observed in living bone tissue. Indeed, we can explain different levels of porosity with different levels/kinds of external load as well as the non-local behavior of the biological phenomenon. For a better insight in the bone adaptive process, the effect of physiological parameters can be examined in detail employing this model and eventually compared with experimental evidences. The results selected show the potentiality of our proposal but further validation of the diffusive model for the stimulus is however necessary.

References

Abali BE, Völlmecke C, Woodward B, Kashtalyan M, Guz I, Müller WH (2012) Numerical modeling of functionally graded materials using a variational formulation. Continuum Mechanics and Thermodynamics 24(4-6):377–390

Abeyaratne R, Knowles JK (2006) Evolution of Phase Transitions. A Continuum Theory. Cambridge University Press, Cambridge

Agerbaek MO, Eriksen EF, Kragstrup J, Mosekilde L, Melsen F (1991) A reconstruction of the remodelling cycle in normal human cortical iliac bone. Bone and mineral 12(2):101–112

Allena R, Cluzel C (2018) Heterogeneous directions of orthotropy in three-dimensional structures: finite element description based on diffusion equations. Mathematics and Mechanics of Complex Systems 6(4):339–351

Altenbach H, Eremeyev V (2009) Eigen-vibrations of plates made of functionally graded material. Computers, Materials, & Continua 9(2):153–178

Altenbach H, Eremeyev V (2015) On the constitutive equations of viscoelastic micropolar plates and shells of differential type. Mathematics and Mechanics of Complex Systems 3(3):273–283

Altenbach H, Eremeyev VA (2014) Vibration analysis of non-linear 6-parameter prestressed shells. Meccanica 49(8):1751–1761

Altenbach H, Eremeyev VA, Naumenko K (2015) On the use of the first order shear deformation plate theory for the analysis of three-layer plates with thin soft core layer. ZAMM-Journal of Applied Mathematics and Mechanics/Zeitschrift für Angewandte Mathematik und Mechanik 95(10):1004–1011

Amar MB, Goriely A (2005) Growth and instability in elastic tissues. Journal of the Mechanics and Physics of Solids 53(10):2284–2319

Ambrosi D, Guillou A (2007) Growth and dissipation in biological tissues. Continuum Mechanics and Thermodynamics 19(5):245–251

Ambrosi D, Preziosi L, Vitale G (2010) The insight of mixtures theory for growth and remodeling. Zeitschrift für angewandte Mathematik und Physik 61(1):177–191

Andreaus U, Giorgio I, Lekszycki T (2014) A 2-D continuum model of a mixture of bone tissue and bio-resorbable material for simulating mass density redistribution under load slowly variable in time. ZAMM-Journal of Applied Mathematics and Mechanics/Zeitschrift für Angewandte Mathematik und Mechanik 94(12):978–1000

Ashby MF, Evans AG, Fleck NA, Gibson LJ, Hutchinson JW, Wadley HNG (2000) Metal Foams: a Design Guid. Butterworth-Heinemann, Boston

Ateshian GA (2007) On the theory of reactive mixtures for modeling biological growth. Biomechanics and Modeling in Mechanobiology 6(6):423–445

Balobanov V, Khakalo S, Niiranen J (2016) Isogeometric analysis of gradient-elastic 1D and 2D problems. In: Altenbach H, Forest S (eds) Generalized Continua as Models for Classical and Advanced Materials, Advanced Structured Materials, vol 42, Springer, Cham

Beaupre GS, Orr TE, Carter DR (1990a) An approach for time-dependent bone modeling and remodeling—application: A preliminary remodeling simulation. Journal of Orthopaedic Research 8(5):662–670

Beaupre GS, Orr TE, Carter DR (1990b) An approach for time-dependent bone modeling and remodeling—theoretical development. Journal of Orthopaedic Research 8(5):651–661

Bednarczyk E, Lekszycki T (2016) A novel mathematical model for growth of capillaries and nutrient supply with application to prediction of osteophyte onset. Zeitschrift für angewandte Mathematik und Physik 67(4):94

Berezovski A, Engelbrecht J, Maugin GA (2008) Numerical Simulation of Waves and Fronts in Inhomogeneous Solids. World Scientific, New Jersey et al.

Berezovski A, Yildizdag ME, Scerrato D (2018) On the wave dispersion in microstructured solids. Continuum Mechanics and Thermodynamics pp 1–20, DOI 10.1007/s00161-018-0683-1

Bonewald LF, Johnson ML (2008) Osteocytes, mechanosensing and Wnt signaling. Bone 42(4):606–615

Bonucci E (2009) The osteocyte: the underestimated conductor of the bone orchestra. Rendiconti Lincei 20(3):237–254

Camar-Eddine M, Seppecher P (2001) Non-local interactions resulting from the homogenization of a linear diffusive medium. Comptes Rendus de l'Academie des Sciences Series I Mathematics 332(5):485–490

Carlen EA, Carvalho MC, Esposito R, Lebowitz JL, Marra R (2009) Droplet minimizers for the Gates–Lebowitz–Penrose free energy functional. Nonlinearity 22(12):2919

Carpentier VT, Wong J, Yeap Y, Gan C, Sutton-Smith P, Badiei A, Fazzalari NL, Kuliwaba JS (2012) Increased proportion of hypermineralized osteocyte lacunae in osteoporotic and osteoarthritic human trabecular bone: Implications for bone remodeling. Bone 50(3):688–694

Cazzani A, Malagù M, Turco E (2016a) Isogeometric analysis of plane-curved beams. Mathematics and Mechanics of Solids 21(5):562–577

Cazzani A, Malagù M, Turco E, Stochino F (2016b) Constitutive models for strongly curved beams in the frame of isogeometric analysis. Mathematics and Mechanics of Solids 21(2):182–209

Chatzigeorgiou G, Javili A, Steinmann P (2014) Unified magnetomechanical homogenization framework with application to magnetorheological elastomers. Mathematics and Mechanics of Solids 19(2):193–211

Chen AE, Ginty DD, Fan CM (2005) Protein kinase A signalling via CREB controls myogenesis induced by Wnt proteins. Nature 433(7023):317

Cluzel C, Allena R (2018) A general method for the determination of the local orthotropic direc-
    tions of heterogeneous materials: application to bone structures using μCT images. Mathemat-
    ics and Mechanics of Complex Systems 6(4):353–367
Colangeli M, De Masi A, Presutti E (2016) Latent heat and the Fourier law. Physics Letters A
    380(20):1710–1713
Colangeli M, De Masi A, Presutti E (2017) Microscopic models for uphill diffusion. Journal of
    Physics A: Mathematical and Theoretical 50(43):435,002
Contrafatto L, Cuomo M (2006) A framework of elastic–plastic damaging model for concrete
    under multiaxial stress states. International Journal of Plasticity 22(12):2272–2300
Cowin SC (1999) Bone poroelasticity. Journal of Biomechanics 32(3):217–238
Cowin SC (ed) (2001) Bone Mechanics Handbook, 2nd edn. CRC Press, Boca Raton
Cuomo M, Contrafatto L, Greco L (2014) A variational model based on isogeometric interpolation
    for the analysis of cracked bodies. International Journal of Engineering Science 80:173–188
Dallas SL, Bonewald LF (2010) Dynamics of the transition from osteoblast to osteocyte. Annals
    of the New York Academy of Sciences 1192(1):437–443
De Masi A, Ferrari PA, Lebowitz JL (1986) Reaction-diffusion equations for interacting particle
    systems. Journal of Statistical Physics 44(3-4):589–644
De Masi A, Gobron T, Presutti E (1995) Travelling fronts in non-local evolution equations. Archive
    for Rational Mechanics and Analysis 132(2):143–205
dell'Isola F, Seppecher P, Madeo A (2012) How contact interactions may depend on the shape of
    Cauchy cuts in Nth gradient continua: approach "à la D'Alembert". Zeitschrift für angewandte
    Mathematik und Physik 63(6):1119–1141
Di Carlo A, Quiligotti S (2002) Growth and balance. Mechanics Research Communications
    29(6):449–456
Diebels S, Steeb H (2003) Stress and couple stress in foams. Computational Materials Science
    28(3–4):714–722
Engelbrecht J, Berezovski A (2015) Reflections on mathematical models of deformation waves in
    elastic microstructured solids. Mathematics and Mechanics of Complex Systems 3(1):43–82
Epstein M, Maugin GA (2000) Thermomechanics of volumetric growth in uniform bodies. Inter-
    national Journal of Plasticity 16(7):951–978
Eremeyev VA, Pietraszkiewicz W (2009) Phase transitions in thermoelastic and thermoviscoelastic
    shells. Archives of Mechanics 61(1):41–67
Eremeyev VA, Pietraszkiewicz W (2011) Thermomechanics of shells undergoing phase transition.
    Journal of Mechanics and Physics of Solids 59(7):1395–1412
Eremeyev VA, Pietraszkiewicz W (2016) Material symmetry group and constitutive equations of
    micropolar anisotropic elastic solids. Mathematics and Mechanics of Solids 21(2):210–221
Eremeyev VA, Lebedev LP, Altenbach H (2013) Foundations of Micropolar Mechanics. Springer,
    Berlin
Eremeyev VA, Lebedev LP, Cloud MJ (2015) The Rayleigh and Courant variational principles in
    the six-parameter shell theory. Mathematics and Mechanics of Solids 20(7):806–822
Eriksen EF (2010) Cellular mechanisms of bone remodeling. Reviews in Endocrine and Metabolic
    Disorders 11(4):219–227
Franciosi P, Spagnuolo M, Salman OU (2019) Mean Green operators of deformable fiber networks
    embedded in a compliant matrix and property estimates. Continuum Mechanics and Thermo-
    dynamics 31(1):101–132
Frost HM (1987) Bone "mass" and the "mechanostat": a proposal. The Anatomical Record
    219(1):1–9
Fung YC (2006) Biomechanics. Mechanical Properties of Living Tissues, 2nd edn. Springer, New
    York
Ganghoffer JF (2012) A contribution to the mechanics and thermodynamics of surface growth. ap-
    plication to bone external remodeling. International Journal of Engineering Science 50(1):166–
    191

Garikipati K, Olberding JE, Narayanan H, Arruda EM, Grosh K, Calve S (2006) Biological remodelling: stationary energy, configurational change, internal variables and dissipation. Journal of
the Mechanics and Physics of Solids 54(7):1493–1515

George D, Allena R, Remond Y (2017) Mechanobiological stimuli for bone remodeling: mechanical energy, cell nutriments and mobility. Computer Methods in Biomechanics and Biomedical
Engineering 20(S1):91–92

George D, Allena R, Remond Y (2018) A multiphysics stimulus for continuum mechanics bone
remodeling. Mathematics and Mechanics of Complex Systems 6(4):307–319

Gibson LJ, Ashby MF (1997) Cellular Solids: Structure and Properties, 2nd edn. Cambridge Solid
State Science Series, Cambridge University Press, Cambridge

Giorgio I, Andreaus U, Scerrato D, dell'Isola F (2016) A visco-poroelastic model of functional
adaptation in bones reconstructed with bio-resorbable materials. Biomechanics and Modeling
in Mechanobiology 15(5):1325–1343

Giorgio I, Andreaus U, dell'Isola F, Lekszycki T (2017a) Viscous second gradient porous materials
for bones reconstructed with bio-resorbable grafts. Extreme Mechanics Letters 13:141–147

Giorgio I, Andreaus U, Lekszycki T, Della Corte A (2017b) The influence of different geometries
of matrix/scaffold on the remodeling process of a bone and bioresorbable material mixture with
voids. Mathematics and Mechanics of Solids 22(5):969–987

Giorgio I, Andreaus U, Scerrato D, Braidotti P (2017c) Modeling of a non-local stimulus for bone
remodeling process under cyclic load: Application to a dental implant using a bioresorbable
porous material. Mathematics and Mechanics of Solids 22(9):1790–1805

Goda I, Assidi M, Ganghoffer JF (2014) A 3D elastic micropolar model of vertebral trabecular
bone from lattice homogenization of the bone microstructure. Biomechanics and Modeling in
Mechanobiology 13(1):53–83

Gong Y, Slee RB, Fukai N, et al (2001) LDL receptor-related protein 5 (LRP5) affects bone accrual
and eye development. Cell 107(4):513–523

Goriely A, Robertson-Tessi M, Tabor M, Vandiver R (2008) Elastic growth models. In: Mondaini
RP, Pardalos PM (eds) Mathematical Modelling of Biosystems, Applied Optimization, vol 102,
Springer, pp 1–44

Hambli R (2014) Connecting mechanics and bone cell activities in the bone remodeling process: an
integrated finite element modeling. Frontiers in Bioengineering and Biotechnology 2(6):1–12

Himeno-Ando A, Izumi Y, Yamaguchi A, Iimura T (2012) Structural differences in the osteocyte
network between the calvaria and long bone revealed by three-dimensional fluorescence mor-
phometry, possibly reflecting distinct mechano-adaptations and sensitivities. Biochemical and
Biophysical Research Communications 417(2):765–770

Holzapfel GA, Ogden RW (eds) (2006) Mechanics of Biological Tissue. Springer, Berlin

van Hove RP, Nolte PA, Vatsa A, Semeins CM, Salmon PL, Smit TH, Klein-Nulend J (2009) Osteocyte morphology in human tibiae of different bone pathologies with different bone mineral
density — Is there a role for mechanosensing? Bone 45(2):321–329

Imatani S, Maugin GA (2002) A constitutive model for material growth and its application to three-
dimensional finite element analysis. Mechanics Research Communications 29(6):477–483

Khalili N, Selvadurai APS (2003) A fully coupled constitutive model for thermo-hydro-mechanical
analysis in elastic media with double porosity. Geophysical Research Letters 30(24)

Komori T (2013) Functions of the osteocyte network in the regulation of bone mass. Cell and
Tissue Research 352(2):191–198

Kühl M, Sheldahl LC, Park M, Miller JR, Moon RT (2000) The Wnt/Ca2+ pathway: a new vertebrate Wnt signaling pathway takes shape. Trends in Genetics 16(7):279–283

Lekszycki T, dell'Isola F (2012) A mixture model with evolving mass densities for describing syn-
thesis and resorption phenomena in bones reconstructed with bio-resorbable materials. ZAMM-
Journal of Applied Mathematics and Mechanics/Zeitschrift für Angewandte Mathematik und
Mechanik 92(6):426–444

Lekszycki T, Bucci S, Del Vescovo D, Turco E, Rizzi NL (2017) A comparison between different approaches for modelling media with viscoelastic properties via optimization analyses.

ZAMM-Journal of Applied Mathematics and Mechanics/Zeitschrift für Angewandte Mathematik und Mechanik 97(5):515–531

Lemaire T, Kaiser J, Naili S, Sansalone V (2010) Modelling of the transport in electrically charged porous media including ionic exchanges. Mechanics Research Communications 37(5):495–499

Li J, Slesarenko V, Rudykh S (2019) Microscopic instabilities and elastic wave propagation in finitely deformed laminates with compressible hyperelastic phases. European Journal of Mechanics-A/Solids 73:126–136

Lu Y, Lekszycki T (2017) Modelling of bone fracture healing: influence of gap size and angiogenesis into bioresorbable bone substitute. Mathematics and Mechanics of Solids 22(10):1997–2010

Lu Y, Lekszycki T (2018) New description of gradual substitution of graft by bone tissue including biomechanical and structural effects, nutrients supply and consumption. Continuum Mechanics and Thermodynamics 30(5):995–1009

Lurie S, Solyaev Y, Volkov A, Volkov-Bogorodskiy D (2018a) Bending problems in the theory of elastic materials with voids and surface effects. Mathematics and Mechanics of Solids 23(5):787–804

Lurie SA, Kalamkarov YO A L and Solyaev, Ustenko AD, Volkov AV (2018b) Continuum microdilatation modeling of auxetic metamaterials. International Journal of Solids and Structures 132:188–200

Madeo A, George D, Lekszycki T, Nierenberger M, Remond Y (2012) A second gradient continuum model accounting for some effects of micro-structure on reconstructed bone remodelling. Comptes Rendus Mécanique 340(8):575–589

Martin RB (1984) Porosity and specific surface of bone. Critical Reviews^{TM} in Biomedical Engineering 10(3):179–222

Menzel A (2005) Modelling of anisotropic growth in biological tissues. Biomechanics and Modeling in Mechanobiology 3(3):147–171

Misra A, Poorsolhjouy P (2015) Identification of higher-order elastic constants for grain assemblies based upon granular micromechanics. Mathematics and Mechanics of Complex Systems 3(3):285–308

Misra A, Marangos O, Parthasarathy R, Spencer P (2013) Micro-scale analysis of compositional and mechanical properties of dentin using homotopic measurements. In: Andreaus U, Iacoviello D (eds) Biomedical Imaging and Computational Modeling in Biomechanics. Lecture Notes in Computational Vision and Biomechanics, vol 4, Springer, Dordrecht, pp 131–141

Misra A, Parthasarathy R, Singh V, Spencer P (2015) Micro-poromechanics model of fluid-saturated chemically active fibrous media. ZAMM-Journal of Applied Mathematics and Mechanics/Zeitschrift für Angewandte Mathematik und Mechanik 95(2):215–234

Mlodzik M (2002) Planar cell polarization: do the same mechanisms regulate Drosophila tissue polarity and vertebrate gastrulation? Trends in Genetics 18(11):564–571

Mullender MG, Huiskes R, Weinans H (1994) A physiological approach to the simulation of bone remodeling as a self-organizational control process. Journal of Biomechanics 27(11):1389–1394

Niiranen J, Niemi AH (2017) Variational formulations and general boundary conditions for sixth-order boundary value problems of gradient-elastic Kirchhoff plates. European Journal of Mechanics-A/Solids 61:164–179

Niiranen J, Balobanov V, Kiendl J, Hosseini SB (2019) Variational formulations, model comparisons and numerical methods for Euler–Bernoulli micro-and nano-beam models. Mathematics and Mechanics of Solids 24(1):312–335

Park HC, Lakes RS (1986) Cosserat micromechanics of human bone: strain redistribution by a hydration-sensitive constituent. Journal of Biomechanics 19(5):385–397

Pinson KI, Brennan J, Monkley S, Avery BJ, Skarnes WC (2000) An LDL-receptor-related protein mediates Wnt signalling in mice. Nature 407(6803):535

Placidi L, Barchiesi E (2018) Energy approach to brittle fracture in strain-gradient modelling. Proceedings of the Royal Society A: Mathematical, Physical and Engineering Sciences 474(2210):20170,878

Placidi L, Barchiesi E, Misra A (2018a) A strain gradient variational approach to damage: a comparison with damage gradient models and numerical results. Mathematics and Mechanics of Complex Systems 6(2):77–100

Placidi L, Misra A, Barchiesi E (2018b) Two-dimensional strain gradient damage modeling: a variational approach. Zeitschrift für angewandte Mathematik und Physik 69(3):56

Prakash C, Singh S, Farina I, Fraternali F, Feo L (2018) Physical-mechanical characterization of biodegradable Mg-3Si-HA composites. PSU Research Review 2(2):152–174

Prendergast PJ, Taylor D (1994) Prediction of bone adaptation using damage accumulation. Journal of Biomechanics 27(8):1067–1076

Roux W (1895) Der Kampf der Teile im Organismus. 1881. Leipzig: Engelmann

Ruimerman R, Hilbers P, van Rietbergen B, Huiskes R (2005) A theoretical framework for strain-related trabecular bone maintenance and adaptation. Journal of Biomechanics 38(4):931–41

Sansalone V, Kaiser J, Naili S, Lemaire T (2013) Interstitial fluid flow within bone canaliculi and electro-chemo-mechanical features of the canalicular milieu. Biomechanics and Modeling in Mechanobiology 12(3):533–553

Scala I, Rosi G, Nguyen VH, Vayron R, Haiat G, Seuret S, Jaffard S, Naili S (2018) Ultrasonic characterization and multiscale analysis for the evaluation of dental implant stability: A sensitivity study. Biomedical Signal Processing and Control 42:37–44

Seppecher P (1993) Equilibrium of a Cahn-Hilliard fluid on a wall: influence of the wetting properties of the fluid upon the stability of a thin liquid film. European Journal of Mechanics Series B Fluids 12:69–69

Seppecher P (2000) Second-gradient theory: application to Cahn-Hilliard fluids. In: Maugin GA, Drouot R, Sidoroff F (eds) Continuum Thermomechanics. Solid Mechanics and Its Applications, vol 76, Springer, Dordrecht, pp 379–388

Spagnuolo M, Barcz K, Pfaff A, dell'Isola F, Franciosi P (2017) Qualitative pivot damage analysis in aluminum printed pantographic sheets: numerics and experiments. Mechanics Research Communications 83:47–52

Spingarn C, Wagner D, Remond Y, George D (2017) Multiphysics of bone remodeling: a 2D mesoscale activation simulation. Bio-medical Materials and Engineering 28(s1):S153–S158

Stern AR, Nicolella DP (2013) Measurement and estimation of osteocyte mechanical strain. Bone 54(2):191–195

Taber LA (1995) Biomechanics of growth, remodeling, and morphogenesis. Applied Mechanics Reviews 48:487–545

Taber LA (2009) Towards a unified theory for morphomechanics. Philosophical Transactions of the Royal Society A: Mathematical, Physical and Engineering Sciences 367(1902):3555–3583

Turner CH (1991) Homeostatic control of bone structure: An application of feedback theory. Bone 12(3):203–217

Turner CH (1998) Three rules for bone adaptation to mechanical stimuli. Bone 23(5):399–407

Vatsa A, Breuls RG, Semeins CM, Salmon PL, Smit TH, Klein-Nulend J (2008) Osteocyte morphology in fibula and calvaria — Is there a role for mechanosensing? Bone 43(3):452–458

Yang JFC, Lakes RS (1982) Experimental study of micropolar and couple stress elasticity in compact bone in bending. Journal of Biomechanics 15(2):91–98

Yeremeyev VA, Zubov LM (1999) The theory of elastic and viscoelastic micropolar liquids. Journal of Applied Mathematics and Mechanics 63(5):755–767

Yildizdag ME, Demirtas M, Ergin A (2018) Multipatch discontinuous Galerkin isogeometric analysis of composite laminates. Continuum Mechanics and Thermodynamics pp 1–14, DOI 10.1007/s00161-018-0696-9

Yildizdag ME, Ardic IT, Demirtas M, Ergin A (2019) Hydroelastic vibration analysis of plates partially submerged in fluid with an isogeometric FE-BE approach. Ocean Engineering 172:316–329

Yoo A, Jasiuk I (2006) Couple-stress moduli of a trabecular bone idealized as a 3D periodic cellular network. Journal of Biomechanics 39(12):2241–2252

Chapter 6
A C_1 Incompatible Mode Element Formulation for Strain Gradient Elasticity

Rainer Glüge

Abstract In the present article, a simple, yet reasonably well working 3rd order 3D tetrahedral incompatible mode finite element with C_1 continuity is presented. It has been implemented into the finite element system Abaqus. To check its capabilities, it is used in elastostatic boundary value problems in conjunction with an elasticity model that includes a classical strain energy and a strain gradient energy contribution. With these examples, we discuss the regularizing effect of the strain gradient contribution on the singularities in classical elasticity in single force indentation tests and bending of an L-shaped sample with a sharp corner. We consider both, large strain gradient contributions and a mesh refinement. We further discuss the transition to the pseudorigid behavior as the strain gradient energy becomes large compared to the strain energy, which results in a material that can only undergo homogeneous deformations, also called pseudorigid. Finally, we propose an improvement of the element formulation, and compare its convergence behavior upon mesh refinement.

Keywords: Continuum mechanics · Finite element method · C_1 continuity · Strain gradient elasticity · Pseudo-rigid

6.1 Introduction

Simple material models involve only the first gradient of the deformation map, named as the deformation gradient $\mathbf{F}$. $\mathbf{F}$ is dimensionless. Consequently, size effects cannot be accounted for in such material models. In most situations, size effects are negligible anyway. There is, however, experimental evidence for a size dependence,

R. Glüge

Otto-von-Guericke–University Magdeburg, Institute of Mechanics, Universitätsplatz 2, 39106 Magdeburg, Germany
e-mail: gluege@ovgu.de

© Springer Nature Switzerland AG 2019
H. Altenbach et al. (eds.), *Higher Gradient Materials and Related Generalized Continua*, Advanced Structured Materials 120,
https://doi.org/10.1007/978-3-030-30406-5_6

in elasticity (Sinclair, 2004) as well as in plasticity (Fleck and Hutchinson, 1993; Peerlings et al, 2001; Wulfinghoff and Böhlke, 2012). In elasticity and plasticity, small samples exhibit a higher relative stiffness and yield stress than larger samples. Consequently, size effects become important when going to smaller scales, which is unavoidable when micro-macro interactions are to be modeled. Thus, as research has been drawn to microscale material modelling during the last decades, taking into account size effects by strain gradient contributions became more popular.

6.1.1 Outline

In this work, we address theoretical as well as practical aspects of a purely elastic strain gradient modelling in the three-dimensional, large strain setting. First, we derive the weak form of the strain-gradient enhanced balance of momentum (Sect. 6.2). Then we propose a simple strain gradient contribution for an ordinary elastic energy (Sect. 6.3). The model is isotropic, and kept as simple as possible. The next section is devoted to details of the finite element formulation of a C_1-continuous, incompatible mode tetrahedral element (Sect. 6.4)[1]. We then examine the behavior of the strain gradient enhanced elastic material at singulatities (Sects. 6.5 and 6.6). Finally, the convergence properties and some improvements are discussed in Sect. 6.7.

6.1.2 Notation

A direct notation is preferred. If the index notation is required, we will make use of implicit summation over the indices 1 to 3. Components are then given w.r.t. orthonormal bases $\{\mathbf{e}_i\}$. Vectors are denoted as bold minuscules (like $\mathbf{u}$), second-order tensors as bold majuscules or bold greek letters (like $\mathbf{F}, \mathbf{H}, \mathbf{T}$), and higher order tensors as blackboard bold majuscules (like $\mathbb{T}, \mathbb{H}$). The dyadic product and scalar contractions are denoted like $(\mathbf{a} \otimes \mathbf{b} \otimes \mathbf{c}) \cdot \cdot (\mathbf{d} \otimes \mathbf{e}) = (\mathbf{b} \cdot \mathbf{d})(\mathbf{c} \cdot \mathbf{e})\mathbf{a}$, with $\cdot$ being the usual scalar product between vectors. The order of the scalar contractions is such that the scalar product properties are maintained, e.g. $\mathbf{A} \cdot \cdot \mathbf{A} > 0$ for real valued tensors $\mathbf{A}$. If only one scalar contraction is carried out, the dot is omitted, e.g., $\mathbf{C} = \mathbf{F}^T \mathbf{F}$. The upper index T denotes the transpose of a second-order tensor, which is $(\mathbf{a} \otimes \mathbf{b})^T := \mathbf{b} \otimes \mathbf{a}$ in terms of base dyads.

An index 0 indicates the reference placement, e.g., $\mathbf{n}_0$ is the surface normal in the reference placement and $\mathbf{n}$ is the surface normal in the current placement. The material and spatial nabla operators are as well distinguished by the index 0. The following symbols are used:

---

[1] Available at https://github.com/Ra-Na/C3D4C1-Abaqus-UEL.

Symbol	Unit	Description
∇_0	m^{-1}	material nabla operator
$\mathbf{b}$	N/kg	body force field in the current placement
$\mathbf{x}_0$	m	location vector in the reference placement
$\mathbf{u}$	m	displacement vector
$\mathbf{x} = \mathbf{x}_0 + \mathbf{u}$	m	location vector the in current placement
$\mathbf{I}$	–	second order identity tensor
$\mathbf{B} = \mathbf{F} \cdot \mathbf{F}^T$	–	left Cauchy-Green tensor
$\mathbf{C} = \mathbf{F}^T \cdot \mathbf{F}$	–	right Cauchy-Green tensor
$\mathbb{C} = \mathbf{F}^T \cdot \mathbb{F}$	–	material strain gradient measure
$\mathbf{H} = \mathbf{u} \otimes \nabla_0$	–	first displacement gradient
$\mathbb{H} = \mathbf{u} \otimes \nabla_0 \otimes \nabla_0$	m^{-1}	second displacement gradient
$w(\mathbf{H}, \mathbb{H})$	Nm/kg	strain energy density
$\mathbf{F} = \mathbf{x} \otimes \nabla_0 = \mathbf{H} + \mathbf{I}$	–	first deformation gradient
$\mathbb{F} = \mathbf{x} \otimes \nabla_0 \otimes \nabla_0 = \mathbb{H}$	m^{-1}	second deformation gradient
$J = \det(\mathbf{F})$	–	determinant of $\mathbf{F}$
ρ_0	kg/m^3	mass density in reference placement
$\rho = \rho_0 / J$	kg/m^3	mass density in current placement
$\mathbf{T} = \rho_0 w_{,\mathbf{H}}$	N/m^2	2nd order first Piola-Kirchhoff stresses
$\mathbb{T} = \rho_0 w_{,\mathbb{H}}$	N/m	3rd order first Piola-Kirchhoff stresses
$\mathbf{n}_0$	–	normal vector in reference placement
$\mathbf{t}_b = (\mathbf{T} - \mathbb{T} \cdot \nabla_0) \cdot \mathbf{n}_0$	N/m^2	boundary tractions
$\mathbf{T}_b = \mathbb{T} \cdot \mathbf{n}_0$	N/m	boundary stresses

6.2 From Local Balance of Momentum to Minimization of the Elastic Potential

We consider the quasistatic setting of a second gradient material. The local balance of momentum (rate of change of momentum per volume) is then

$$\mathbf{o} = \mathbf{T} \cdot \nabla_0 - \mathbb{T} \cdot\cdot \nabla_0 \otimes \nabla_0 + \rho \mathbf{b}, \quad [\text{N/m}^3] \tag{6.1}$$

see, e.g., Sect. 2 in Bertram and Forest (2014). The local balance of angular momentum is implied by the symmetry of the Cauchy stress tensor $\boldsymbol{\sigma} = \frac{1}{J} \mathbf{T}\mathbf{F}^T$. Here, the symmetry of $\boldsymbol{\sigma}$ is ensured through the potential relation $\mathbf{T} = w_{,\mathbf{F}}$ and the independence of w on the polar part of $\mathbf{F}$. No requirements on the higher order stress tensor $\mathbb{T}$ emerge. Nevertheless, only the part of $\mathbb{T}$ that is symmetric in the second and third entry/index enters the local balance of momentum (Eq. (6.1)). The unknown function is the displacement field $\mathbf{u}$ in terms of the reference coordinates $\mathbf{x}_0$. >From $\mathbf{u}$ we have the strain deformation measures $\mathbf{F} = \mathbf{u} \otimes \nabla_0 + \mathbf{I}$ and $\mathbb{F} = \mathbf{F} \otimes \nabla_0$, which enter the material model via the strain energy w and give the stresses $\mathbf{T}$ and $\mathbb{T}$.

Let $\overline{\mathbf{u}}$ be the solution of the latter differential equation (Eq. (6.1)) in V, where $\overline{\mathbf{u}}^*$, $\overline{\mathbf{H}}^*$, $\overline{\mathbf{t}}_b^*$ and $\overline{\mathbb{T}}_b^*$ denote the boundary conditions that are satisfied by $\overline{\mathbf{u}}$. Scalar multiplication of the latter differential equation at the solution with a test function $\delta\mathbf{u}$ and integration over V gives

$$0 = \int \delta\mathbf{u} \cdot (\overline{\mathbf{T}} \cdot \nabla_0 - \overline{\mathbb{T}} \cdot\cdot \nabla_0 \otimes \nabla_0 + \rho\mathbf{b})\mathrm{d}V. \quad [\mathrm{Nm}] \tag{6.2}$$

We interpret the test function $\delta\mathbf{u}$ as a displacement deviation from the solution, i.e. $\tilde{\mathbf{u}} = \overline{\mathbf{u}} + \delta\mathbf{u}$. Due to the linearity of $\mathbf{H}$ and $\mathbb{H}$ in $\mathbf{u}$, this additive deviation is inherited to $\mathbf{H}$ and $\mathbb{H}$, i.e. $\tilde{\mathbf{H}} = \overline{\mathbf{H}} + \delta\mathbf{H}$ and $\tilde{\mathbb{H}} = \overline{\mathbb{H}} + \delta\mathbb{H}$. With the product rule backwards and Gauß's theorem we get

$$0 = \int \delta\mathbf{u} \cdot \rho\mathbf{b}\mathrm{d}V + \int \delta\mathbf{u} \cdot (\overline{\mathbf{T}} - \overline{\mathbb{T}} \cdot \nabla_0) \cdot \mathbf{n}_0 \mathrm{d}A+ \tag{6.3}$$

$$- \int \delta\mathbf{H} \cdot\cdot \overline{\mathbf{T}}\mathrm{d}V + \int \delta\mathbf{H} \cdot\cdot \overline{\mathbb{T}} \cdot \nabla_0 \mathrm{d}V. \quad [\mathrm{Nm}] \tag{6.4}$$

Repeating this with the last integral gives

$$0 = \int \delta\mathbf{u} \cdot \rho\mathbf{b}\mathrm{d}V + \int \delta\mathbf{u} \cdot \underbrace{(\overline{\mathbf{T}} - \overline{\mathbb{T}} \cdot \nabla_0) \cdot \mathbf{n}_0}_{\overline{\mathbf{t}}_b} \mathrm{d}A+ \tag{6.5}$$

$$- \int \delta\mathbf{H} \cdot\cdot \overline{\mathbf{T}}\mathrm{d}V + \int \delta\mathbf{H} \cdot\cdot \underbrace{\overline{\mathbb{T}} \cdot \mathbf{n}_0}_{\overline{\mathbb{T}}_b} \mathrm{d}A - \int \delta\mathbb{H} \cdot\cdot\cdot \overline{\mathbb{T}}\mathrm{d}V. \quad [\mathrm{Nm}] \tag{6.6}$$

We assume that the test function is compliant to the displacement boundary conditions $\overline{\mathbf{u}}^*$ and the displacement gradient boundary conditions $\overline{\mathbf{H}}^*$. Consequently, the surface integrals vanish where $\mathbf{u}$ or $\mathbf{H}$ are prescribed, since $\delta\mathbf{u}$ and $\delta\mathbf{H}$ vanish there. We therefore restrict the surface integration to where $\mathbf{t}_b$ or $\mathbb{T}_b$ is prescribed, referred to as A_{dyn}. We presume that the solution $\overline{\mathbf{u}}$ satisfies the traction and the stress boundary conditions, so we can replace $\overline{\mathbf{t}}_b$ by $\mathbf{t}_b^*$ and $\overline{\mathbb{T}}_b$ by $\mathbb{T}_b^*$. We further assume hyperelastic material behavior, such that the potential relations $\mathbf{T} = w_{,\mathbf{H}}$ and $\mathbb{T} = w_{,\mathbb{H}}$ hold for the first Piola-Kirchhoff stresses and a similar stress tensor of third order, where w is the stored elastic energy. We can write

$$0 = \int \delta\mathbf{u} \cdot \mathbf{b}\,\mathrm{d}m + \int \delta\mathbf{u} \cdot \mathbf{t}_b^* + \delta\mathbf{H} \cdot\cdot \mathbb{T}_b^* \,\mathrm{d}A_{\mathrm{dyn}}+ \tag{6.7}$$

$$- \int (\delta\mathbf{H} \cdot\cdot w_{,\mathbf{H}} |_{\overline{\mathbf{H}}} + \delta\mathbb{H} \cdot\cdot\cdot w_{,\mathbb{H}} |_{\overline{\mathbb{H}}}) \,\mathrm{d}m. \quad [\mathrm{Nm}] \tag{6.8}$$

Note that w is mass-specific, i.e. the factor ρ_0 is contained in w. When we now consider arbitrary fields $\mathbf{H}$ and $\mathbb{H}$ instead of $\overline{\mathbf{H}}$ and $\overline{\mathbb{H}}$, we see that the latter equation is the stationarity condition to the following variational problem:

Find the extremum of

$$\Pi := - \int \mathbf{u} \cdot \mathbf{b} dm - \int (\mathbf{u} \cdot \mathbf{t}_b^* + \mathbf{H} \cdot \cdot \mathbf{T}_b^*) \, dA_{\mathrm{dyn}} + \int w(\mathbf{H}, \mathbb{H}) dm \quad [\mathrm{Nm}]$$

(6.9)

over $\mathbf{u}$, where $\mathbf{H} = \mathbf{u} \otimes \nabla_0$ and $\mathbb{H} = \mathbf{u} \otimes \nabla_0 \otimes \nabla_0$,

and $\mathbf{u}$ has to satisfy some BC and continuity requirements. We presume that w is positive or zero, thus the extremal must be a minimum. In conclusion, the equilibrium of the local balance of momentum in V corresponds to the minimization of the functional Π over $\mathbf{u}$ in V. This is the classical minimum of the elastic potential principle extended to strain gradients, as one can see when dropping the dependence on $\mathbb{H}$ and setting $\mathbb{T} = \mathbb{O}$.

The solution to the variational problem is unique if w satisfies some convexity requirements. In case of nonlinear elasticity of simple materials, this condition is termed *quasiconvexity*. It is, in words, that the minimization of the latter problem gives a unique result for all possible domains and boundary conditions. We will presume the same for the gradient elasticity, i.e. the latter minimization is unique for all domains and boundary conditions that one can think of. We will see later that our choice of w is safe in this regard.

6.3 Ciarlets Elastic Energy

Ciarlets strain energy (Ciarlet, 1988) is close to the St.-Venant-Kirchhoff strain energy, but in contrast to the latter it is polyconvex in $\mathbf{F}$,

$$w(\mathbf{F}) = \frac{\lambda}{4}(\mathit{III} - \ln \mathit{III}) + \frac{\mu}{2}(I - \ln \mathit{III}).$$

(6.10)

The roman numbers I and III denote the first and third principle invariants of the Cauchy-Green stretch tensors $\mathbf{B}$ and $\mathbf{C}$, λ and μ are Lamé's constants. Ciarlets strain energy serves as the starting point for our elastic strain gradient extension.

6.3.1 Strain Gradient Extension

To comply with the principle of invariance under rigid body motions, w is best formulated in terms of material deformation measures. While we use $\mathbf{C} = \mathbf{F}^T \mathbf{F}$ for the first gradient, we need to define a similar material measure for the second gradient $\mathbb{F} = \mathbf{x} \otimes \nabla_0 \otimes \nabla_0$. One way to do so is

$$\mathbb{C} = \mathbb{F}^T \mathbb{F}.$$

(6.11)

Unlike $\mathbf{C}$, $\mathbb{C}$ can not be interpreted as a metric.

There are no popular invariant definitions for third order tensors, since there is no eigenvalue problem for tensors of uneven order. One simple invariant is the norm, which we term g to indicate the strain gradient contribution,

$$g = \sqrt{\mathbb{C} \cdots \mathbb{C}} = \sqrt{C_{ijk} C_{ijk}}. \tag{6.12}$$

We are now ready to extend w by a summand that contains a strain gradient contribution. We choose

$$w_g(\mathbf{F}) = \frac{\lambda}{4}(I\!I\!I - \ln I\!I\!I) + \frac{\mu}{2}(I - \ln I\!I\!I) + \frac{\alpha}{2}g^2, \tag{6.13}$$

where α is a new material parameter. The contribution $\alpha g^2/2$ should be sufficiently convex close to $\mathbb{C} = \mathbb{O}$, i.e. we do not expect a loss of convexity and hence problems with the uniqueness of the solution as long as the strains and the strain gradient are not too large.

6.3.2 Stress-strain Relations

We consider the first Piola–Kirchhoff stresses $\mathbf{T}$ as most practical for our purpose. Especially from an implementation point of view, we can write down the potential Π (Eq. (6.9)) using the material nabla operator ∇_0. Thus we need in the FE implementation the same operators ∇_0 and $\nabla_0 \otimes \nabla_0$ for calculating the first and second gradient of $\mathbf{u}$ and for calculating the residuals, which simplifies the implementation considerably. We refer to $\mathbb{T}$ thus as the first Piola–Kirchhoff stress tensor of third order, which is derived from w similar to the usual first Piola–Kirchhoff stresses,

$$\mathbf{T} = \frac{\partial w}{\partial \mathbf{F}}, \qquad \mathbb{T} = \frac{\partial w}{\partial \mathbb{F}}. \tag{6.14}$$

The derivatives are

$$\mathbf{T} = \mu\mathbf{F} + \left[\frac{\lambda}{2}(I\!I\!I - 1) - \mu\right]\mathbf{F}^{-T} + \alpha F_{mjl}C_{njl}\mathbf{e}_m \otimes \mathbf{e}_n,$$
$$\mathbb{T} = \alpha\mathbf{F}\mathbb{C} = \alpha\mathbf{B}\mathbb{F}. \tag{6.15}$$

The number of indices distinguishes $\mathbf{F}, \mathbb{F}$ and $\mathbf{C}, \mathbb{C}$ in the index notation. Additional to the stress-strain (gradient) relation, we need the linearisations. The indices at the brackets indicate the order of the base vectors in the tensor product basis, e.g., $(T_{ik}F_{kj})_{ij} = T_{ik}F_{kj}\mathbf{e}_i \otimes \mathbf{e}_j$.

$$\frac{\partial \mathbf{T}}{\partial \mathbf{F}} = \mu\mathbb{I} - \left[\frac{\lambda}{2}(I\!I\!I - 1) - \mu\right](F_{jk}^{-1}F_{il}^{-T})_{ijkl} +$$
$$+ \lambda I\!I\!I\mathbf{F}^{-T} \otimes \mathbf{F}^{-T} + \alpha(F_{ixy}F_{kxy}\delta_{jl})_{ijkl}, \tag{6.16}$$

$$\frac{\partial \mathbf{T}}{\partial \mathbb{F}} = \alpha (\delta_{ik} C_{jlm} + F_{kj} F_{ilm})_{ijklm},$$

$$\frac{\partial \mathbb{T}}{\partial \mathbf{F}} = \alpha (\delta_{il} C_{mjk} + F_{im} F_{ljk})_{ijklm}, \qquad (6.17)$$

$$\frac{\partial \mathbb{T}}{\partial \mathbb{F}} = \alpha \mathbf{B} \mathbb{I}^6 = \alpha (B_{il} \delta_{jm} \delta_{kn})_{ijklmn}.$$

As mentioned before, the expressions are quite simple thanks to the specific choice of $\mathbb{C} = \mathbf{F}^T \mathbb{F}$.

6.3.3 Material parameters

For our study, we choose to fix the classical material parameters to

$$\lambda = \frac{E\nu}{(1+\nu)(1-2\nu)}, \qquad \mu = \frac{E}{2(1+\nu)}, \qquad (6.18)$$

$$E = 200 \text{ GPa}, \qquad \nu = 0.3, \qquad (6.19)$$

and vary α. To put α into context, we need to have a closer look at the physical units. The free energy is usually given specific to mass or volume in the reference placement,

$$[w] = \text{J/kg}, \qquad (6.20)$$

$$[\rho_0 w] = \text{J/m}^3 = \text{N/m}^2. \qquad (6.21)$$

Mostly, the latter is preferred, which manifests in the units for the elastic constants in MPa. Thus, E takes a value of $2 \cdot 10^{11}$ N/m^2, and ν remains dimensionless. The term αg^2 in Eq. (6.13) must therefore take the unit N/m^2. Since $[g] = \text{m}^{-1}$, we have $[\alpha] = N$. In conclusion, α and E can only be related meaningful to each other with a squared length, or more reasonable, with a squared curvature. The physical unit of α can be interpreted as energy [Nm] per curvature squared [1/m^2] per volume [m^3], i.e. it reflects the stored energy per volume due to the bending of material line elements.

6.4 Element Formulation

6.4.1 Overview

For an explicit account to a strain gradient extension, we need C_1-continuous elements. In three dimensions, this is a rather complicated requirement. There are only few formulations, see for example Papanicolopulos et al (2009) for a hexahedral element or Alfeld (1984); Alfeld and Sorokina (2009); Walkington (2014) for tetrahedral elements. Most formulations have some drawbacks. The element of Papanicolopulos et al (2009) requires a regular meshing and some preprocessing, the element of Alfeld (1984) has a transfinite interpolant, i.e. the usual linear relation between polynomial coefficients and the nodal degrees of freedom (DOF) is lost. Also, often high gradients of the field are used as DOF, which do not appear in the theory.

Another strategy is the mixed element formulation or implicit strain gradient extension, for example employed in Zybell et al (2012), who proposed a hexahedral element. This approach introduces the strains as independent DOF. A term in the energy density that involves Lagrange multipliers is responsible for enforcing the differential relation between the kinematic DOF. Hence, it involves a larger number of DOF, 162 in case of Zybell et al (2012). Peerlings et al (2001) suggest this approach, compared to an explicit strain gradient extension in a one-dimensional case. It is very popular in strain gradient plasticity, since the plastic strains are not directly the gradient of the displacement, but more complicated relation needs to be enforced, like the dislocation density tensor in crystal plasticity (Forest et al, 2014), see e.g. Dimitrijevic and Hackl (2008); Wulfinghoff and Böhlke (2012). Programmatic finite element environments like the FEniCS-project[2] make such multi-field formulations relatively easy, see for example, Abali et al (2017); Phunpeng and Baiz (2015), the same holds for implementations in Comsol (Reiher et al, 2017). These implementations require no special purpose elements, the only drawback is the larger number of DOF.

One may also enforce the constraints by the penalty method instead of the Lagrange method. A common error is then to treat the penalty parameters as fitting parameters.

6.4.2 Incompatible Mode Element Formulation

Here, the goal is a formulation that allows for a free, unstructured mesh with the smallest possible number of degrees of freedom. In order to satisfy C_1 continuity at the complete interface between two neighboring elements, Zenísek (1973) showed that one needs a complete polynomial of order

$$o = 2^d + 1 \tag{6.22}$$

when a triangular ($d = 2$) or a tetrahedral ($d = 3$) element shape are used. The number of monomials m that such a polynomial has is given by the binomial coefficient

$$m = \binom{o + d}{o}. \tag{6.23}$$

---

[2] https://fenicsproject.org/

For $d = 3$ we have $m = 220$. In general, the number of monomials corresponds to the number of DOF, where 220 is unusually large. One way to reduce the number of DOF is the composite element approach. It consists in relaxing the C_∞-continuity inside the element to C_1-continuity, and use the additional functional freedom to obtain C_1-continuity at the element interface. Walkington (2014) was able to develop a C_1 continuous tetrahedral element with 45 DOF. Other such elements are listed in Alfeld and Sorokina (2009). This is still a relatively large number of DOF. It is therefore much more common to implement incompatible mode (non-conforming) elements, i.e. to relax the requirement that the interpolating polynomial is C_1 continuous at every point of an element interface.

Thus, we choose a different starting point. We aim for the simplest possible tetrahedral element, and see how many monomial coefficients we can adopt to the minimum set of degrees of freedom (DOF) that we have. The simplest possible tetrahedron has four nodes. Assuming that the value of the field and its gradient are DOF at each node, we end up with 16 DOF per element and scalar field. Looking at the number of monomials of complete polynomials of different order in 3D, we note that we lie between 10 (quadratic) and 20 (cubic). Thus, one might think that one may discard four of the 20 monomials of the cubic polynomial in 3D, and adopt the 16 remaining coefficients to the 16 DOF. Unfortunately, there is no good choice for which of the 20 monomials should be discarded. When one discards four monomials in a symmetric manner (e.g. x^3, y^3, z^3 and xyz), it turns out that the remaining coefficients do not allow for interpolating any given set of DOF. When one discards four monomials in a cyclic symmetric manner (e.g. xy^2, yz^2, zx^2 and xyz), one can uniquely determine the remaining coefficients to each set of DOF. Unfortunately, the axis of cyclic symmetry (i.e. the 111-direction) becomes apparent in the results. Thus, one needs to put some more effort in. In case of the tetrahedral elements it is common practice to additionally tie the face centers of neighboring elements to each other, i.e. to make sure that the value of the field or the normal gradient is the same in the face center of two neighboring elements. For a tetrahedral element, these four additional requirements complement the number of DOF such that a complete third order polynomial can be uniquely determined from the set of 20 DOF.

Still, it is not very practical to have mixed nodes, namely some with the value and the gradient of the field, and some only with the value or one gradient component in a finite element mesh. Therefore we employed a serendipity-like formulation, such that only 16 primary DOF are present in one element, but a full third order polynomial with 20 monomials is used as a shape function. The four additional field values ϕ_{qi} (see Fig. 6.1) are calculated as intermediate quantities. Such elements have been studied by Ming and Xu (2007), who attests them a good performance.

These four additional quasi-DOF are determined in the following manner: At the three adjacent corner points to each element face, there is the value of the field plus its gradient given. Consequently, we can calculate linear extrapolations of the field from these points. Additionally, we can interpolate linearly between the three values of the field at the three adjacent corner points. The following approach has been used for the four face-centers:

Fig. 6.1 Sketch of the tetrahedral element that has been
implemented.

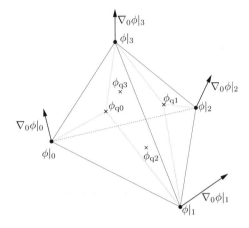

$$
\begin{aligned}
\phi_{\text{q}ijk} = \frac{1}{6}\Big(& \phi|_i + \phi|_j + \phi|_k + \\
& + \phi|_i + \nabla\phi|_i \cdot (\mathbf{x}_{ijk} - \mathbf{x}_i) + \\
& + \phi|_j + \nabla\phi|_j \cdot (\mathbf{x}_{ijk} - \mathbf{x}_j) + \\
& + \phi|_k + \nabla\phi|_k \cdot (\mathbf{x}_{ijk} - \mathbf{x}_k)\Big) \\
& i, j, k \in \{0, 1, 2, 3\}, \qquad i < j < k.
\end{aligned}
\tag{6.24}
$$

$\phi|_i$ denotes the value of the field ϕ at the ith node, while the $\phi_{\text{q}ijk}$ and $\mathbf{x}_{ijk}$ denote the quasi-DOF-value of the field and the location vector at the face center between the ith, j^{th} and k^{th} node.

For simplicity, we restrict ourselves to have the quasi-DOF as linear inter- and extrapolations from the DOF, but some arbitrariness remains in this choice. Different approaches can be taken to eliminate this arbitrariness. The given choice is such that in case of a quadratic $\phi(\mathbf{x})$-field, the $\phi_{\text{q}ijk}$-values are calculated exactly[3]. The reason is that we will later extend this approach to interpolate also the gradient at the face centers, for which a linear interpolation is used, see Sect. 6.7. Thus, the interpolation of the function and its gradient are related by differentiation up to quadratic functions at the points of the quasi-DOF.

Having established an approach, we now look at the element formulation. Our 16 DOF (see Fig. 6.1) are collected in a vector [DOF]. Since the described extrapolation from the 16 DOF to the four quasi-DOF (QDOF) is linear, we can write it as a matrix-vector product

---

[3] See https://github.com/Ra-Na/C3D4C1-Abaqus-UEL.

$$\phi_{q0} = [L_0(\mathbf{x}_0^{\mathrm{ele}})] \cdot [\phi|_0, \phi|_1, \phi|_3, \nabla_0\phi|_0, \nabla_0\phi|_1, \nabla_0\phi|_3],$$
$$\phi_{q1} = [L_1(\mathbf{x}_0^{\mathrm{ele}})] \cdot [\phi|_1, \phi|_2, \phi|_3, \nabla_0\phi|_1, \nabla_0\phi|_2, \nabla_0\phi|_3],$$
$$\phi_{q2} = [L_2(\mathbf{x}_0^{\mathrm{ele}})] \cdot [\phi|_0, \phi|_1, \phi|_2, \nabla_0\phi|_0, \nabla_0\phi|_1, \nabla_0\phi|_2],$$
$$\phi_{q3} = [L_3(\mathbf{x}_0^{\mathrm{ele}})] \cdot [\phi|_0, \phi|_2, \phi|_3, \nabla_0\phi|_0, \nabla_0\phi|_2, \nabla_0\phi|_3],$$

(6.25)

where the vectors $[L_i(\mathbf{x}_0^{\mathrm{ele}})]$ depend only on the coordinates of the corner nodes. The 20 DOF and quasi-DOF are given through a linear mapping

$$[QDOF]_{20} = [M_1(\mathbf{x}_0^{\mathrm{ele}})]_{20\times16} \cdot [DOF]_{16}. \tag{6.26}$$

M_1 stands simply for the first big elemental matrix that we need. Its upper 16×16 matrix is the identity matrix, since the 16 elemental DOF are merely conducted to the first 16 entries of the QDOF vector. The last four rows correspond to the $[L_i(\mathbf{x}_0^{\mathrm{ele}})]$-vectors.

Denoting the vector of polynomial coefficients by $[C]_{20}$, we can relate them linearly to the 20 quasi-DOF by

$$[QDOF]_{20} = [M_2(\mathbf{x}_0^{\mathrm{ele}})]_{20\times20} \cdot [C]_{20}, \tag{6.27}$$

where the matrix M_2 again depends only on the nodal coordinates $\mathbf{x}_0^{\mathrm{ele}}$ and the interpolating polynomial. By inverting M_2, we can determine a mapping

$$[C]_{20} = [M_2(\mathbf{x}_0^{\mathrm{ele}})]_{20\times20}^{-1} \cdot [M_1(\mathbf{x}_0^{\mathrm{ele}})]_{20\times16} \cdot [DOF]_{16}. \tag{6.28}$$

that relates the 16 elemental DOF uniquely to the 20 monomial coefficients.

Note that we did not use a reference element. Also, the product $[M_2(\mathbf{x}_0^{\mathrm{ele}})]_{20\times20}^{-1} \cdot [M_1(\mathbf{x}_0^{\mathrm{ele}})]_{20\times16}$ is calculated only once numerically for each element in a preprocessing step, and is then stored, such that one can directly compute the coefficients $[C]_{20}$ for any $[DOF]_{16}$-vector quite efficiently. With the coefficients at hand, one can then easily evaluate the shape functions and their derivatives at any given coordinate. We subject the implementation of the element to a patch test (Sect. 6.4.4) and the element itself to a convergence study (Sect. 6.8).

6.4.3 Numerical Integration

To fully integrate a third order polynomial over a tetrahedral domain, 5 integration points are sufficient. However, Babuska et al (2011) recommend an overintegration when increased regularity requirements are present. A list of quadrature rules can be found in Jinyun (1984) and at the accompanying web page[4]. We observed indeed a better performance when using 29 integration points with only positive weights, which are sufficient to fully integrate a sixth order polynomial. The integration point

---

[4] http://nines.cs.kuleuven.be/ecf/mtables.html

Fig. 6.2 The 29 integration points, with the relative
weight represented by the
spheres diameters.

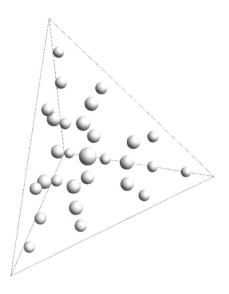

coordinates and their weight factors can be found at the Github repository[5] for the
reference element (corner points at 0,0,0; 1,0,0; 0,1,0; 0,0,1), as depicted in Fig. 6.2.

It remains to discuss the determination of integration points in the element as
it is embedded in the meshing. With only four nodes, one can uniquely identify
a linear mapping that maps line elements from the reference element, for which
the integration point coordinates are given, to the element as it is embedded in the
meshing. Thus, the integration point coordinates $\mathbf{g}$ are given by

$$\mathbf{g} = \mathbf{x}_0^{\text{ele}} + \left[\left(\mathbf{x}_1^{\text{ele}} - \mathbf{x}_0^{\text{ele}} \right) \otimes \mathbf{e}_1 + \left(\mathbf{x}_2^{\text{ele}} - \mathbf{x}_0^{\text{ele}} \right) \otimes \mathbf{e}_2 + \left(\mathbf{x}_3^{\text{ele}} - \mathbf{x}_0^{\text{ele}} \right) \otimes \mathbf{e}_3 \right] \cdot \widetilde{\mathbf{g}}$$

in terms of the integration point coordinates $\widetilde{\mathbf{g}}$ in the reference element and the
location vectors of the four corner nodes $\mathbf{x}_i^{\text{ele}}$ of the element as embedded in the
meshing.

6.4.4 Testing of the Implementation

The element has been implemented as a user element in the FE system Abaqus and
then subjected to a patch test, employing the material model given in Sect. 6.3. A
cubic domain has been meshed by a large number of elements. The boundary conditions where such that the body is subjected to a homogeneous displacement gradient

---

[5] https://github.com/Ra-Na/C3D4C1-Abaqus-UEblob/master/2_Mathematica/numerical_integration.p

$\overline{\mathbf{H}}$, i.e. the displacements at the boundary are given by $\overline{\mathbf{H}}\mathbf{x}_i^{\text{ele}}$, and the displacement gradient $\overline{\mathbf{H}}$ itself was prescribed. The resultant displacement field was linear according to $\mathbf{u} = \overline{\mathbf{H}} \cdot \mathbf{x}_0$. This is a universal solution of $\mathbf{T} \cdot \nabla_0 - \mathbb{T} \cdot \cdot \nabla_0 \otimes \nabla_0 = \mathbf{o}$, since the homogeneity of $\mathbf{H}$ is conducted to $\mathbf{T}$, which is hence divergence free, while $\mathbb{T} = \mathbb{O}$ due to $\mathbb{F} = \mathbb{O}$ due to the homogeneity of $\mathbf{H}$. It would be nice to have nontrivial universal solutions to subject the element to a more thorough testing. Unfortunately, there are none at hand for a strain-gradient elastic material so far.

6.5 Single Force Indentation Simulations

To examine the effect of the strain gradient contribution, we subjected a disk to an in-plane, centric single force, according to Fig. 6.3. The deformation is prescribed to be plane, thus we used only one element for the thickness in direction of $\mathbf{e}_3$.

6.5.1 The Boundary Conditions

$\mathbf{t}_b$ and $\mathbf{T}_b$ have been prescribed as stress free everywhere except at the two force application points at the top of the plane, which correspond to both sides of the plate in $\mathbf{e}_3$ direction. At these points, an indentation force of 1 N has been applied. At the bottom, displacements are only allowed parallel to $\mathbf{e}_1$ except at the origin, which is displacement free. Everywhere in the model u_3 is zero and

$$H_{33} = H_{13} = H_{31} = H_{23} = H_{32} = 0 \tag{6.29}$$

due to the plane strain requirement.

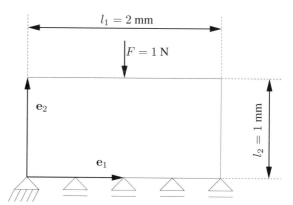

Fig. 6.3 Elastostatic boundary value problem for the examination of the single force singularity.

6.5.2 Meshing

We examined meshes of three degrees of coarseness, namely meshes with 4704, 22021 and 72594 elements. The mesh density is locally increased near the force application point. The smallest element edge length is halfed at each step of mesh refinement.

6.5.3 Results

6.5.3.1 Transition Behavior as $\alpha = 0 \ldots \infty$

Keeping $E = 200.000$ MPa and $\nu = 0.3$ fixed and increasing α from 0 (classical theory) to infinity can physically be interpreted as equipping the elastic material with an increasing resistance against inhomogeneous strain fields. For the described test, the displacement in e_1 and e_2 directions is plotted in Fig. 6.4 for the finest of the three meshes. One can see that the displacement field becomes more linear, i.e. less curved, as α is increased.

6.5.3.2 Singularity

Choosing $\alpha = 0$ deactivates the strain gradient contribution. As expected, the equivalent von Mises stress σ_v, as well as the displacement, diverge when the mesh is refined, see Tables 6.1 and 6.2. Interpreting these stresses is futile, since they

Table 6.1 Largest equivalent stress near the point of load application.

α in N	max. σ_V in MPa for Mesh 1	max. σ_V in MPa for Mesh 2	max. σ_V in MPa for Mesh 3
0	812.0	2015.0	3397.5
100	152.4	158.7	169.1
200	109.1	106.0	117.8
1000	47.73	45.03	49.83
10000	17.86	17.37	19.5
100000	13.32	13.38	14.96

Table 6.2 Indentation displacement for different meshes and parameters α.

α in N	max. $|u_y|$ in mm for Mesh 1	max. $|u_y|$ in mm for Mesh 2	max. $|u_y|$ in mm for Mesh 3
0	0.00026	0.000299	0.000338
100	0.000143	0.00013	0.000143
200	0.0001209	0.000117	0.0001222
1000	0.0000806	0.0000793	0.0000819
10000	0.000052	0.0000507	0.000052
100000	0.0000455	0.0000455	0.0000455

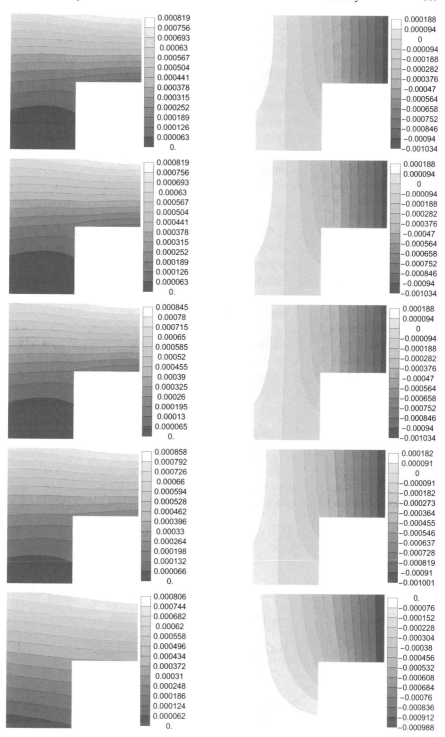

Fig. 6.4 Displacement in $\mathbf{e}_1$ (left) and $\mathbf{e}_2$ (right) directions for different values of α ($\alpha = \{0, 100, 1000, 10000, 100000\}$ N from top to bottom for the finest meshing.

are controlled purely numerically, as the analytical solution has a singularity there (Boussinesq (1885), see also Bertram and Glüge (2015) problem 16). Workarounds have been developed in order to make sense of such stress singularities, specifically at crack tips. Griffith (1921) firstly observed that a volume integral of the strain energy enclosing the crack tip remains bounded, which led to the development of the J-integral (Rice, 1968). This underlines that the interpretation and use of a singular solution requires an integral evaluation. Stresses and strains cannot be accounted for near singularities in classical elasticity. Consequently, the bulk of classical yield-, flow-, fracture or phase transformation criteria, all formulated in terms of stresses or strains, can only be used in sufficient distance from singularities, according to the de Saint-Venant's principle. Although we have grown used to this, it is not a satisfactory situation.

The strain gradient extension appears to remove the classical stress singularity. For large values of α, the equivalent von Mises stress depends only weakly on the FE discretization. The very same holds for the indentation displacement, which classically diverges, but converges if a strain gradient contribution is included.

Considering the maximum strain gradient energy, we see that the value does not converge: Even for $\alpha = 100000$ N, we find the values $1.698 \ 10^{-3}$, $3.726 \ 10^{-3}$ and $19.72 \ 10^{-3}$ as the element edge length is halved. Thus, we can confirm that the strain gradient extension pushes the stress singularity to the higher order stresses (Reiher et al, 2017). Consequently, if one can give a physical interpretation to the strain gradient energy and measure the higher order elasticities, one is able to make sense of the classical stresses even at singularities.

6.5.4 A Comment on Pseudorigid Bodies

Steigmann (2006); Casey (2006) consider materials that can only undergo homogeneous deformations. Such materials are called pseudorigid. While rigid bodies can only undergo motions of the form

$$\mathbf{x}(t) = \mathbf{F}(t) \cdot \mathbf{x}_0 + \mathbf{c}(t), \qquad \mathbf{F} \in Orth^+, \tag{6.30}$$

pseudorigid bodies obey

$$\mathbf{x}(t) = \mathbf{F}(t) \cdot \mathbf{x}_0 + \mathbf{c}(t), \qquad \mathbf{F} \in Inv^+. \tag{6.31}$$

These global constraints can well be written as local constraints, namely

$$\mathbf{F} \in Orth^+, \qquad \mathbf{F} \otimes \nabla_0 = \mathbb{O} \tag{6.32}$$

and

$$\mathbf{F} \in Inv^+, \qquad \mathbf{F} \otimes \nabla_0 = \mathbb{O}. \tag{6.33}$$

In the rigid case, the restriction of $\mathbf{F}$ to $Orth^+$ may be written as $\mathbf{F}^T\mathbf{F} = \mathbf{I}$. There is a well established methodology due to Truesdell and Noll (1965) that allows to deal with constraints on $\mathbf{F}$. It results in an additive split of the stresses into a reactive part and a constitutive part. The constitutive part depends on the motion alone, while the reactive part cannot be determined from the motion alone. In the rigid case, the total of six constraints on $\mathbf{F}$ renders the Cauchy stresses completely reactive.

Casey (2006) points out that in the pseudorigid case, no constraints on $\mathbf{F}$ exist, and a different approach to determine the stresses is required, around which a dispute with Steigmann evolves. The strain gradient elasticity is a versatile tool to examine the pseudorigid case by considering the limits $\alpha \to \infty$ and $\mathbb{F} \to \mathbb{O}$ (Bertram and Glüge, 2016). Note that for the strain energy at hand we have $g = 0 \Leftrightarrow \mathbb{F} = \mathbb{O}$. For the latter boundary value problem, we have seen that the displacement field becomes linear as α is increased, producing a homogeneous (pseudorigid) deformation. Thus, by taking $\alpha \to \infty$, we should be able to tell what is going on. Having a look at our stress-strain relations Eq. (6.15), we see that both stresses are linear in α but $\mathbf{T}$ is quadratic in $\mathbb{F}$ while $\mathbb{T}$ is linear in $\mathbb{F}$. Thus, if the limit $\alpha\mathbb{F}$ with $\alpha \to \infty$ and $\mathbb{F} \to \mathbb{O}$ leaves $\mathbb{T}$ undetermined but finite, we have to conclude that the strain gradient contribution to $\mathbf{T}$ vanishes as it is quadratic in $\mathbb{F}$. This appears to be a point in favor of Steigmann (2006), who argues that the pseudorigid constraint does not affect $\mathbf{T}$.

6.6 Sharp Corner Simulations

In this section, we study the effect of a different kind of singularity. Instead of a singular stress at the boundary, we consider a singularity in the geometry of the boundary, namely an L-shaped bending sample, see Fig. 6.5. Again, the deformation is prescribed to be plane, thus we used only one element for the thickness in direction of $\mathbf{e}_3$.

6.6.1 The Boundary Conditions

Where no displacement or displacement gradient boundary conditions have been prescribed, $\mathbf{t}_b$ and $\mathbf{T}_b$ have been prescribed to be zero. At the bottom, displacements are only allowed parallel to $\mathbf{e}_1$ except at the origin, which is displacement free. At the vertical upper right edge, the displacement in direction of $\mathbf{e}_1$ is free, while the displacement in the $\mathbf{e}_2$ direction is confined to -0.001 mm. Everywhere in the sample u_3 is zero, and due to the plane strain requirement we have also

$$H_{33} = H_{13} = H_{31} = H_{23} = H_{32} = 0 \tag{6.34}$$

everywhere.

Fig. 6.5 Elastostatic boundary value problem for the
examination of the sharp corner singularity.

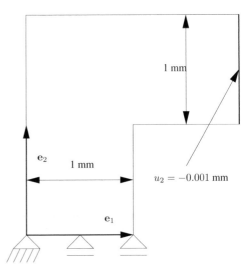

6.6.2 Meshing

We examined meshes of five degrees of coarseness, listed in Table 6.3.

Table 6.3 Mesh densities for the sharp corner simulations.

Mesh no.	No. of elements
mesh 1	2372
mesh 2	3977
mesh 3	14248
mesh 4	28611
mesh 5	58902

6.6.3 Results

6.6.3.1 Transition Behavior as $\alpha \to \infty$

Again, we kept $E = 200\,000$ MPa and $\nu = 0.3$ constant and increased α from 0 (classical theory) to a large value. For the described test, the displacement in the e_1 and e_2 directions is plotted in Fig. 6.6 for mesh 4. Again, one can see that the displacement field becomes smoother, i.e. less curved, as α is increased. Note that the displacement field does not become linear, since this is not possible due to the displacement boundary conditions. Likewise, the equivalent von Mises stress distribution becomes smoother as α is increased.

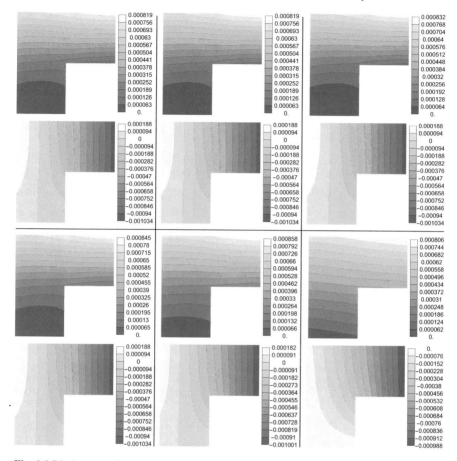

Fig. 6.6 Displacement field in mm in $\mathbf{e}_1$ (upper subfigure) and $\mathbf{e}_2$ (lower subfigure) directions for $\alpha = \{0, 100, 200, 1000, 10000, 100000\}$ N, row-wise from top left to bottom right.

For small values of α, one recognizes the typical linear bending stress distribution in the lower section of the L. The neutral axis is shifted slightly to the left due to the superimposed normal force, see the first subfigure of Fig. 6.7. As α is increased, the bending stress distribution becomes smoother and eventually fades away, and the stress concentration changes its position and its maximum value. It moves from the sharp corner to the right bottom corner. This can be seen in Table 6.4 by the nonmonotonic behavior of max. σ_V as α is increased. For small α, σ_V diverges at the sharp corner as the mesh is refined. For large alpha, the stress maximum is in the bottom right corner, but it converges. For intermediate α, the stresses distribute between these two locations, resulting in a decrease of the maximum σ_V.

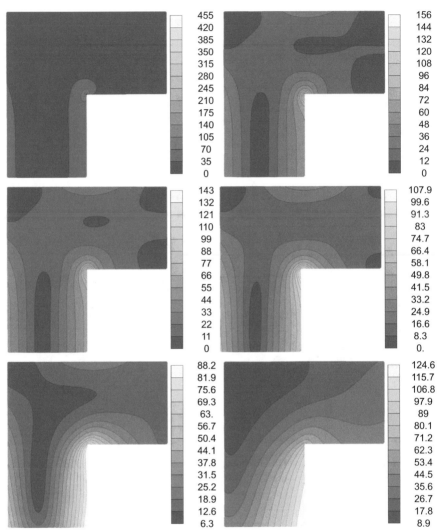

Fig. 6.7 Equivalent von Mises stress distribution in MPa for different values of α
($\alpha = \{0, 100, 200, 1000, 10000, 100000\}$ N, row-wise from top left to bottom right.

6.6.3.2 Convergence on Mesh Refinement

Considering the strain energies, we observe that the classical part of the strain energy (and hence the classical stresses) converge, while the gradient part does not. In Fig. 6.8, one can see that for $\alpha = 1000$ N and the 5 different meshes, the global energy and the classical strain energy density converge, but the strain gradient energy density diverges as the mesh is refined. This underlines the capability of the

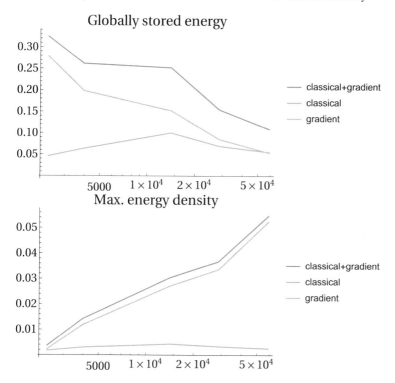

Fig. 6.8 Globally stored energy and maximum energy density in V for the sharp corner simulations with $\alpha = 1000$ N for different meshes.

strain gradient extension to push the singularities to higher order stresses (Reiher et al, 2017). One notices that the classical strain energy firstly increases as the mesh is refined, which is not expected from the FE method: As the mesh is refined, the minimum of Π (Eq. (6.9)) should be approached from above. The reason for this behavior is the incompatible mode element formulation, discussed in Sect. 6.8.

Table 6.4 Largest von Mises stress for different mesh densities and parameters α.

α in N	max(σ_V) for Mesh 1	max(σ_V) for Mesh 2	max(σ_V) for Mesh 3	max(σ_V) for Mesh 4
0	125.9	229.9	300.2	465.1
100	102.3	140.7	161.4	166.9
200	95.51	140.1	141.8	146.1
1000	92.89	152.1	107.8	109.4
10000	122.0	173.3	89.86	89.97
100000	166.4	180.6	129.4	129.4

6.7 Improvement of the Element Formulation

As discussed in Sect. 6.4.4, the implemented element passes the patch test. Nevertheless, it has some disadvantageous properties. It is constructed such that the C_1-continuity is only met at the corner points, but not along the faces and edges of the element. There is not even C_0 continuity everywhere at the element surface. Consequently, there are incompatible deformation modes, which reduce the energy below the analytic solution. These incompatibilities can be imagined as cleavages between adjacent elements. By refining the mesh, the incompatible modes are suppressed as the density continuous points is increased.

The presented element formulation ensures C_0-continuity only at the face centers. In order to reduce the incompatibility, we may introduce more connecting points or higher order continuity. One possibility it to blow up the 16 primary DOF up by 40 quasi-DOF, which consist in the function value and the gradient (four quasi-DOF) at the four face centers and the six edge centers. The choice is such that a complete 3D fifth order polynomial can be determine from the total of 56 supporting values. The quasi-DOF at the additional edge centers are obtained by

$$
\phi_{\mathrm{q}ij} = \frac{1}{4} \left(\phi|_i + \phi|_j + (\phi|_i + \nabla\phi|_i \cdot (\mathbf{x}_{ij} - \mathbf{x}_i)) + (\phi|_j + \nabla\phi|_j \cdot (\mathbf{x}_{ij} - \mathbf{x}_j)) \right)
$$
$$
\nabla\phi_{\mathrm{q}ij} = \frac{1}{2} (\nabla\phi|_i + \nabla\phi|_j)
$$
$$
i, j \in \{0, 1, 2, 3\}, \qquad i < j,
$$

(6.35)

where $\phi|_i$ and $\nabla\phi|_i$ denote the value and the gradient of the field ϕ at the ith node, while the $\phi_{\mathrm{q}ij}$ and $\mathbf{x}_{ij}$ denote the quasi-DOF-value of the field and the location vector at the face center between the ith and the jth node. The gradient at the face centers is the linear interpolation of the gradient at the adjacent corner nodes,

$$
\nabla\phi_{\mathrm{q}ijk} = \frac{1}{3} (\nabla\phi|_i + \nabla\phi|_j + \nabla\phi|_k)
$$
$$
i, j, k \in \{0, 1, 2, 3\}, \qquad i < j < k.
$$

(6.36)

6.8 Convergence Study

We compared the performance of both element formulations by checking the convergence upon mesh refinement in a tension test, with rigidly displaced end faces (see Fig. 6.9) on a cubic sample. The strain gradient part has been removed from the material, i.e. $\alpha = 0$ N. Four different meshes have been considered with 211, 1122, 8136 and 57101 elements.

As can be seen in Fig. 6.10, both element types approximate the result from below, which is because of the incompatible modes. For minimization principles, one would expect convergence from above. The incompatible modes allow a reduc-

Fig. 6.9 Classical strain energy in a tensile test with rigidly displaced ends on a cubic sample.

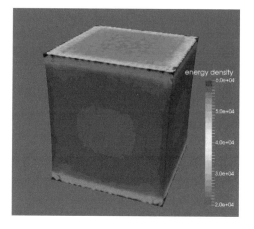

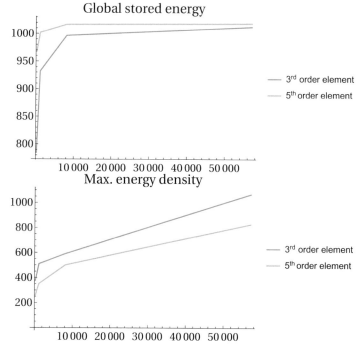

Fig. 6.10 Globally stored energy and maximum energy density in V for a tension test with $\alpha = 0$ N for different meshes.

tion of the potential through discontinuous deformations which are actually not in the function space of admissible solutions. Nevertheless, as the mesh density is increased, the incompatible deformations are suppressed through the increasing number of coupling points. We observe that the element with 10 C_1 continuous couple points (fifth order element) converges faster than the element with four C_0 couple points. This is clear: The incompatible deformation is suppressed better in this case, hence the convergence is faster. At the same time, the maximum strain gradient energy diverges slower for the fifth order element. The classical strain energy density converges in both cases.

6.9 Conclusion

It remains a challenge to construct 3D C_1-continuous finite elements. One can distinguish several strategies, namely

- high order polynomial interpolation,
- transfinite interpolation (Alfeld, 1984), which uses transfinite interpolants from few nodal degrees of freedom,
- macro-elements or composite elements, which remove geometric constraints on the interpolation by splitting the element into sub-elements (Alfeld and Sorokina, 2009), which leads to lower order interpolants on the sub-elements,
- incompatible mode or non-conforming elements, which allow for lower-order interpolants by enforcing C_1 continuity only at specific points.

Of these, the incompatible mode elements yield the simplest interpolants and nodal data structure. Unfortunately, they are not strictly C_1-continuous. Nevertheless, from an engineering point of view, they seem to yield reasonable results. We have used such an approach to implement a strain gradient elasticity, and checked its behavior in problematic boundary value problems. For single force indentation tests, we found that the indentation displacement and von Mises stress remain regular at the indentation point, instead of diverging upon mesh refinement as in classical elasticity. Instead, the higher-order stresses display singular behavior. The same statements hold for a sharp corner bending test. These results are in line with theoretical findings that the discontinuity is shifted to the highest gradient, see, e.g. Reiher et al (2017). We were also able to examine the case that the strain gradient contribution dominates the classical strain energy, which results in a pseudorigid material behavior. To summarize, the incompatible mode elements seem to work nicely and behave as expected.

There is however one drawback: The relaxation of C_1-continuity allows for kinematic incompatibilities at the element faces, like gaps and overlaps. These give enough functional freedom to reduce the global energy *below* the actual solution, such that convergence upon mesh refinement takes place from below. This implies that one needs a fine mesh everywhere to suppress these kinematic incompatibilities, even where the interpolated field has no steep gradients. In contrast, in usual finite element formulations the mesh can be coarser in regions of homogeneous deformation.

References

Abali BE, Müller WH, dell'Isola F (2017) Theory and computation of higher gradient elasticity theories based on action principles. Archive of Applied Mechanics 87(9):1495–1510

Alfeld P (1984) A discrete C^1 interpolant for tetrahedral data. The Rocky Mountain Journal of Mathematics 14(1):5–16

Alfeld P, Sorokina T (2009) Two tetrahedral C^1 cubic macro elements. Journal of Approximation Theory 157(1):53–69

Babuska I, Banerjee U, Li H (2011) The effect of numerical integration on the finite element approximation of linear functionals. Numerische Mathematik 117(1):65–88

Bertram A, Forest S (2014) The thermodynamics of gradient elastoplasticity. Continuum Mechanics and Thermodynamics 26:269–286

Bertram A, Glüge R (2015) Solid Mechanics- Theory, Modeling, and Problems. Springer

Bertram A, Glüge R (2016) Gradient materials with internal constraints. Mathematics and Mechanics of Complex Systems 4:1–15

Boussinesq J (1885) Application des potentiels à l'étude de l'équilibre et du mouvement des solides élastiques. Gauthier-Villars, Paris

Casey J (2006) The ideal pseudo-rigid continuum. Proceedings of the Royal Society of London Series A: Mathematical, Physical and Engineering Sciences 462(2074):3185–3195

Ciarlet P (1988) Mathematical Elasticity Volume 1: Three-Dimensional Elasticity. Studies in Mathematics and its Applications, Volume 20, Elsevier Science Publishers B.V.

Dimitrijevic B, Hackl K (2008) A method for gradient enhancement of continuum damage models. Technische Mechanik 28(1):43–52

Fleck N, Hutchinson J (1993) A phenomenological theory for strain gradient effects in plasticity. Journal of the Mechanics and Physics of Solids 41(12):1825–1857

Forest S, Ammar K, Appolaire B, Cordero N, Gaubert A (2014) Micromorphic approach to crystal plasticity and phase transformation. In: Schröder J, Hackl K (eds) Plasticity and Beyond: Microstructures, Crystal-Plasticity and Phase Transitions, Springer, Vienna, CISM International Centre for Mechanical Sciences, Courses and Lectures, vol 550, pp 131–198

Griffith AA (1921) The phenomena of rupture and flow in solids. Philosophical Transactions of the Royal Society of London Series A, Containing Papers of a Mathematical or Physical Character 221(582–593):163–198

Jinyun Y (1984) Symmetric gaussian quadrature formulae for tetrahedronal regions. Computer Methods in Applied Mechanics and Engineering 43:349–353

Ming W, Xu J (2007) Nonconforming tetrahedral finite elements for fourth order elliptic equations. Mathematics of Computation 76(257):1–18

Papanicolopulos SA, Zervos A, Vardoulakis I (2009) A three-dimensional C^1 finite element for gradient elasticity. International Journal for Numerical Methods in Engineering 77(10):1396–1415

Peerlings R, Geers M, de Borst R, Brekelmans W (2001) A critical comparison of nonlocal and gradient-enhanced softening continua. International Journal of Solids and Structures 38(44–45):7723–7746

Phunpeng V, Baiz P (2015) Mixed finite element formulations for strain-gradient elasticity problems using the fenics environment. Finite Elements in Analysis and Design 96:23–40

Reiher J, Giorgio I, Bertram A (2017) Finite-element analysis of polyhedra under point and line forces in second-strain gradient elasticity. Journal of Engineering Mechanics 143(2)

Rice J (1968) A path independent integral and the approximate analysis of strain concentration by notches and cracks. Transaction of ASME Journal of Applied Mechanics 35:379–386

Sinclair G (2004) Stress singularities in classical elasticity—I: Removal, interpretation, and analysis. Applied Mechanics Reviews 57(4):251–298

Steigmann D (2006) On pseudo-rigid bodies. Proceedings of the Royal Society of London Series A: Mathematical, Physical and Engineering Sciences 462(2066):559–565

Truesdell CA, Noll W (1965) The non-linear field theories of mechanics. In: Flügge S (ed) Encyclopedia of Physics / Handbuch der Physik, Springer, Berlin, vol III/1: Principles of Classical Mechanics and Field Theory / Prinzipien der Klassischen Mechanik und Feldtheorie

Walkington N (2014) A C^1 tetrahedral finite element without edge degrees of freedom. SIAM Journal on Numerical Analysis 52:330–342

Wulfinghoff S, Böhlke T (2012) Equivalent plastic strain gradient enhancement of single crystal plasticity: theory and numerics. Proceedings of the Royal Society of London Series A: Mathematical, Physical and Engineering Sciences 468:2682–2703

Zeníšek A (1973) Polynomial approximation on tetrahedrons in the finite element method. Journal of Approximation Theory 7(4):334–351

Zybell L, Mühlich U, Kuna M, Zhang Z (2012) A three-dimensional finite element for gradient elasticity based on a mixed-type formulation. Computational Materials Science 52(1):268–273

Chapter 7
A Comparison of Boundary Element Method and Finite Element Method Dynamic Solutions for Poroelastic Column

Leonid A. Igumnov, Aleksandr A. Ipatov, Andrey N. Petrov, Svetlana Yu. Litvinchuk, Aron Pfaff, and Victor A. Eremeyev

Abstract Boundary element approach for solving boundary-value problems of saturated poroelastic solid dynamics is presented. Our boundary element approach is based on step scheme for numerical inversion of Laplace transform. Biot's model of poroelastic media with four base functions is employed in order to describe wave propagation process, base functions are skeleton displacements and pore pressure of the fluid filler. The problem of the load acting on a poroelastic prismatic solid is solved by means of developed software based on boundary element approach. A comparison of obtained BEM solution with numerical-analytical solution and also with FEM solution from ANSYS is presented.

Keywords: Boundary element method · Finite element method · Dynamic solution · Boundary integral equation · Poroelastic column

7.1 Introduction

Porous medium is a solid with pore system, filled with a liquid or gas. Wide range of natural and artificial materials can be treated as a porous media, for example

L.A. Igumnov, A.A. Ipatov, A.N. Petrov, S.Yu. Litvinchuk, V.A. Eremeyev
Research Institute for Mechanics, Nizhny Novgorod Lobachevsky State University, 23, Gagarin av. 603950 Nizhny Novgorod, Russia
e-mail: igumnov@mech.unn.ru; ipatov@mech.unn.ru; andrey.petrov@mech.unn.ru

A. Pfaff
Fraunhofer Institute for High-Speed Dynamics, Ernst-Mach-Institut, Ernst-Zermelo-Str. 4, Freiburg, 79104, Germany
e-mail: Aron.Pfaff@emi.fraunhofer.de

V.A. Eremeyev
Faculty of Civil and Environmental Engineering, Gdańsk University of Technology, ul. Gabriela Narutowicza 11/12, Gdańsk, 80-233, Poland
e-mail: eremeyev.victor@gmail.com

© Springer Nature Switzerland AG 2019
H. Altenbach et al. (eds.), *Higher Gradient Materials and Related Generalized Continua*, Advanced Structured Materials 120,
https://doi.org/10.1007/978-3-030-30406-5_7

121

rocks, soils, biological tissues, foams, ceramics, etc. Since poroelastic materials extensively used in many branches of industry, theoretical and experimental researches of its dynamic behavior is a great interest of many disciplines. It is important for development of ideas about the processes accompanying the use of modern technologies connected with porous media. Satisfactory accuracy of the results of such studies cannot be achieved using elastic or viscoelastic models of the material and requires the development of adequate mathematical, methodological and software tools.

Studying of the dynamic processes in porous media began from the discovery of the experimental law of liquid filtration in a porous medium, made by Darcy (1956). Model of poroelastic medium and effective stress principal for taking into account an influence of pore filler on quasistatic deformation of soils first was introduced by Terzaghi (1923). In 1935 by generalizing of Terzaghi's approach on three dimensional case, Biot introduced theory of classical three dimensional quasistatic poroelasticity (Biot, 1935). In 1944 J. Frenkel developed full set of dynamic equations that describes acoustics of isotropic poroelastic media (Frenkel, 2005). Thereafter M. Biot introduced the set of differential equations for porous media motion based on generalized relations between stresses and strains of Frenkel's approach (Biot, 1956a, 1962, 1956b). Biot theory is a significant contribution in development of continuum mechanics, it is a generalization of classical theory of elasticity on saturated porous medium.

Biot's model correctly describes processes of deformation of an elastic porous medium and fluid flow in it. It is assumed that the space containing poroelastic medium is filled with a two-phase material, and one phase corresponds to the elastic skeleton, and the another one to the fluid in pores. Both phases are present at each point of the physical space, and the phase distribution in space is described by macroscopic quantities such as porosity. The fundamental property of a poroelastic saturated medium, following from Biot's theory, is the existence of two longitudinal waves in such media, fast and slow, and also a shear wave. The fast longitudinal wave and the transversal wave are similar in their nature to waves in an elastic medium, whereas the slow longitudinal wave is characteristic of a porous medium. The slow wave is more difficult to detect, as its amplitude is considerably smaller than that of the fast longitudinal wave.

Nowadays the analyzing and optimization of oil and gas fields development, earthquake proof construction and bioengineering became more and more complicated and multidisciplinary, that leads to increasing of experimental research cost for such problems. Therefore, the use of mathematical modeling in such cases as an additional research tool to reduce the expenditure of funds and resources is not only preferable, but also necessary. Due to the active development of computing in the last decade, various numerical methods are widely used in many engineering applications. The most commonly used methods for solving problems of deformable solids mechanics are the finite element method (FEM) and boundary element method (BEM). A detailed review of studies devoted to the modelling of liquid-saturated porous media using BEM and FEM is presented in Schanz (2009).

7.2 Mathematical Model

The constitutive equations for a poroelastic medium in terms of stress according to works by Biot can be formulated as follows:

$$\sigma_{ij} = 2G\varepsilon_{ij}^{s} + \left(K - \frac{2}{3}G\right)\varepsilon_{kk}^{s} - \alpha\delta_{ij}p , \qquad (7.1)$$

where K and G are the elastic moduli of the porous material, p is the porous pressure of the filling material. Coefficient α is determined as

$$\alpha = 1 - \frac{K}{K^{s}} , \qquad (7.2)$$

where K^{s} is bulk modulus of the skeleton grains. The stress tensor components of the solid body are ε_{ij}^{s} . The kinematic relations, in the assumption of small strain gradients, have the following form:

$$\varepsilon_{ij}^{s} = \frac{1}{2}\left(u_{i,j}^{s} + u_{j,i}^{s}\right) , \qquad (7.3)$$

where u_i^s are solid body displacements.

In addition to net stress σ_{ij} , change of the liquid volume per unit of initial volume is introduced as

$$\zeta = \alpha\varepsilon_{kk}^{s} + \frac{\phi^2}{R}p , \qquad (7.4)$$

where porosity is designated as

$$\phi = \frac{V^{f}}{V} , \qquad (7.5)$$

The parameter characterizing the relation between the solid body and the liquid:

$$R = \frac{\phi^2 K^{f}K^{s2}}{K^{f}(K^{s} - K) + \phi K^{s}(K^{s} - K^{f})} . \qquad (7.6)$$

In (7.5), V^{f} is the volume of interconnected pores in a specimen with volume V, in (7.6), K^{f} is the bulk modulus of the liquid. Change of the liquid is described by the continuity equation:

$$\dot{\zeta} + q_{i,i} = 0 . \qquad (7.7)$$

In (7.7), the notion of specific flow, or filtration vector is introduced as

$$q_i = \phi\left(\dot{u}_i^{f} - \dot{u}_i^{s}\right) , \qquad (7.8)$$

where u_i^f is displacement of the liquid. To describe liquid transfer, a dynamic version of Darcy's law is used:

$$q_i = -\kappa\left(p_{,i} + \frac{\varrho_a}{\phi}\left(\ddot{u}_i^{f} - \ddot{u}_i^{s}\right) + \varrho_f\right) \qquad (7.9)$$

where κ is permeability, ϱ_f is density of the filling material. To describe the dynamic interaction between the liquid and the skeleton, Biot introduced an additional density:

$$\varrho_a = B\phi\varrho_f , \qquad (7.10)$$

where B is a coefficient depending on the pore geometry and excitation frequency. In what follows, the coefficient B is taken equal to 0.66.

7.2.1 u_i^s-p-formulation in Laplace Domain

Transformation into Laplace region makes it possible to write an expression for q_i, keeping in mind (7.8) and (7.9), as follows:

$$\hat{q}_i = -\frac{\beta}{s^2\varrho_f}(\hat{p}_{,i} + s^2\varrho_f\hat{u}_i^s) ,$$

$$\beta = \frac{\kappa\varrho_f\phi^2 s^2}{\phi^2 s + s^2\kappa(\varrho_a + \phi\varrho_f)} . \qquad (7.11)$$

Symbol $\hat{}$ denotes Laplace transform with complex variable s. The momentum balance equation for a mixture is formulated in Biot's work. Dynamic equilibrium, with the account of (7.11), is defined as:

$$\hat{\sigma}_{ij,j} + \hat{F}_i + s^2\varrho\hat{u}_i^s = \beta\left(\hat{p}_{,i} - s^2\varrho_f\hat{u}_i^s\right) , \qquad (7.12)$$

where $\varrho = \varrho_s(1 - \phi) + \phi\varrho_f$, ϱ_s is density of the skeleton grains, $\hat{F}_i$ is volume force density.

The differential equation system in Laplace representations for displacements $\hat{u}_i^s$ and pore pressure $\hat{p}$ in the absence of volume forces has the following form:

$$G\hat{u}_{i,jj}^s + \left(K + \frac{1}{3}G\right)\hat{u}_{j,ij}^s - (\alpha - \beta)\hat{p}_{,i} - s^2(\varrho - \beta\varrho_f)\hat{u}_i^s = 0 , \qquad (7.13)$$

$$\frac{\beta}{s\varrho_f} - \frac{\phi^2 s}{R}\hat{p}_i - (\alpha - \beta)s\hat{u}_{i,i}^s = 0, \; \mathbf{x} \in \Omega, \; \Omega \subset \mathbf{R}^3 . \qquad (7.14)$$

A generalized displacement vector and a generalized force vector are additionally introduced as

$$\mathbf{u}(\mathbf{x}, s) = (\hat{u}_1^s, \hat{u}_2^s, \hat{u}_3^s, \hat{p}) , \qquad (7.15)$$

$$\mathbf{t}(\mathbf{x}, s) = \left(\hat{t}_1^s, \hat{t}_2^s, \hat{t}_3^s, \hat{q}\right) , \qquad (7.16)$$

where $\hat{t}_i = \hat{\sigma}_{ij}n_i$ and $\hat{q} = \hat{q}_in_i$, n_i are components of the vector of the normal to the boundary of region Ω. Equations (7.13) and (7.14), supplemented with boundary conditions:

$$\mathbf{u}(\mathbf{x}, s) = \tilde{\mathbf{u}}, \; \mathbf{x} \in \Gamma^u , \qquad (7.17)$$

$$\mathbf{t}(\mathbf{x}, s) = \widetilde{\mathbf{t}}, \ \mathbf{x} \in \Gamma^{\sigma} \ , \tag{7.18}$$

where Γ^{u} is the Dirichlet's boundary and Γ^{σ} is the Neumann's boundary, fully describe the boundary problem in the representations of the 3D isotropic dynamic theory of poroelasticity.

7.2.2 u_i^{s}–p-formulation in Time Domain

Engineering software ANSYS provides tools for modeling poroelastic medium based on Biot model. Finite element implementation is based on motion equations for variables u_i^{s} and p with some simplifications. In particular, the second derivative of the relative displacements of the filler is omitted from dynamic equilibrium equation, and also the static Darcy's law is used. The corresponding equations in this case read as follows:

$$\sigma_{ij,j} + F_i = \varrho \ddot{u}_i^{\mathrm{s}} \ , \tag{7.19}$$

$$q_i = -\kappa p_{,i} \ , \tag{7.20}$$

final differential equation system takes the following form:

$$G u_{i,jj}^{\mathrm{s}} + \left(K + \frac{G}{3} \right) u_{j,ij}^{\mathrm{s}} - \alpha p_{,i} - \varrho \ddot{u}_i^{\mathrm{s}} = F_i \ , \tag{7.21}$$

$$\kappa p_{,ii} - \frac{\phi}{R} \dot{p} - \alpha \dot{u}_{i,i}^{\mathrm{s}} = 0 \ . \tag{7.22}$$

7.3 Boundary Integral Equation and Boundary Element Methodology

Boundary-value problem (7.13)–(7.18) is solved using the direct boundary element method (BEM), based on a combined use of integral Laplace transform and boundary integral equation (BIE) of the 3D isotropic theory of poroelasticity:

$$\mathbf{C}(\mathbf{y})\mathbf{u}(\mathbf{y}, s) + \int_{\Gamma} \mathbf{T}(\mathbf{x}, \mathbf{y}, s)\mathbf{u}(\mathbf{x}, s) \Delta\Gamma = \int_{\Gamma} \mathbf{U}(\mathbf{x}, \mathbf{y}, s)\mathbf{t}(\mathbf{x}, s)\Delta\Gamma \ , \mathbf{x} \, , \mathbf{y} \in \Gamma \ ,$$
$$\tag{7.23}$$

where $\mathbf{U}(\mathbf{x}, \mathbf{y}, s)$ and $\mathbf{T}(\mathbf{x},\mathbf{y},s)$ are matrices of fundamental and singular solutions, respectively, $\mathbf{x}$ is an integration point, $\mathbf{y}$ is an observation point. The values of the coefficients of matrix $\mathbf{C}$ are defined by the geometry of boundary Γ . A procedure for obtaining BIE's, based on the weighted residual method can be found in Schanz (2001). Some of the problems of the arising kernels of BIE's are discussed in Amenitsky et al (2009).

Equations (7.23) comprise singular integrals in the sense of Cauchy, which are quite difficult to compute. Use of the boundary properties of retarded potentials makes it possible, based on Igumnov et al (2016), to write down a regular representation of Eq. (7.23):

$$\int_{\Gamma} \Big(\mathbf{T}(\mathbf{x}, \mathbf{y}, s)\mathbf{u}(\mathbf{x}, s) - \mathbf{T}^0(\mathbf{x}, \mathbf{y}, s)\mathbf{u}(\mathbf{y}, s) - $$
$$- \mathbf{U}(\mathbf{x}, \mathbf{y}, s)\mathbf{t}(\mathbf{x}, s) \Big) \Delta\Gamma = 0 \ , \mathbf{x} \ , \mathbf{y} \in \Gamma \ , \tag{7.24}$$

where $\mathbf{T}^0(\mathbf{x}, \mathbf{y}, s)$ is singularity matrix. Using Eq. (7.24), it is possible to construct a boundary-element solution of the BIE.

In the result of spatial discretization, boundary Γ is represented with a set of K_E quadrangular eight-node boundary elements. The geometry of each element E_K is defined by biquadratic functions of form N_m and the global coordinates of nodes $\mathbf{x}_m^k$, related as (Bazhenov and Igumnov, 2008)

$$\mathbf{x}(\zeta) = \sum_{m=1}^{8} N_m(\xi)\mathbf{x}_m^k \ , k = 1..K \tag{7.25}$$

where $\xi = (\xi_1, \xi_2) \in (-1, 1)^2$ are local coordinates. According to correlated interpolation model in Goldshteyn (1978), displacements are described using bilinear elements with the related bilinear functions of form $R_l(\xi)$, and surface generalized forces are described with constant boundary elements:

$$\mathbf{u}(\xi) = \sum_{l=1}^{4} R_l(\xi)\mathbf{u}_l^k \ , \tag{7.26}$$

$$\mathbf{t}(\xi) = \mathbf{t}^k \ , \tag{7.27}$$

where $\mathbf{u}_m^k$ and $\mathbf{t}^k$ are nodal values of displacements and tractions, respectively, over element E_k. A discrete representation of BIE written at the nodes of the approximation of boundary functions $\mathbf{y}^i$, using the collocation method and accounting for (7.25)–(7.27), is of the following form:

$$\sum_{k=1}^{K_E}\sum_{m=1}^{4} \Delta\mathbf{T}_{mi}^k \mathbf{u}_m^k = \sum_{k=1}^{K} \Delta\mathbf{U}_{mi}^k \mathbf{t}^k \ ,$$

$$\Delta\mathbf{U}_{mi}^k = \int_{-1}^{1} \mathbf{U}(\mathbf{x}^k(\xi), \mathbf{y}^i, s)J^k(\xi)\Delta\xi \ ,$$

$$\Delta\mathbf{T}_{mi}^k = \int_{-1}^{1} \Big(R_m(\xi)\mathbf{T}(\mathbf{x}^k(\xi), \mathbf{y}^i, s) - \mathbf{I} \cdot \mathbf{T}^0(\mathbf{x}^k(\xi), \mathbf{y}^i)) \, J^k(\xi)\Delta\xi \ , \tag{7.28}$$

where $\mathbf{I}$ is unit matrix, J^k is Jacobean of the local coordinates into global ones.

The elements of matrices ΔU_{mi}^k, ΔT_{mi}^k are computed using numerical integration schemes depending on the kind of integral (nonsingular or singular). Nonsingular integrals arise, when the collocation point does not belong to the element. Here, standard Gaussian-type formula is used in combination with a hierarchical subdivision of the elements (Lachat and Watson, 1976). Singular integrals arise, when the collocation point is situated on the element being integrated over. In this case, new local coordinates are introduced, making it possible to avoid a singularity in the integrand and to use Gaussian integration.

7.4 Laplace Transform Inversion

Consider a method based on the fundamental integration theorem – the stepped method of numerical inversion of Laplace transform. Consider the following integral:

$$y(t) = \int_0^1 f(\tau)\Delta\tau \tag{7.29}$$

Integral (7.29) gives rise to Cauchy problem for an ordinary differential equation:

$$\frac{\Delta}{\Delta t}x(t) + sxt + C \, , x(0) = 0 \, . \tag{7.30}$$

Integral (7.29) is substituted for by a quadrature sum, weighting factors of which are determined using Laplace representation and the linear multistep method (Igumnov and Petrov, 2016). Further derivation is based on the results of those works. The traditional stepped method of integrating the original is that integral (7.29) is calculated using the following relation:

$$y(0) = 0 \, , y(n\Delta t) = \sum_{k=1}^n w_k(\Delta t) \, , n = 1..N \, ,$$

$$w_n(\Delta t) = \frac{\theta^{-n}}{L} \sum_{l=0}^{L-1} \hat{f}\left(\frac{\gamma(\theta e^{il\frac{2\pi}{L}})}{\Delta t}\right) e^{-inl\frac{2\pi}{L}} \, , \tag{7.31}$$

where Δt is time step, $\gamma(z) = 3/2 - 2z + z^2/2$, N is number of time steps, θ is the parameter of the method.

7.5 Numerical Example

The 3D poroelastic column loaded by a Heaviside-type function is considered as example to study the behavior of transformation method. The width of the column is 1 m, the height – 3 m. The column has zero displacements on one end and prescribed normal force on the other end. A boundary-element mesh of 2744 quadrangular elements is employed in the computations. Finite element mesh is consist of 12288 8-noded elements CPT215. The problem statement is presented in Fig. 7.1. The parameters of the fully saturated porous material corresponds to the sandstone:

$$\phi = 0.23, K = 4.8 \cdot 10^9 \, \text{N/m}^2, G = 7.2 \cdot 10^9 \, \text{N/m}^2,$$
$$K^s = 3.6 \cdot 10^{10} \, \text{N/m}^2, K^f = 3.3 \cdot 10^9 \, \text{N/m}^2,$$
$$\rho = 2458 \, \text{kg/m}^3, \rho_f = 1000 \, \text{kg/m}^3, \kappa = 1.9 \cdot 10^{-10} \, \text{m}^4/(\text{N} \cdot \text{s}).$$

Time domain solution from BEM is obtained with time-step method parameter : $\theta = 0.997$, $\Delta t = 5 \cdot 10^{-6}$ s, $L = 2650$. When solving a problem using ANSYS, the Newmark method is used with the step size $\Delta t = 5 \cdot 10^{-3}$ s. A macro was written in the ANSYS parametric design language to carry out the procedure automatically, the details are given in the Appendix. Both numerical solutions are compared with the 1D numerical-analytical solution in Schanz (2009). Plots for displacements on loaded end and pore pressure on free end, and also in the middle point of the column are presented on Figs. 7.2–7.6.

Figures 7.2 and 7.3 demonstrates that solutions obtained by BEM and FEM with static Darcy's law differ little. The greatest differences are observed in the pore pressure plots and consist in the magnitude of the oscillations at the points of the wave front.

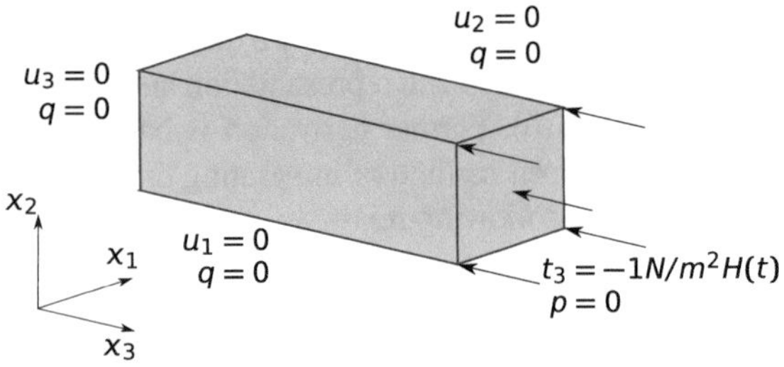

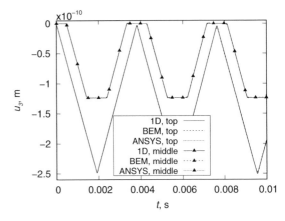

Fig. 7.1 Geometry and boundary conditions of the problem

Fig. 7.2 Comparison of BEM, FEM and numerical-analytical solutions: displacement at the loaded end and at the middle of the column

Fig. 7.3 Comparison of BEM, FEM and numerical-analytical solutions: pore pressure at the middle and at free end of the column

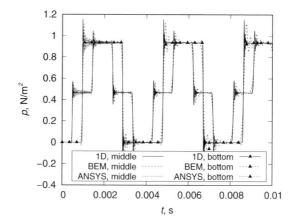

Fig. 7.4 Displacement at the top and the middle of the column for the increased permeability

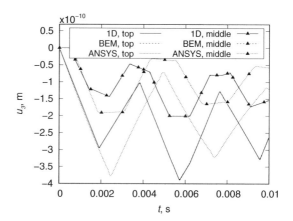

The principal differences may be observed when slow longitudinal wave is modeled. This wave has a large dispersion and is difficult to identify, so for its detection by numerical-analytical methods, it is necessary to artificially reduce the viscosity of the filler, setting a high permeability coefficient. On Figs. 7.4–7.6 calculation results with permeability coefficient $\kappa = 1.9 \cdot 10^{-10}\,\mathrm{m^4/(N \cdot s)}$ are presented.

Figure 7.4 shows that increasing of permeability coefficient value does not allow to achieve investigated effect and identify slow longitudinal wave on displacement responses. Herewith both numerical solutions demonstrate similar behavior, but differ significantly in quantitative terms. However an effect of slow wave appearance is clearly observed on dynamic responses of pore pressure obtained by means of BEM. Dynamic response of pore pressure obtained by means of FEM based on simplified equations and static Darcy's law demonstrates nonphysical behavior.

Fig. 7.5 Pore pressure at the middle of the column for the increased permeability

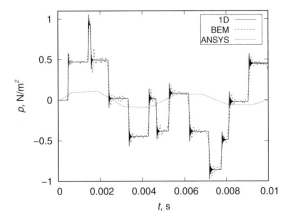

Fig. 7.6 Pore pressure at the bottom of the column for the increased permeability

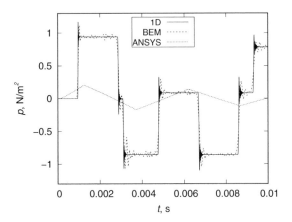

7.6 Conclusion

The results of numerical modeling of the fully saturated poroelastic medium are presented. The Biot model is used as a mathematical model of the porous medium. The application of the BEM and FEM is performed, in both cases the u_i^s–p-formulation is used. The boundary-element scheme is based on boundary integral equations for dynamic poroelasticity applied with time-step method of numerical Laplace transform inversion. Finite element results are obtained using ANSYS. Numerical results comparison provided on the example of the problem about Heaviside-type load acting on the poroelastic column. Significant differences between numerical results are observed in case of slow longitudinal wave modeling. In this case, the simplified equations of motion usage in finite element method software leads to non-physical results.

Acknowledgements This work was supported by a grant from the Government of the Russian Federation (contract No. 14.Y26.31.0031).

Appendix

```
 1  /CLEAR,NOSTAR
 2  /PREP7
 3                              ! Material properties
 4  K = 4.8E+9
 5  G = 7.2E+9
 6  E = 9*K*G/(3*K+G)           ! Young's modulus
 7  NU = (3*K-2*G)/2/(3*K+G)    ! Poisson's ratio
 8  RHO_B = 2458                ! Density of porous media
 9  PHI = 0.19                  ! Porosity
10  FPX =   1.9E-10             ! Permeability
11  K_S = 3.6E+10               ! Bulk modulus of solid
12  K_F = 3.3E+9                ! Bulk modulus of fluid
13
14  TT = 0.01                   ! Time limit
15  NTSTEP = 1000               ! Number of time steps
16
17  DE = 3                      ! Depth of column
18  W = 1                       ! Width of column
19  H = 1                       ! Heigth of column
20  R = 1
21  NELEM =   16                !
22  ET,1,CPT215                 ! 3D 8-node coupled pore
23                              !pressure element
24  KEYOPT,1,12,1               ! Pressure degree
25                              ! of freedom is enabled
26  BLC5, 0, 0, W, H, DE        ! Creating the block
27  AESIZE, ALL, W/NELEM        ! Mapping of mesh
28  MSHKEY, 1
29  VMESH,ALL
30
31  MP,EX,1,E                   ! Set Young's modulus
32  MP,NUXY,1,NU                ! Set Poisson's ratio
33  MP,DENS,1,RHO_B             ! Density of porous media
34  TB,PM,1,,,PERM
35  TBDATA,1,FPX,FPX,FPX        ! Permeability coefficients
36  TB,PM,1,,,SP
37  TBDATA,1,K_S                ! Set bulk modulus of solid
38  TB,PM,1,,,FP
39  TBDATA,1,K_F                ! Set bulk modulus of fluid
40  TBDATA,3,PHI                ! Set porosity
41
42  NSEL,S,LOC,Z,0              ! Boundary conditions
43  D,ALL,UX,0
44  D,ALL,UY,0
45  D,ALL,UZ,0
46  SF,ALL,FFLX,0
47  ALLSEL,ALL
48  NSEL,S,LOC,Z,DE
49  D,ALL,PRES,0
50  ALLSEL,ALL
51  NSEL,S,LOC,X,W/2
```

```
52   D , ALL , UX , 0
53   SF , ALL , FFLX , 0
54   ALLSEL , ALL
55   NSEL , S , LOC , X , -W / 2
56   D , ALL , UX , 0
57   SF , ALL , FFLX , 0
58   ALLSEL , ALL
59   NSEL , S , LOC , Y , H / 2
60   D , ALL , UY , 0
61   SF , ALL , FFLX , 0
62   ALLSEL , ALL
63   NSEL , S , LOC , Y , -H / 2
64   D , ALL , UY , 0
65   SF , ALL , FFLX , 0
66   ALLSEL , ALL
67
68   FINISH
69   / SOLU
70   ANTYPE , TRANSIENT
71   AUTOTS , OFF
72   OUTRES , NSOL , ALL
73   TIME , TT
74   NSUBST , NTSTEP
75   KBC , 1                            ! Stepped  loading
76   NSEL , S , LOC , Z , DE
77   SF , ALL , PRES , 1
78   ALLSEL , ALL
79   SOLVE
80   FINISH
81
82   / POST26
83   NSEL , S , LOC , Z , 3 , 3
84   NSEL , R , LOC , X , 0 , 0
85   NSEL , R , LOC , Y , 0 , 0
86   *GET , NUMBER , NODE , , NUM , MAX
87   ALLSEL , ALL
88   NSOL , 2 , NUMBER , U , Z , UZ_TOP
89   STORE , MERGE
90   *GET , SIZE , VARI , , NSETS
91   NSEL , S , LOC , Z , 1 . 5 , 1 . 5
92   NSEL , R , LOC , X , 0 , 0
93   NSEL , R , LOC , Y , 0 , 0
94   *GET , NUMBER , NODE , , NUM , MAX
95   ALLSEL , ALL
96   NSOL , 3 , NUMBER , U , Z , UZ_MIDDLE
97   STORE , MERGE
98   ANSOL , 4 , NUMBER , S , Z , SZ_MIDDLE
99   STORE , MERGE
100  NSOL , 5 , NUMBER , PRES , , PRES_MIDDLE
101  STORE , MERGE
102  ALLSEL , ALL
103  NSEL , S , LOC , Z , 0 , 0
104  NSEL , R , LOC , X , 0 , 0
105  NSEL , R , LOC , Y , 0 , 0
```

```
106  *GET,NUMBER,NODE, ,NUM,MAX
107  ALLSEL,ALL
108  ANSOL,6 ,NUMBER, S ,Z ,SZ_BOTTOM
109  STORE,MERGE
110  NSOL,7 ,NUMBER,PRES , ,PRES_BOTTOM
111  STORE,MERGE
112
113  *DIM,TIME_H ,ARRAY, SIZE
114  *DIM,UZ_TOP_DATA ,ARRAY, SIZE
115  *DIM,UZ_MIDDLE_DATA ,ARRAY, SIZE
116  *DIM,SZ_MIDDLE_DATA ,ARRAY, SIZE
117  *DIM,PRES_MIDDLE_DATA ,ARRAY, SIZE
118  *DIM,SZ_BOTTOM_DATA ,ARRAY, SIZE
119  *DIM,PRES_BOTTOM_DATA ,ARRAY, SIZE
120  VGET,TIME_H(1) ,1
121  VGET,UZ_TOP_DATA(1) ,2
122  VGET,UZ_MIDDLE_DATA(1) ,3
123  VGET,SZ_MIDDLE_DATA(1) ,4
124  VGET,PRES_MIDDLE_DATA(1) ,5
125  VGET,SZ_BOTTOM_DATA(1) ,6
126  VGET,PRES_BOTTOM_DATA(1) ,7
127  !Write  array  in  given  format  to  file
128  *CFOPEN, 'SOLUTION ' ,DAT
129  *VWRITE,TIME_H(1) ,UZ_TOP_DATA(1) ,UZ_MIDDLE_DATA(1) ,
130  SZ_MIDDLE_DATA(1) ,  PRES_MIDDLE_DATA(1) ,
131  SZ_BOTTOM_DATA(1) ,PRES_BOTTOM_DATA(1)
132  (e15.7 ,1x ,e15.7 ,e15.7 ,1x ,e15.7 ,e15.7 ,1x ,e15.7 ,e15.7)
133  *CFCLOSE
```

References

Amenitsky AV, Belov AA, Igumnov LA, Karelin IS (2009) Granichnye integral'nye uravneniya dlya resheniya dinamicheskikh zadach trekhmernoi teorii porouprugosti (Boundary integral equations for analyzing dynamic problems of 3-D porouselasticity, in Russ.). Problems of Strength and Plasticity 71:164–171

Bazhenov VG, Igumnov LA (2008) Boundary Integral Equations and Boundary Element Methods in Treating the Problems of 3D Elastodynamics with Coupled Fields (in Russ.). Fizmatlet, Moscow

Biot MA (1935) Le problème de la consolidation des matières argileuses sous une charge. Annales de la Société Scientifique de Bruxelles, série B 55:110–113

Biot MA (1956a) Theory of propagation of elastic waves in a fluid-saturated porous solid. I. low-frequency range. The Journal of the Acoustical Society of America 28(2):168–178, DOI 10.1121/1.1908239

Biot MA (1956b) Theory of propagation of elastic waves in a fluid-saturated porous solid. II. higher frequency range. The Journal of the Acoustical Society of America 28(2):179–191, DOI 10.1121/1.1908241

Biot MA (1962) Generalized theory of acoustic propagation in porous dissipative media. The Journal of the Acoustical Society of America 34(9A):1254–1264, DOI 10.1121/1.1918315

Darcy H (1956) Les fontaines publiques de la ville de Dijon: Exposition et application des principes a suivre et des formules a employer dans les questions de distribution d'eau; ouvrage terminé par un appendice relatif aux fournitures d'eau de plusieurs villes au filtrage des eaux et a la fabrication des tuyaux de fonte, de plomb, de tole et de bitume. Dalmont, Paris

Frenkel J (2005) On the theory of seismic and seismoelectric phenomena in a moist soil. Journal of Engineering Mechanics 131(9):879–887, DOI 10.1061/(ASCE)0733-9399(2005)131:9(879)

Goldshteyn RV (1978) Boundary Integral Equations Method: Numerical Aspects and Application in Mechanics. Mir, Moskva

Igumnov LA, Petrov AN (2016) Modelirovanie dinamiki chastichno nasyshchennykh poro-uprugikh tel na osnove metoda granichno-vremennykh elementov (Dynamics of partially saturated poroelastic solids by boundary-element method (in Russ.). PNRPU Mechanics Bulletin 3:47–61, DOI 10.15593/perm.mech/2016.3.03

Igumnov LA, Litvinchuk SY, Petrov AN, Ipatov AA (2016) Numerically analytical modeling the dynamics of a prismatic body of two- and three-component materials. In: Advanced Materials, Springer Proceedings in Physics, vol 175, pp 505–516, DOI 10.1007/978-3-319-26324-3-35

Lachat JC, Watson JO (1976) Effective numerical treatment of boundary integral equations: A formulation for three-dimensional elastostatics. International Journal for Numerical Methods in Engineering 10(5):991–1005, DOI 10.1002/nme.1620100503

Schanz M (2001) Wave Propagation in Viscoelastic and Poroelastic Continua: A Boundary Element Approach. Springer-Verlag, Berlin

Schanz M (2009) Poroelastodynamics: Linear models, analytical solutions, and numerical methods. Applied Mechanics Reviews 62(3):030,803, DOI 10.1115/1.3090831

Terzaghi K (1923) Die Berechnung der Durchlassigkeitsziffer des Tones aus dem Verlauf der Hydrodynamichen Spannungserscheinungen. Akademie der Wissenschaften in Wien, Math Naturwiss Kl, Sitzungsberichte Abteilung II 132:125–138

Chapter 8
From Generalized Theories of Media with Fields of Defects to Closed Variational Models of the Coupled Gradient Thermoelasticity and Thermal Conductivity

Sergey Lurie and Petr Belov

Abstract A gradient theory of the coupled theory of elasticity, thermoelasticity and thermal conductivity based on a generalized model of media with fields of defects is developed. Defectiveness is defined only by free dilatation and the tensor of incompatible distortions is determined by the spherical tensor. In the general case, the unified model includes scale parameters that are responsible for mechanical and temperature scale effects. In the particular case the proposed model describes the gradient thermoelasticity, in which effects are controlled by a mechanical scale parameter, and in the limit case, it describes the classical thermoelasticity, when this parameter tends to zero. The analysis of boundary problems of the general model is given. Particular cases are considered, and it is shown that gradient thermal conductivity and thermoelasticity make it possible to simulate thermal resistance and size effects.

Keywords: Generalized Mindlin's medium · Free dilatations · Gradient thermoelasticity · Reversible thermal conductivity · Cohesive interactions · Scale effects

8.1 Introduction

For classical thermoelasticity, there are excellent monographs and review studies (Landau and Lifshitz, 1986; Nowacki, 1986; Coleman and Noll, 1963; Carlson, 1972; Müller and Ruggeri, 1993). However, the classical theory becomes insufficient for the description of distribution of heat at low temperatures (Joseph and Preziosi, 1989) for understanding the thermoelastic properties of small-sized systems (Lepri et al, 2003; Dhar, 2008), as well as for the development of nanostructured microelectronic components and various thermoelectric devices (Zhang and

S. Lurie · P. Belov
Institute of Applied Mechanics of Russian Academy of Sciences, Moscow, Russia
e-mail: salurie@mail.ru; belovpa@yandex.ru

© Springer Nature Switzerland AG 2019
H. Altenbach et al. (eds.), *Higher Gradient Materials and Related Generalized Continua*, Advanced Structured Materials 120,
https://doi.org/10.1007/978-3-030-30406-5_8

Li, 2010). To take into account the scale effects that manifest themselves in hetero-geneous materials within micro nanostructures, thermoelasticity and thermal con-ductivity should be carefully analyzed and, if possible, modified. For example, it is known that the wave properties of heat transfer can manifest themselves as conse-quences of scale effects (Gusev and Lurie, 2013; Gendelman and Savin, 2010).

Coupled effects can be significant, they are usually ignored, i.e. it is believed that changes in elastic deformation in classical uncoupled thermoelasticity do not affect the temperature distribution and vice versa. In Biot (1956) it was formulated the theory of coupled thermoelasticity, which allows to eliminate this disadvantage of classical uncoupled theory in describing the dynamic behavior of structures. Unfor-tunately, the equations describing the heat transfer process for both coupled and un-coupled theories, although being different, are of the diffusion type, predicting infi-nite speeds of heat distribution, which contradicts both physical intuition and exper-imental observations. This contradiction was removed as a result of the development of thermodynamics and the generalization of the theory of thermal conductivity and coupled thermoelasticity, originally due to Maxwell (1865); Vernotte (1958); Catta-neo and Hebd (1958), and then due to basic researches (Lord and Shulman, 1967; Green and Lindsay, 1972; Bahar and Hetnarski, 1978; Sherief and Helmy, 1999; Sherief, 1993), as well as studies of coupled thermoelasticity and thermal conduc-tivity directly for inhomogeneous structures (Filopoulos et al, 2014). In Green and Lindsay (1972); Bahar and Hetnarski (1978); Sherief and Helmy (1999); Sherief (1993), the coupled thermoelasticity was extended by including thermal relaxation in the governing relations. A version of extended thermodynamics for microme-chanics of inhomogeneous media is developed in Filopoulos et al (2014), based on generalized gradient models, so the result is a variant of a non-local theory with two scale parameters. It is necessary to note Sobolev (1991), where the hyperbolic-ity of the transfer processes is discussed. Extended thermodynamics constructed by invoking generalized relations for heat conduction of the Maxwell–Catttaneo type leads to the wave nature of heat distribution and can be interpreted as hyperbolic thermo-elasticity.

In Mahan (1988); Iesan (1991); Williams (1989), models of extended thermody-namics of deformation processes are considered in the framework of the space-time theory of elasticity including both thermomechanics and thermal conductivity ap-plied to homogeneous bodies and mixtures. It should be noted that perhaps for the first time the gradient thermal conductivity was proposed in Mahan (1988), and the thermodynamics of two-phase structures was constructed in Iesan (1991). These studies were continued in the works, with reference to the specific problems of ther-momechanics, not only of homogeneous bodies (Jou et al, 2010; Barone and Sell-eri, 2012; Kienzler and Maugin, 2001), but also of inhomogeneous materials (Iesan, 1983; Martinez and Quintanilla, 1998; Antoci and Mihich, 1999). In Iesan (2004, 2009); Knyazeva and Evstigneev (2010), attempts were made to link the problems of thermomechanics with the structural characteristics of materials to explain the coupled effects associated with heat transfer in inhomogeneous materials.

The universal variable principle of possible displacements was used for con-struction of a covariant model of reversible dynamic thermoelasticity in Lurie and

Belov (2000, 2001, 2013, 2018); Belov et al (2006); Belov and Lurie (2012); Lurie et al (2019), variational closed models of heat transfer using variational approaches have been constructed for the reversible and irreversible processes. In Lurie and Belov (2000, 2001); Belov et al (2006) dynamic thermoelasticity was developed as generalized space-time continuum with symmetrical generalized stresses. Herein, it was demonstrated that the formulation of any thermodynamic model (reversible or irreversible) as 4D continuum model contradicts the traditional notion about 4D continua physical isotropy. To explain the difference of physical properties of heat flow and impulse requires the cancellation of 4D stress tensor symmetry hypotheses (Belov and Lurie, 2012; Lurie and Belov, 2013, 2018). As it turned out, the space-time elasticity includes traditional thermoelasticity and also entropy and temperature, which are the diagonal components of the space-time representation of the 4D strain tensors and the stress tensors, respectively. In Lurie and Belov (2018) it was shown that the Maxwell–Cattaneo law and Fourier law can be defined for the reversible processes. We realize, however, that the presented space–time framework can also be employed for studying other coupled phenomena such as thermoelectricity or thermomagnetism, as well as for describing large-strain deformations of solids, and fluid mechanics (see as example Lurie et al, 2019).

Size effects in the thermal conductivity and thermoelasticity were investigated in theoretical and experimental studies (Stewart and Norris, 2000; Molaro et al, 2015; Zenkour et al, 2015). In Kavner and Panero (2004); Cardona and Sievert (2000); Challamel et al (2016) it was shown that for inhomogeneous structures, gradient effects and coupled effects and thermal relaxation effects are very important as in assessing the effective thermal conductivity, effective thermomechanical properties as in the study of their dynamic behavior. In Challamel et al (2016), it was shown that nonlocality is also essential for thermal and mechanical behavior with the same scale length for both phenomena. In this case, non-local models (scale parameter) are identified by the characteristic lattice length of the lattix structure and the scale effect can be decisive in the presence of strong microstructural effects.

Recently, in thermomechanics, as well as earlier in the 80s in the mechanics of inhomogeneous media there occurred an interest in the development of generalized applied gradient theories, which allow us to take into account not only strain gradients, but also temperature gradients of a higher order than in classical models. An example of such a continuum thermomechanical theory is the generalized Aifantis theory. In Forest and Aifantis (2010), this theory extends to thermomechanical problems; it introduces the concepts of hypertemperature and hyperentropy associated with gradientness. We note some monographs containing informative reviews of current studies of nonequilibrium and irreversible thermodynamic processes of deformation, thermomechanics of composites, molecular dynamics, physical processes occurring in micro/nanoscale inhomogeneous media, methods for modeling heat transfer of micro/nanoscale heat, accounting of the interface in problems of thermomechanics (Tzou, 2014; Povstenko, 2015; Harry et al, 2012; Gudlur, 2010).

Specially, it is worth to note the interface effects characteristic for heterogeneous materials, especially if the density of the phase boundaries is high. Such effects are associated with thermo-barrier effects at the interfaces. To describe the thermal barrier properties of micro/nanoscale structures, methods of atomistic modeling are usually used (Gaughey and Kaviany, 2006; Turney et al, 2009; Zhou et al, 2007; Qiu et al, 2012; Xu and Li, 2009; Sellan et al, 2010). There are continuum models that take into account the electron and phonon conductivity of metals and are used for thin-layer coatings with the metal-semiconductor structure (Majumdar and Reddy, 2004; Ordonez-Miranda et al, 2011; Hopkins et al, 2009; Anisimov et al, 1974; Kaganov et al, 1957; Tzou, 2014; Rawat and Sands, 2006; Hopkins et al, 2009). In this case, in fact, the existence of two types of temperature fields is postulated, for each of which the process of heat conduction is determined by the Fourier law. Consequently, the Fourier law extends to both types of conductivity.

The effect of the thermal resistance of the internal boundaries of the medium was for the first time confirmed by Kapitza (1971). The Kapitza effect, in essence, consists in the proportionality of the heat flux and the temperature jump during the crossing of the boundary, and, in fact, establishes the existence of its own thermal resistance of the boundary of the media and the validity of the Fourier law for the boundaries. When the effects of thermal resistance of boundaries in continual models is described, it is set, on the contact of media, the temperature jump (Majumdar and Reddy, 2004; Ordonez-Miranda et al, 2011), the value of which is related to the coefficient of thermal resistance of the boundaries, which, in turn, is a physical characteristic of the boundary of the contacting materials under study. A large number of works is devoted to the study of thermal barrier effects in thermal conductivity for specific inhomogeneous structures. Thus, in Chen et al (2012), thermo-barrier effects (thermal resistance of interface boundaries) are studied together with scale effects for a specific two-phase nano-structured material of silicon oxide-selikon, which is used in electronics. In Dong and Wen (2014), it is given an analysis of the effects of thermal resistance on the phase boundaries in nanocrystalline materials, multilayer coatings, micro/nano mechanical devices, etc.

The use of variational methods is also promising from the point of view of both formulation of correct boundary problems of coherent gradient thermoelasticity and thermal conductivity, allowing to take into account the interfacial thermomechanical interactions by invoking the continuous theory of adhesion. However, we note that similar methods in coupled heat conduction have not been fully implemented to date. Partially, these issues will be addressed below, where it is proposed to model scale effects and coupled effects using generalized gradient dilatation models for stationary problems.

8.2 Kinematics of Gradient Continuous Media and Gradient Media with Fields of Defects

The kinematic variables of a gradient continuous medium are: displacement vector R_i, second-rank tensor D_{ij} – distortion tensor, third-rank tensor D_{ijk} – curvature tensor. By definition, if there are no dislocations in continuous media, therefore distortions are integrable. The integrability conditions are:

$$D_{im,n}e_{mnj} = 0, \tag{8.1}$$

where e_{mnj} is the Levi-Civita pseudo-tensor. Otherwise dislocations take place

$$D_{im,n}e_{mnj} = \Xi_{ij}, \tag{8.2}$$

where $\Xi_{ij} \neq 0$ is the De Vit dislocation density tensor
    From (8.1) it follows that

$$D_{ij} = R_{i,j}. \tag{8.3}$$

We can also introduce a third rank curvature tensor D_{ijn}, which is a gradient of
the distortion tensor $D_{ijk} = D_{ij,k}$. For continuous defect-free media, there must be
satisfied the integrability conditions of the curvature tensor, that are the conditions
for the existence of a curvilinear integral in determining the distortion tensor D_{in}
through the curvature tensor D_{ijn}:

$$D_{ijm,n}e_{mnk} = 0. \tag{8.4}$$

Otherwise there are disclinations $D_{ijm,n}e_{mnk} = \Omega_{ijk}$, where the pseudo-tensor
$\Omega_{ijk} \neq 0$completely coincides with the definition of the Frank disclinations density
tensor. From (8.4) and (8.3) it follows that

$$D_{ijk} = R_{i,jk}. \tag{8.5}$$

Thus, the kinematic model of gradient continuous media is the Cauchy relations
(8.3) and the corresponding relations for constrained curvatures (8.5). The arguments of the Lagrange functional, respectively, are tensor of the first rank R_i is the
displacement tensor; tensor of the second rank $D_{ij} = R_{i,j}$ is the distortion tensor;
tensor of the third rank $D_{ijk} = R_{i,jk}$ is the curvature tensor.
    Consider a particular case of media in which there are no declinations and the
relations (8.4) are valid, and there are non-zero fields of conserved dislocations (see
(8.2)). The general solution of relations (8.2) is the sum of the particular solution
D_{ij}^2 of the inhomogeneous equation (8.2) and the general solution $R_{i,j}$ of the homogeneous equation (8.1):

$$D_{ij} = R_{i,j} + D_{ij}^2. \tag{8.6}$$

Formal integration (8.6) allows to determine the non-integrable part of the displacements:

$$D_i^2 = \int_{M_0}^{M_x} D_{ij}^2 dx_j = \int_{M_0}^{M_x} D_{ij}^2 s_j ds, \tag{8.7}$$

where the discontinuous, non-integrable part of the vector of generalized displacements (dislocation vector) by virtue of (8.2); $s_j = dx_j/ds$ meaning the tangent
vector of the integration path. We can write

$$D_i = R_i + D_i^2, \tag{8.8}$$

here R_i is the continuous part of the vector of generalized displacements D_i. The kinematic model of gradient defective media (media with conserved dislocations) is the generalized Cauchy relations (8.6) and the corresponding relations for constrained curvatures (8.5).

The arguments for the Lagrange functional are respectively: tensor of the first rank R_i which is the continuous part of the generalized displacements; tensors of the second rank $D_{ij}^1 = R_{i,j}$, D_{ij}^2; distortions of the first and second grade (integrable and non-integrable distortions); tensors of the third rank $D_{ij,k}^1 = R_{i,jk}$, $D_{ij,k}^2$ are curvatures of the first and second grade. If there are no dislocations $D_{ij}^2 = 0$, then the defective media degenerates into a defect-free, continuous media.

8.3 Variational Statement of Generalized Gradient Media with Fields of Defects

We use the variational method for constructing models of media, according to which the model of a linear media is completely determined by the kinematic model and the quadratic form in the general representation of potential energy. We propose here the following Lagrange functional, which defines a generalized gradient model of a medium with a field of conserved dislocations:

$$L = A - U,$$

$$A = \iiint P_i^V R_i \, dV + \oiint P_i^F R_i \, dF,$$

$$U = \iiint U_V \, dV + \oiint U_F \, dF,$$

$$2U_V = C_{ijmn}^{pq} D_{ij}^p D_{mn}^q + C_{ijkmnl}^{pq} D_{ij,k}^p D_{mn,l}^q, p, q = 1, 2,$$

$$2U_F = A_{ijmn}^{pq} D_{ij}^p D_{mn}^q + A_{ijkmn}^{pq} D_{ij,k}^p D_{mn}^q + A_{ijkmnl}^{pq} D_{ij,k}^p D_{mn,l}^q, \quad p, q = 1, 2,$$

$$\tag{8.9}$$

here C_{ijmn}^{pq} is fourth-rank moduli tensor, C_{ijmn}^{11} is fourth-rank moduli tensor of classical elasticity, C_{ijmn}^{pq}, $p \neq q$ are the elastic moduli that determine the coupled effects of the interaction of deformation fields and dislocation fields $C_{ijmn}^{pq} = C_{mnij}^{qp}$, C_{ijkmnl}^{pq} is the tensor of gradient modules of the sixth rank, C_{ijkmnl}^{11} is the tensor of gradient Mindlin-Toupin modules of the defect-free media and, C_{ijkmnl}^{pq}, $p \neq q$ are the gradient moduli that determine the coupled effects, $C_{ijkmnl}^{pq} = C_{mnlijk}^{qp}$, C_{ijkmnl}^{22} is the tensor of gradient Mindlin's modulus.

The adhesion properties of the body surface are determined by the transversal-isotropic tensors of the adhesive modules of the fourth, fifth and sixth ranks A_{ijmn}^{pq}, A_{ijkmn}^{pq}, A_{ijkmnl}^{pq}, $A_{ijmn}^{pq} = A_{mnij}^{qp}$. In the following, we will consider a model of the media that takes into account scale effects up to the second order. Therefore, we will not take into account the tensor modulus of the sixth rank $A_{ijkmnl}^{pq} = 0$. In order for the potential adhesion energy to remain positively defined, the components of the fifth rank adhesive modulus tensor should be set to zero $A_{ijkmn}^{pq} = 0$.

Since in temperature problems of mechanics of isotropic continuous media, temperature causes volume deformations, we will further consider models of media in which the fields of defect-dislocations are determined only by the spherical tensor of free deformations. Thus, in the kinematic model we will consider only free dilatations θ^2 and their curvatures $D_{ij,k}^2 = \theta_{,k}^2 \delta_{ij}/3$. Accordingly, the potential energy of free dilatations curvatures in (8.9) will take on a simpler form:

$$C_{ijkmnl}^{pq} D_{ij,k}^p D_{mn,l}^q = C^{pq}\theta_{,k}^p \theta_{,k}^q, \qquad C^{pq}\delta_{kl} = C_{ijkmnl}^{pq}\delta_{ij}\delta_{mn}/9.$$

In the result we can introduce the following Lagrangian for generalized media

$$L = A - \iiint U_V dV - \iint U_F dF,$$
$$2U_V = G_{ijmn}^{11} R_{i,j} R_{m,n} + K^{11}\theta^1\theta^1 + 2K^{12}\theta^1\theta^2 + K^{22}\theta^2\theta^2 + $$
$$+ C^{11}\theta_{,a}^1\theta_{,a}^1 + 2C^{12}\theta_{,a}^1\theta_{,a}^2 + C^{22}\theta_{,a}^2\theta_{,a}^2,$$
$$2U_F = A^{11}\theta^1\theta^1 + 2A^{12}\theta^1\theta^2 + A^{22}\theta^2\theta^2,$$

(8.10)

where $G_{ijmn}^{pq} = 2\mu^{pq}(\delta_{im}\delta_{jn}/2 + \delta_{in}\delta_{jm}/2 - \delta_{ij}\delta_{mn}/3)$, K^{11}, K^{22} are classical bulk modulus and free dilatation bulk modulus respectively, K^{12} is the bulk modulus which describe the coupled effects, C^{11} is the gradient modulus, C^{12} and C^{22} are coupled gradient modulus and gradient free dilatation modulus.

Let us formulate the physical model for the generalized media (8.10). Physical relationships between generalized stresses and generalized kinematic variables (deformations, deformation curvatures, free deformations and their curvatures) in the body volume could be written based on the Green formulas. Cauchy stresses σ_{ij}^1, moments m_a^1, dislocation stresses σ_{ij}^2, and dislocation moments m_a^2 are the following view:

$$\begin{cases} \sigma_{ij}^1 = \dfrac{\partial U_V}{\partial R_{i,j}} = 2\mu(R_{i,j}/2 + R_{j,i}/2 - R_{r,r}\delta_{ij}/3) + (K^{11}R_{r,r} + K^{12}\theta^2)\delta_{ij}, \\[2mm] \sigma_{ij}^2 = \dfrac{\partial U_V}{\partial(\delta_{ij}\theta^2)} = (K^{12}R_{r,r} + K^{22}\theta^2)\delta_{ij}, \end{cases}$$

(8.11)

as well as

$$\begin{cases} m_a^1 = \dfrac{\partial U_V}{\partial \theta_a^1} = (C^{11}R_{r,r} + C^{12}\theta^2)_{,a}, \\[2mm] m_a^2 = \dfrac{\partial U_V}{\partial \theta_a^2} = (C^{12}R_{r,r} + C^{22}\theta^2)_{,a}. \end{cases}$$

On the surface of the media under consideration, the force model allows the existence of two types of adhesive "forces" that are scalars:

$$\begin{cases} a^1 = \dfrac{\partial U_F}{\partial \theta^1} = A^{11}\theta^1 + A^{12}\theta^2 \\[2mm] a^2 = \dfrac{\partial U_F}{\partial \theta^2} = A^{12}\theta^1 + A^{22}\theta^2 \end{cases}$$

(8.12)

Let us formulate variational model for the generalized media damaged by the defects field connected with free dilatation. For this one let us write the variational equation corresponding to the stationary condition of the Lagrangian (8.10). Taking into account Eqs. (8.11), (8.12) we receive the following variational equation in force terms $(\sigma^2 = \sigma_{ij}^2 \delta_{ij}/3)$:

$$
\begin{aligned}
\delta L &= \iiint P_i^V \delta R_i dV + \oiint P_i^F \delta R_i dF - \\
&\quad - \iiint (\sigma_{ij}^1 \delta R_{i,j} + m_a^1 \delta\theta_{,a}^1 + \sigma^2 \delta\theta^2 + m_a^2 \delta\theta_{,a}^2) dV = \\
&= \iiint \left\{ [(\sigma_{ij}^1 - m_{a,a}^1 \delta_{ij})_{,j} + P_i^V]\delta R_i + (m_{a,a}^2 - \sigma^2)\delta\theta^2 \right\} dV + \\
&\quad + \oiint \left\{ [P_i^F - (\sigma_{ij}^1 - m_{a,a}^1 \delta_{ij})n_j]\delta R_i \right. \\
&\quad \left. - (m_a^1 n_a + a^1)\delta R_{r,r} - (m_a^2 n_a + a^2)\delta\theta^2 \right\} dF = 0.
\end{aligned}
$$

The last equation can be represented in kinematic variables as follows:

$$
\begin{aligned}
\delta L &= \iiint \left\{ [\mu \Delta R_i + (\mu + \lambda)R_{j,ji} - C^{11}\Delta R_{r,ri} + \right. \\
&\quad + K^{12}\theta_{,i}^2 - C^{12}\Delta\theta_{,i}^2 + P_i^V]\delta R_i + \\
&\quad \left. + (C^{22}\Delta\theta^2 - K^{22}\theta^2 - K^{12}R_{r,r} + C^{12}\Delta R_{r,r})\delta\theta^2 \right\} dV + \\
&\quad + \oiint \left\{ [P_i^F - (\mu(R_{i,j} + R_{j,i}) + \lambda R_{r,r}\delta_{ij} - \right. \\
&\quad - C^{11}\Delta R_{r,r}\delta_{ij} + K^{12}\theta^2\delta_{ij} - C^{12}\Delta\theta^2\delta_{ij})n_j]\delta R_i - \\
&\quad - [(C^{11}R_{j,j} + C^{12}\theta^2)_{,a}n_a + A^{11}R_{j,j} + A^{12}\theta^2]\delta R_{r,r} - \\
&\quad \left. - [(C^{12}R_{j,j} + C^{22}\theta^2)_{,a}n_a + A^{12}R_{j,j} + A^{22}\theta^2]\delta\theta^2 \right\} dF = 0. \quad (8.13)
\end{aligned}
$$

8.4 Mathematical Statement for Generalized Gradient Dilatation Model

Variational equation (8.13) allows us to formulate a closed formulation of the boundary value problem, including the governing equation as the Euler equations with variations of generalized displacements in the integral over the volume and the whole spectrum of natural boundary conditions in the surface integral. By equating to zero the first variation of (8.13) with respect to admissible kinematic fields δR_i and $\delta\theta^2$, we obtain the following equilibrium equations:

$$
\mu \Delta R_i + (\mu + \lambda)R_{j,ji} - C^{11}\Delta R_{r,ri} + K^{12}\theta_{,i}^2 - C^{12}\Delta\theta_{,i}^2 + P_i^V = 0, \quad (8.14)
$$
$$
C^{22}\Delta\theta^2 - K^{22}\theta^2 - K^{12}R_{r,r} + C^{12}\Delta R_{r,r} = 0. \quad (8.15)
$$

The interpretation of two groups of equilibrium equations (8.14), (8.15) is quite obvious. The system of equations (8.14) should be considered as a system of three-dimensional thermoelasticity equilibrium equations, written in displacements, where the terms containing $\theta_{,i}^2$ determine the volume change associated with the temperature field. The scalar equilibrium equation (8.15) remains to be interpreted as the "heat conduction" equation. So, the system of equations (8.14), (8.15) determines the system of equations of coupled thermoelasticity-thermal conductivity. The following natural/essential boundary conditions are derived by such variational procedure. Accordingly, the terms in the surface integral (8.13) give three pairs of alternative almost "classical" boundary conditions:

$$\oiint [P_i^F - (\mu(R_{i,j} + R_{j,i}) + \lambda R_{r,r}\delta_{ij} - C^{11}\Delta R_{r,r}\delta_{ij} +$$
$$+ K^{12}\theta^2\delta_{ij} - C^{12}\Delta\theta^2\delta_{ij})n_j]\delta R_i dF = 0. \tag{8.16}$$

one pair of alternative "non-classical" gradient boundary conditions:

$$\oiint [(C^{11}R_{r,r} + C^{12}\theta^2)_{,a}n_a + A^{11}\theta^1 + A^{12}\theta^2]\delta R_{r,r} dF = 0 \tag{8.17}$$

and, finally, a pair of scalar "thermodynamic" boundary conditions:

$$\oiint [(C^{12}R_{r,r} + C^{22}\theta^2)_{,a}n_a + A^{12}\theta^1 + A^{22}\theta^2]\delta\theta^2 dF = 0 \tag{8.18}$$

8.5 Identification of Generalized Stress Factors of the Model

Let us briefly discuss the interpretation of the constitutive equations (8.11). The physical meaning of the stress tensor σ_{ij}^1 is obvious – this is the Cauchy stress tensor in the thermoelasticity problem, if the value θ^2 is considered equal to the free volume change (non-integrable or inconsistent volume change – dilatation). We dwell in more detail on the interpretation of the physical meaning of the force and kinematic factors related to the thermomechanics of deformations and try to give an interpretation of all the physical parameters of a coupled model under consideration through the known thermomechanical characteristics of isotropic media.

Let us consider the defining relations for the spherical part of the "classical" $\sigma^1 = \sigma_{ij}^1\delta_{ij}/3$ and "non-classical" $\sigma^2 = \sigma_{ij}^2\delta_{ij}/3$ stresses. Excluding from these equations of Hooke's law an inconsistent change in volume, i.e. the value θ^2, we obtain the connection between two force factors: the spherical "classical" tensor σ^1 and the spherical tensor σ^2:

$$\sigma^1 = \left(K^{11} - \frac{K^{12}K^{12}}{K^{22}}\right)\theta^1 + \frac{K^{12}}{K^{22}}\sigma^2, \quad \theta^2 = \frac{1}{K^{22}}\sigma^2 - \frac{K^{12}}{K^{22}}\theta^1. \tag{8.19}$$

The relation (8.19) can be considered as a mathematical formulation of the Duhamel–Neumann hypothesis, if the force factor σ^1 is interpreted as pressure p, and the force factor σ^2 is defined as the difference between local and global temperatures T. In this case, Eqs. (8.19) can be written in the form of the Duhamel–Neumann equations:

$$p = \left(K^{11} - \frac{K^{12}K^{12}}{K^{22}} \right) \theta^1 + \frac{K^{12}}{K^{22}} T = K_T(R_{k,k} - \alpha T),$$

$$\theta^2 = \frac{1}{K^{22}} T - \frac{K^{12}}{K^{22}} \theta^1. \tag{8.20}$$

The interpretations proposed above lead to the identification of non-classical modules K^{12} and K^{22}, through the known thermomechanical parameters of the media: adiabatic bulk module $K^{11} = K = (2\mu/3) + \lambda$, the isothermal bulk module K_T and the coefficient of thermal expansion α:

$$K^{11} = K = \frac{2\mu}{3} + \lambda, \quad \left(K^{11} - \frac{K^{12}K^{12}}{K^{22}} \right) = K_T, \quad \frac{K^{12}}{K^{22}} = -K_T\alpha \tag{8.21}$$

or

$$K^{11} = K, \quad K^{12} = -\frac{K - K_T}{K_T\alpha}, \quad K^{22} = \frac{1}{K_T\alpha}\frac{K - K_T}{K_T\alpha}. \tag{8.22}$$

From Eq. (8.21) we can immediately implies the well-known thermodynamic inequality:

$$K - K_T > 0$$

as

$$K - \frac{K^{12}K^{12}}{K^{22}} = K_T \quad \text{and} \quad \frac{K^{12}K^{12}}{K^{22}} > 0.$$

Let us return to the equations of Hooke's law for σ^1 and σ^2. We write them using the above interpretations ($\sigma^1 = p$, $\sigma^2 = T$) in the following form:

$$p = K\theta^1 - \frac{K - K_T}{K_T\alpha}\theta^2 \quad T = -\frac{K - K_T}{K_T\alpha}\theta^1 + \frac{1}{K_T\alpha}\frac{K - K_T}{K_T\alpha}\theta^2$$

or

$$\theta^1 = \frac{1}{K_T}[p + K_T\alpha T], \quad \theta^2 = \frac{K_T\alpha}{K_T}\left[p + \frac{K(K_T\alpha)}{(K - K_T)} T \right]. \tag{8.23}$$

We now consider the constitutive equations for the vector of double stresses ("moment" stresses) m_k^2. Substituting (8.23) into (8.11) we obtain governing equations in terms stress factors:

$$m_k^2 = \left[\frac{C^{12}}{K_T} + C^{22}\alpha \right] p_{,k} + \left[C^{12} + C^{22}\frac{KK_T\alpha}{(K - K_T)} \right] \alpha T_{,k}. \tag{8.24}$$

Let us consider the case when $R_{k,k} = 0$. Then, we receive directly from Duhamel–Neumann equations (8.20):

$$p_{,k} = -(K_T\alpha)T_{,k}.$$

Excluding from (8.24) the value $p_{,k}$ using above equation we obtain the equality which can be interpreted as Fourier law for the heat flow

$$m_a^2 = \frac{C^{22}}{K^{22}} T_{,a}.$$ (8.25)

Then we obtain treatment for the double stress

$$q_a = -k_V T_{,a}, \qquad m_a^2 = -q_a$$

and

$$\frac{C^{22}}{K^{22}} = k_V,$$ (8.26)

k_V is the thermal conductivity coefficient at constant volume.

Let us consider the case when $p = 0$. Then directly from (8.24) we obtain:

$$m_a^2 = \left[C^{12} + C^{22}(K_T\alpha)\frac{K}{(K - K_T)} \right] \alpha T_{,a} \quad \text{or} \quad q_a = -k_P T_{,a}.$$ (8.27)

Using (8.27) we found the second equation for C^{12}, C^{22} treatment:

$$\left[C^{12} + C^{22}(K_T\alpha)\frac{K}{(K - K_T)} \right] \alpha = k_P,$$ (8.28)

where k_P is the thermal conductivity coefficient at constant pressure.

Thus, following (8.25) and (8.27) we can see double stress m_a^2 is proportional to the temperature gradient, similar to the heat flux vector and proposed identification of m_a^2 from $(-q_a)$ is quite adequate. Moreover, the coefficients C^{12}, C^{22} can be identified through heat conductivity coefficient at constant volume k_V and heat conductivity coefficient at constant pressure k_P:

$$C^{12} = \frac{k_P K_T - k_V K}{K_T\alpha}, \qquad C^{22} = k_V \frac{(K - K_T)}{(K_T\alpha)^2}.$$ (8.29)

As a result the generalized Hooke's law for the "temperature" double stress (heat flux) takes the form:

$$q_a = -m_a^2 = -\frac{(k_P - k_V)}{K_T\alpha} p_{,a} - k_P T_{,a}.$$

Note, that the similar equation for the heat flux was recieved in the recent work by Lurie and Belov (2018) in the frame work of coupled extended thermodynamic. As a result, a reversible theory of coupled gradient thermoelasticity and thermal conductivity was formulated, in which non-classical modules $K^{11}, K^{12}, K^{22}, C^{12}, C^{22}$ are explicitly defined by relations (8.22) and (8.26), (8.29) through the well-known classical thermomechanical parameters of the media K, K_T, α, k_V, k_P.

Gradient physical modulus C^{11} defines scale effects in the gradient model of the media in the absence of thermodynamic processes. It is convenient to introduce the following definitions of this modulus $C^{11} = l^2 K_T$ where the parameter l determines the length of cohesive interactions and the scale effect associated with them.

In conclusion, taking into account Eqs. (8.11), (8.12), (8.27), we obtained the variational equation of the coupled gradient thermoelasticity and thermal conductivity in a form:

$$
\begin{aligned}
\delta L = \iiint \Big\{ & \Big[\mu \Delta R_i + (\mu + \lambda) R_{j,ji} - l^2 K_T \Delta R_{r,ri} - \\
& - \frac{(K - K_T)}{(K_T \alpha)} \theta_{,i}^2 - \frac{(k_p K_T - k_V K)}{(K_T \alpha)} \Delta \theta_{,i}^2 + P_i^V \Big] \delta R_i + \\
& + \frac{(K - K_T)}{(K_T \alpha)^2} \Big[k_V \Delta \theta^2 - \theta^2 + K_T \alpha R_{r,r} + \\
& + (K_T \alpha) \frac{(k_p K_T - k_V K)}{(K - K_T)} \Delta R_{r,r} \Big] \delta \theta^2 \Big\} dV + \\
& + \oiint \Big\{ \Big[P_i^F - [\mu(R_{i,j} + R_{j,i}) + \lambda R_{r,r} \delta_{ij} - l^2 K_T \Delta R_{r,r} \delta_{ij} - \\
& - \frac{(K - K_T)}{(K_T \alpha)} \theta^2 \delta_{ij} - \frac{(k_p K_T - k_V K)}{(K_T \alpha)} \Delta \theta^2 \delta_{ij}] n_j \Big] \delta R_i - \\
& - \Big[(l^2 K_T R_{j,j} + \frac{(k_p K_T - k_V K)}{(K_T \alpha)} \theta^2)_{,a} n_a + A^{11} R_{j,j} + A^{12} \theta^2 \Big] \delta R_{r,r} - \\
& - \Big[(\frac{(k_p K_T - k_V K)}{(K_T \alpha)} R_{j,j} + k_V \frac{(K - K_T)}{(K_T \alpha)^2} \theta^2)_{,a} n_a + \\
& + A^{12} R_{j,j} + A^{22} \theta^2 \Big] \delta \theta^2 \Big\} dF = 0.
\end{aligned}
$$

(8.30)

The variation equation (8.30) completely defines the boundary value problem for the considered coherent theory of thermoelasticity and heat conduction. Note that the Euler equations, which are in (8.30) with the variation of the components displacement vectors are equilibrium equations of gradient thermoelasticity, if we assume that the temperature field is given. Similarly, the classic natural boundary conditions (static conditions):

$$
\begin{aligned}
P_i^F - \Big[& \mu(R_{i,j} + R_{j,i}) + \lambda R_{r,r} \delta_{ij} - l^2 K_T \Delta R_{r,r} \delta_{ij} - \\
& - \frac{(K - K_T)}{(K_T \alpha)} \theta^2 \delta_{ij} - \frac{(k_p K_T - k_V K)}{(K_T \alpha)} \Delta \theta^2 \delta_{ij} \Big] n_j = 0
\end{aligned}
$$

and non-classical gradient natural condition

$$
\Big(l^2 K_T R_{j,j} + \frac{(k_p K_T - k_V K)}{(K_T \alpha)} \theta^2 \Big)_{,a} n_a + A^{11} R_{j,j} + A^{12} \theta^2 = 0
$$

could be written exclusively in terms of displacements and temperatures, using the relation $\theta^2 = (1/K^{22})T - (K^{12}/K^{22}) R_{k,k}$, see Eq. (8.20). Consequently, both the equilibrium equations of the gradient thermoelasticity and the boundary value problem as a whole give here a closed boundary value problem in the case of a given temperature field.

On the other hand, the task is connected when the general formulation is considered and the temperature field is not known a priori. Indeed, it is sufficient to consider the equations of equilibrium for thermoelasticity written above, together with the Euler equation for thermal conductivity, see Eq. (8.30),

$$\frac{(K - K_T)}{(K_T \alpha)^2} \left[k_V \Delta \theta^2 - \theta^2 + K_T \alpha R_{r,r} + (K_T \alpha) \frac{(k_p K_T - k_V K)}{(K - K_T)} \Delta R_{r,r} \right] = 0$$

and take into account thermodynamic boundary conditions, see Eq. (8.30),

$$\oiint \left\{ \left[\frac{(k_p K_T - k_V K_T)}{(K_T \alpha)} R_{j,j} + k_V T \right]_{,a} n_a + \right.$$
$$\left. + [A^{12} + A^{22}(K_T \alpha)] R_{j,j} + A^{22} \frac{(K_T \alpha)^2}{(K - K_T)} T \right\} \delta \left(T + \frac{K - K_T}{K_T \alpha} R_{r,r} \right) dF = 0$$

during formulating a closed boundary value problem.

8.6 Particular Model: Gradient Thermoelasticity

Let us assume that temperature field is known. Then, directly from variation equation (8.30) we can obtain the system of equilibrium equations for the gradient thermoelasticity. Taking in to account (8.20) can write these equations in the terms of displacements and temperature:

$$\mu \Delta R_i + (\mu + \lambda_T) R_{j,ji} - (C^{11} + k_p K_T - k_V K) \Delta R_{r,ri} -$$
$$- (K_T \alpha) T_{,i} - \frac{(k_p K_T - k_V K)}{(K - K_T)} (K_T \alpha) \Delta T_{,i} + P_i^V = 0. \qquad (8.31)$$

Accordingly, similar the three pare of the classical boundary conditions could be written as follows:

$$\oiint \left\{ P_i^F - \mu(R_{i,j} + R_{j,i}) n_j - \lambda_T R_{r,r} n_i + [(C^{11} - k_V K + k_p K_T) \Delta R_{r,r}] n_i + \right.$$
$$\left. + (K_T \alpha) \left[T - \frac{(k_V K - k_p K_T)}{(K - K_T)} \Delta T \right] n_i \right\} \delta R_i dF = 0. \qquad (8.32)$$

Finally, the terms with variation of $\delta \theta^2$ in Eq. (8.30) gives after accounting (8.20) the pare of the non-classical gradient boundary conditions

$$\iint\left\{\left[(C^{11}+k_pK_T-k_VK)R_{j,j}+\frac{(k_pK_T-k_VK)}{(K-K_T)}(K_T\alpha)T\right]_{,a}n_a+\right.$$

$$\left.+(A^{11}+A^{12}(K_T\alpha))R_{j,j}+A^{12}\frac{(K_T\alpha)^2}{(K-K_T)}T\right\}\delta R_{r,r}dF=0. \qquad (8.33)$$

Note that in a general gradient theory, a variational statement leads not to a scalar nonclassical boundary condition, but to a vector one. However, in this case, we obtain a scalar pair of boundary conditions, since we are considering a gradient generalized dilatation theory.

Gradient theory of thermoelasticity (8.31)–(8.33) allows to take into account more high temperature field gradients and more accurately investigate edge effects along the contact boundaries, ensuring not only the continuity of displacements, but also the continuity of their normal derivatives. As a result, the gradient thermoelasticity refines the localized stress state in the vicinity of the contact boundaries, which is extremely important, for example, for the estimation of the the strength and destruction of thermal protective coatings.

8.7 Particular Model: Gradient Thermal Conductivity Model

Let us consider again the governing system of equations for coupled thermoelasticity and thermal conductivity. Eliminating from the heat conduction equation $\Delta\theta^1$, we obtain the governing equation for temperature; after a few transformations, we obtain the generalized biharmonic equation for temperature:

$$l_1^2l_2^2\Delta\Delta T-(l_1^2+l_2^2)\Delta T+T=\frac{(C^{12}-K^{12}/K^{22}C^{22})}{[2\mu+\lambda-(K^{12}K^{12})/K^{22}]}P_{r,r}^V, \qquad (8.34)$$

where

$$\begin{cases}\dfrac{(C^{11}-2C^{12}(K^{12}/K^{22})+C^{22}(2\mu+\lambda)/K^{22}}{[2\mu+\lambda-(K^{12}K^{12})/K^{22}]}=l_1^2+l_2^2,\\[3mm]\dfrac{(C^{22}/K^{22})(C^{11}-C^{12}C^{12}/C^{22})}{[2\mu+\lambda-(K^{12}K^{12})/K^{22}]}=l_1^2l_2^2.\end{cases}$$

Assume that the gradient temperature field is determinative

$$\{\Delta\Delta T,(l_1^{-2}+l_2^{-2})\Delta T\}\gg(T+\tilde{P})/l_1^2l_2^2,$$

$$\tilde{P}^V=-\frac{(C^{12}-C^{22}K^{12}/K^{22})}{[2\mu+\lambda-(K^{12}K^{12})/K^{22}]}P_{r,r}^V,$$

then from Eq. (8.34), we obtain the following gradient thermal conductivity equation

$$l_T^2\Delta\Delta T-\Delta T=f^V, \qquad (8.35)$$

where $l_T^2 = l_1^2 l_2^2 / (l_1^2 + l_2^2)$.

The value f^V in (8.35) specifies the field of heat sources in the volume and we can accept that the source field is determined by the value $-(T + \tilde{P})/l_1^2 l_2^2 + f_\theta^V$, where f_θ^V is the "clear temperature" source field, which could be added in the variational equation (8.30) as a possible work of this force factor on the kinematic parameter $\delta\theta^2$. Considering (8.30) for the coupled problem in absence of the source field f_θ^V for the "clear temperature", thermal sources can be approximately estimated by an expression $-\tilde{P}/l_1^2 l_2^2$.

Thus, we have identified the gradient equation of thermal conductivity. The corresponding boundary value problem is contained both pairs of alternative non-classical boundary conditions (8.17), (8.18). It can be shown that from the general representation (8.30) for a Lagrangian of a generalized media under certain assumptions it is also possible to obtain a variational formulation of the problem of stationary gradient thermal conductivity:

$$\delta L = \delta(A - U),$$

$$A \equiv \int_V f^V T \, dV + \int_F f^F T \, ds,$$

$$U = \int_V U^V \, dV + \int_F U^F \, ds, \qquad U^V = \frac{1}{2} k_V (T_{,i} T_{,i} + l_T^2 \Delta T \Delta T), \tag{8.36}$$

$$U^F = \frac{1}{2} (aTT + bT\dot{T} + R_s \dot{T}\dot{T}), \qquad \dot{T} = \partial T / \partial n.$$

Here, f^V, f^F designate in the volume V and on the surface F scalar fields that define the density of heat sources; a, b, R_s are the physical constant that define the surface properties, R_s is the parameter that determines the thermal barrier properties of the boundary, n is external unit normal to the boundary of the media.

Equation (8.36) following the variational procedure gives the closed boundary value problem for the gradient stationary heat conduction

$$\delta L = \int_V k_V (l_T^2 \Delta \Delta T - \Delta T) \delta T \, dV +$$

$$+ \int_F [(-k_V l_T^2 \Delta \dot{T} + k_V \dot{T} + f^F - aT - b\dot{T}) \, \delta T + \tag{8.37}$$

$$+ (k_V l_T^2 \Delta T - bT - R_s \dot{T}) \, \delta \dot{T}] dF.$$

The variational equality (8.37) defines the gradient heat conduction equation and contact conditions at the boundary of two phases in the case of a contact heat conduction problem.

Let us consider an example illustrating that the gradient model of thermal conductivity is capable to describe a non-classical temperature distribution in the vicinity of the phase boundaries in inhomogeneous materials. In Schelling et al (2002) the molecular dynamics method was used to study interfacial effects in heat conduction, associated with a boundary thermal resistance. A periodic structure was considered where on the fragments of a small length the heat sources and sinks were specified.

In Schelling et al (2002) non-classical interfacial effects were studied in the vicinity of the contact boundaries of the middle part of the representative fragment with the region where the heat sources are specified. Sillinger–Weber silicon as a model system was considered as a material, in which the so-called electronic thermal conductivity is actually absent. The size of the regions for specifying the source and heat sink δ was chosen much smaller than the length of the representative fragment. The characteristic distribution of heat in silicon is presented in Fig. 8.1 at an average temperature of $500°$ K.

The source is localized at $z = 39$ nm, and sink at $z = 117$ nm. In the middle of the fragment, the classical Fourier heat propagation law is realized. However, in the vicinity of the interphase boundaries, there is a sharp change in temperature, which is characteristic for the interphase boundaries. This effect is usually associated with the effect of thermal resistance.

We considered the same representative fragment in a one-dimensional formulation for the gradient thermal conductivity (8.37) and obtained the solution of the problem of heat distribution in a conventionally composite periodic structure similar to the one studied in Schelling et al (2002). In the composite medium, select a representative fragment of a length $2l$ (Fig. 8.1). In a periodic structure (with periodic boundary conditions) along the edges of a fragment in a region with a length 2δ, it is given a uniformly distributed heat source with a density $f^V = q_0$, and, accordingly, the sink $f^V = -q_0$.

The results of solving the boundary value problem (8.37), are presented in Figure 8.2a), b). It was assumed that $Q = q_0/k_V$, k_V is the coefficient of thermal conductivity. To obtain qualitative results, the following parameters were set as $Q = 10$, $\delta = 0.14$ nm, $l = 80$ nm, $l^{-2} = 5$ nm^{-2}.

Modeling the interfacial thermal resistance, it was assumed that $R_S/k_V = 7, 24$, (Fig. 8.2b). R_S is the parameter characterizing the thermal resistance of the boundary. In the case of an ideal thermal contact Fig. 8.2a) it was supposed that $R_S = 0$. The results of solution derived from the gradient thermal conductivity in the Fig. 8.2 show that the gradient model provides a qualitatively adequate description of the nonclassical effects of the interfacial boundary layer in thermal conductivity and allows treating the adhesive parameter in the boundary conditions as a parameter of the thermal resistance. Note that the formal introduction of adhesive effects of the

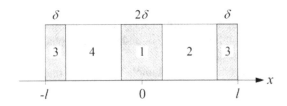

Fig. 8.1 Representative fragment of the periodic structure.

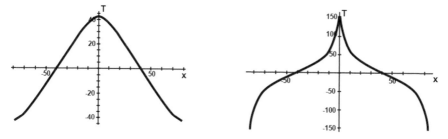

Fig. 8.2 Temperature distribution profiles for a periodic structure in the interval $-l < x < l$: a) $R_S = 0$ ideal contact without interfacial thermal resistance, b) $R_S/k_V = 7, 24$ interface with the thermal resistance.

Gurtin–Murdoch type (parameters a, b in (8.37)) does not provide adequate modeling of thermal resistance. Away from the contact zones, the solution corresponds exactly to the classical Fourier law, which follows directly from the structure of the solution of the gradient thermal conductivity. On the other hand, the curve presented in Fig. 8.2 a) gives a smooth temperature distribution with a smooth contact over the boundaries. Such a situation is likely to occur in metal-to-metal structures when the heat transfer process is controlled by electronic thermal conductivity.

8.8 Conclusion

A reversible theory of coupled gradient thermoelasticity and thermal conductivity was formulated, in which all non-classical modules are explicitly defined through the well-known classical thermomechanical parameters of the media. In the proposed, model there are also gradient physical moduli which describe the scale effects. Adhesion mechanical and temperature interactions also take into account. We show that the general representation for the generalized media allows to give the separate variational formulation of the problems of gradient thermoelastisity and stationary gradient thermal conductivity. The solution of the problem of heat distribution in a conventionally composite periodic structure was used for an explanation of effect of Kapitza.

Acknowledgements The authors are deeply grateful to Professor Holm Altenbach for professional and useful discussions of our research. This work was carried out with support from the Russian Government Foundation of Institute of Applied Mechanics of RAS, Project -AAAA-A-19-119012290177-0 and particularly supported by the Russian Foundation for Basic Research grant 18-01-00553-A.

References

Anisimov SI, Kapeliovich BL, Perel'man TL (1974) Electron emission from metal surfaces exposed to ultrashort laser pulses. Sov Phys-JETP 39(2):375–377

Antoci S, Mihich L (1999) A four-dimensional Hooke's law can encompass linear elasticity and inertia. Il Nuovo Cimento B 8:873–880

Bahar LY, Hetnarski RB (1978) State space approach to thermoelasticity. J Therm Stress 1:135–145

Barone M, Selleri F (eds) (2012) Frontiers of Fundamental Physics. Springer Science & Business Media, New York

Belov PA, Lurie SA (2012) Ideal nonsymmetric 4D-medium as a model of invertible dynamic thermoelasticity. Mech Solids 47(5):580–590

Belov PA, Gorshkov AG, Lurie SA (2006) Variational model of nonholonomic 4D-media. Mech Solids 41(6):22–35

Biot MA (1956) Thermoelasticity and irreversible thermo-dynamics. J Appl Phys 27:240–253

Cardona JM, Sievert R (2000) Thermoelasticity of second-grade media. In: Maugin GA, Drouot R, Sidoroff F (eds) Continuum hermomechanics, the Art and Science of Modelling Material Behaviour, Kluwer Academic Publishers, Dordrecht, pp 163–176

Carlson DE (1972) Linear thermoelasticity. In: Flügge S (ed) Handbuch der Physik, Springer, Berlin, vol VIa/2: Linear Theories of Elasticity and Thermoelasticity (ed. by C. Truesdell), pp 297–345

Cattaneo C, Hebd CR (1958) Sur une forme de l'équation de la chaleur éliminant le paradoxe d'une propagation instantanée. C R Acad Sci 247:431–432

Challamel N, Grazide C, Picandet V, Perrot A, Zhang Y (2016) A nonlocal fourier's law and its application to the heat conduction of one-dimensional and two-dimensional thermal lattices. Comptes Rendus Mécanique 344(4):388–401

Chen J, Zhang G, Li B (2012) Thermal contact resistance across nanoscale silicon dioxide and silicon interface. Journal of Applied Physics 112(6):064,319

Coleman BD, Noll W (1963) The thermodynamics of elastic materials with heat conduction and viscosity. Arch Ration Mech Anal 13:167–178

Dhar A (2008) Heat transport in low-dimensional systems. Adv Phys 57:457

Dong H, Wen R B Melnik (2014) Relative importance of grain boundaries and size effects in thermal conductivity of nanocrystalline materials. Scientific Reports 4(7037)

Filopoulos SP, Papathanasiou TK, Markolefas SI, Tsamaphyros GJ (2014) Generalized thermoelastic models for linear elastic materials with micro-structure Part II: Enhanced Lord–Shulman model. J Therm Stress 37:642–659

Forest S, Aifantis EC (2010) Some links between recent gradient thermo – elasto – plasticity theories and the thermomechanics of generalized continua. Int J Solids Struct 47:3367–3376

Gaughey AJH, Kaviany M (2006) Phonon transport in molecular dynamics simulations: Formulation and thermal conductivity prediction. Advances in Heat Transfer 39:169–254

Gendelman OV, Savin AV (2010) Nonstationary heat conduction in one-dimensional chains with conserved momentum. Phys Rev E 81:0201,103

Green AE, Lindsay KA (1972) Thermoelasticity. J Elast 2:1–7

Gudlur P (2010) Thermoelastic Properties of Particle Reinforced Composites at the Micro and Macro Scales. Texas, A & M University

Gusev A, Lurie S (2013) Wave-relaxation duality of heat propagation in fermi–pasta–ulam chains. Modern Phys Lett B 26(22):1250,145

Harry G, Bodiya TP, DeSalvo R (eds) (2012) Optical Coatings and Thermal Noise in Precision Measurement. Cambridge University Press, Cambridge

Hopkins PE, Kassebaum JL, Norris PM (2009) Effects of electron scattering at metal-nonmetal interfaces on electron-phonon equilibration in gold films. J Appl Phys 105(2):023,710

Iesan D (1983) Thermoelasticity of nonsimple materials. J Therm Stress 6:385–397

Iesan D (1991) On the theory of mixtures of thermoelastic solids. J Therm Stress 13(4):389–408

Iesan D (2004) Thermoelastic Models of Continua. Kluwer Acad. Publ., Dordrecht
Iesan D (2009) Classical and Generalized Models of Elastic Rods. Chapman & Hall CRC Press, New York
Joseph DD, Preziosi L (1989) Heat waves. Rev Mod Phys 61:41–73
Jou D, Casa-Vásquez J, Lebon G (2010) Extended Irreversible Thermodynamics. Springer, New York
Kaganov MI, Lifshitz IM, Tanatarov LV (1957) Relaxation between electrons and crystalline lattices. Sov Phys JETP 4(2):173–178
Kapitza PL (1971) The study of heat transfer in Helium II. In: Helium 4 - The Commonwealth and International Library: Selected Readings in Physics, Elsevier, pp 114–153
Kavner A, Panero W (2004) Temperature gradients and evaluation of thermoelastic properties in the synchrotron-based laser-heated diamond cell. Physics of the Earth and Planetary Interiors 143–144:527–539
Kienzler R, Maugin GA (eds) (2001) Configurational Mechanics of Materials, CISM International Centre for Mechanical Sciences, vol 427. Springer, Vienna
Knyazeva AG, Evstigneev NK (2010) Interrelations between heat and mechanical processes during solid phase chemical conversion under loading. Procedia Computer Science 1(1):2613–2622
Landau LD, Lifshitz EM (1986) Course of Theoretical Theory of Elasticity. Theory of Elasticity. Vol. 7. Pergamon Press, Oxford
Lepri S, Livi R, Politi A (2003) Thermal conduction in classical lowdimensional lattices. Phys Rep 377:1–80
Lord HW, Shulman YA (1967) Generalized dynamical theory of thermoelasticity. J Mech Phys Solids 15:299–309
Lurie SA, Belov PA (2000) Mathematical Models of Continuum Mechanics and Physical Fields. Izd-vo VTs, Moscow
Lurie SA, Belov PA (2001) A variational model for nonholonomic media. Mekh Komp Mater Konstr [J Comp Mech Design (Engl Transl)] 7(2):266–276
Lurie SA, Belov PA (2013) Theory of space–time dissipative elasticity and scale effects. Nanoscale Systems MMTA 2(1):66–178
Lurie SA, Belov PA (2018) On the nature of the relaxation time, the Maxwell–Cattaneo and Fourier law in the thermodynamics of a continuous medium and the scale effects in thermal conductivity. Continuum Mechanics and Thermodynamics DOI doi.org/10.1007/s00161-018-0718-7
Lurie SA, Belov PA, Solyaev YO (2019) Mechanistic model of generalized non-antisymmetrical electrodynamics. In: Altenbach H, Belyaev A, Eremeyev V, Krivtsov A, Porubov A (eds) Dynamical Processes in Generalized Continua and Structures, Springer, Cham, Advanced Structured Materials, vol 103
Mahan GD (1988) Nonlocal theory of thermal conductivity. Phys Rev B 38(3):1963–1969
Majumdar A, Reddy P (2004) Role of electron–phonon coupling in thermal conductance of metal–nonmetal interfaces. Appl Phys Lett 84(23):4768–4770
Martinez F, Quintanilla R (1998) On the incremental problem in thermoelasticity of nonsimple materials. Z Angew Math Mech 78:703–710
Maxwell JC (1865) A dynamical theory of the electromagnetic field. Philosophical Transactions of the Royal Society of London 155:459–512
Molaro JL, Byrne S, Langer SA (2015) Grain-scale thermoelastic stresses and spatiotemporal temperature gradients on airless bodies, implications for rock breakdown. J Geophys Res: Planets 120(2):255–277
Müller I, Ruggeri T (1993) Extended Thermodynamics, Springer Tracts in Natural Philosophy, vol 37. Springer, Berlin
Nowacki W (1986) Thermoelasticity. Pergamon Press, Oxford
Ordonez-Miranda J, Alvarado-Gil JJ, Yang R (2011) The effect of the electron-phonon coupling on the effective thermal conductivity of metal-nonmetal multilayers. J Appl Phys 109(9):094,310
Povstenko Y (2015) Fractional Thermoelasticity. Springer

Qiu B, Bao H, Zhang G, Wu Y, Ruan X (2012) Molecular dynamics simulations of lattice thermal conductivity and spectral phonon mean free path of PbTe: Bulk and nanostructures. Computat Mater Sci 53(1):278–285

Rawat V, Sands TD (2006) Growth of TiN/GaN metal/semiconductor multilayers by reactive pulsed laser deposition. J Appl Phys 100(6):064,901

Schelling PK, Phillpot SR, Keblinski P (2002) Comparison of atomic-level simulation methods for computing thermal conductivity. Phys Rev B 65:144,306

Sellan DP, Turney JE, McGaughey AJH, Amon CH (2010) Cross-plane phonon transport in thin films. J Appl Phys 108(11):113,524

Sherief HH (1993) State space formulation for generalized thermoelasticity with one relaxation time including heat sources. J Therm Stress 16:163–180

Sherief HH, Helmy K (1999) A two dimensional generalized thermoelasticity problem for a half-space. J Therm Stresses 22:897–910

Sobolev SL (1991) Transport processes and traveling waves in systems with local nonequilibrium. Sov Phys Usp 34(3):217–229

Stewart DA, Norris PM (2000) Size effects on the thermal conductivity of thin metallic wires: microscale implications. Microscale Thermophysical Engineering 4(2):89–101

Turney JE, Landry ES, McGaughey AJH, Amon CH (2009) Predicting phonon properties and thermal conductivity from anharmonic lattice dynamics calculations and molecular dynamics simulations. Phys Rev B 79(6):064,301

Tzou DY (2014) Macro- to Microscale Heat Transfer: the Lagging Behavior, 2nd edn. Series in Chemical and Mechanical Engineering, Taylor & Francis, Washington, D.C.

Vernotte MP (1958) Les paradoxes de la théorie continue de l'équation de la chaleur. C R de l'Académie des Sci 246(3):154–3155

Williams D (1989) The elastic energy-momentum tensor in special relativity. Ann of Phys 196(2):345–360

Xu Y, Li G (2009) Strain effect analysis on phonon thermal conductivity of two-dimensional nanocomposites. J Appl Phys 106(11):114,302

Zenkour AM, Abouelregal AE, Alnefaie KA, Zhang X, Aifantis EC (2015) Nonlocal thermoelasticity theory for thermal-shock nanobeams with temperature-dependent thermal conductivity. J Therm Stress 38(9):1049–1067

Zhang G, Li BW (2010) Impacts of doping on thermal and thermoelectric properties of nanomaterials. NanoScale 2:1058–1068

Zhou Y, Anglin B, Strachan A (2007) Phonon thermal conductivity in nanolaminated composite metals via molecular dynamics. J Chem Phys 127(8):184,702–11

Chapter 9
Mathematical Modeling of Elastic Thin Bodies with one Small Size

Mikhail Nikabadze and Armine Ulukhanyan

Abstract Some questions on parametrization with an arbitrary base surface of a thin-body domain with one small size are considered. This parametrization is convenient to use in those cases when the domain of the thin body does not have symmetry with respect to any surface. In addition, it is more convenient to find the moments of mechanical quantities than the classical. Various families of bases (frames) and the corresponding families of parameterizations generated by them are considered. Expressions for the components of the second rank unit tensor are obtained. Representations of some differential operators, the system of motion equations, and the constitutive relation (CR) of the micropolar theory of elasticity are given for the considered parametrization of a thin body domain. The main recurrence formulas of system of orthogonal Legendre polynomials are written out and some additional recurrence relations are obtained, which play an important role in the construction of various variants of thin bodies. The definitions of the moment of the kth order of a certain value with respect to an arbitrary system of orthogonal polynomials and system of Legendre polynomials are given. Expressions are obtained for the moments of the kth order of partial derivatives and some expressions for the system of Legendre polynomials. Various representations of the system of motion equations and CR in the moments for the theory of thin bodies are given. Boundary conditions are derived. The CR of the classical and micropolar theory of the zero approximation and approximation of order r in the moments are obtained. The boundary conditions of physical and thermal contents on the front surfaces are given. The statements of dynamic problems in moments of the approximation (r, M) of a micropolar thermomechanics of a deformable thin body, as well as a non-stationary temperature

M. Nikabadze
Lomonosov Moscow State University
Bauman Moscow State Technical University, Moscow, Russia
e-mail: nikabadze@mail.ru

A. Ulukhanyan
Bauman Moscow State Technical University, Moscow, Russia
e-mail: armine_msu@mail.ru

© Springer Nature Switzerland AG 2019
H. Altenbach et al. (eds.), *Higher Gradient Materials and Related Generalized Continua*, Advanced Structured Materials 120,
https://doi.org/10.1007/978-3-030-30406-5_9

problem in moments are given. It should be noted that using the considered method of constructing a theory of thin bodies with one small size, we obtain an infinite system of equations, which has the advantage that it contains quantities depending on two variables, the base surface Gaussian coordinates x^1, x^2. So, to reduce the number of independent variables by one we need to increase the number of equations to infinity, which of course has its obvious practical inconveniences. In this connection, the reduction of the infinite system to the finite is made.

Keywords: Micropolar theory · Thin body · Constitutive relations · Boundary value problems

9.1 Introduction

Nowadays, there are several methods for constructing theories of rods, plates, shells, and multilayered structures. For example, in the case of single-layer plates these methods are based on (Altenbach, 1991; Ambartsumyan, 1987; Grigolyuk and Selezov, 1973; Naghdi, 1972; Reissner, 1985; Wunderlich, 1973):

1. hypotheses about stressed and/or deformed states;
2. expansion of all geometrical and mechanical quantities in the series;
3. asymptotic integration;
4. concepts of two-dimensional deformable surfaces;
5. method of sequential differentiation.

These methods differ in their use in practical calculations, the level of mathematical rigor, etc. At the same time, they reduce three-dimensional systems of partial differential equations, which describe the mechanical behavior of a real structure, to two-dimensional ones. A similar picture occurs when considering multilayered structures, i.e. when calculating multilayered structures as equivalent single-layer structures, we solve again a system of two-dimensional differential equations.

The first method, which is also called the method of hypotheses (Filin, 1987), is the closest to engineering concepts. The original problem is simplified after making some assumptions (hypotheses). Such hypotheses are primarily associated with Kirchhoff (1850); Reissner E. (1944); Hencky (1947); Ambartsumyan (1958, 1970, 1974, 1987); Levinson (1980); Pelekh (1973, 1978); Khoroshun (1978, 1985); Chernykh (1986, 1988); Tvalchrelidze et al (1984); Tvalchrelidze (1984, 1986, 1994); Nikabadze (1988a,b, 1989, 1990a,b,c,d, 1991, 1998a,b, 1999a,b,c,d, 2000a,b, 2001a,b,d,c, 2002a,b, 2003, 2005, 2014a) among others. Various kinematic models are presented, for example, in Lewiński (1987).

The second method is related to the expansion in power series with respect to the transverse coordinate (Lo et al, 1977a,b; Kienzler, 1982; Preußer, 1984), to the expansion via Legendre polynomials (Mindlin and Medick, 1959; Medick, 1966; Hertelendy, 1968; Soler, 1969; Fellers and Soler, 1970; Vekua, 1955, 1964, 1965, 1970, 1972, 1982; Meunargiya, 1987; Nikabadze, 2004a,b, 2006, 2014a; Nikabadze and Ulukhanyan, 2005a,b, 2008, 2016; Ulukhanyan, 2011; Kantor et al, 2013; Pelekh and Sukhorolskii, 1977, 1980; Pelekh et al, 1988; Chepiga, 1976, 1977, 1986a,b,c; Alekseev, 1994, 1995, 2000; Alekseev et al, 2001; Alekseev and Annin, 2003; Alekseev and Demeshkin, 2003; Dergileva, 1976; Volchkov and Dergileva, 1977; Volchkov et al, 1994; Volchkov and Dergileva, 1999, 2004, 2007; Volchkov, 2000; Vajeva and Volchkov, 2005; Ivanov, 1976, 1977, 1979, 1980; Egorova et al, 2015; Kuznetsova et al, 2018; Zhavoronok, 2014, 2017, 2018; Zozulya and Saez, 2014, 2016; Zozulya, 2017a,b,c), to the expansion in series in a system of given functions (Vasiliev and Lurie, 1990a,b), to the expansion via Chebyshev polynomials (Nikabadze, 2007a,b,c, 2014a; Nikabadze and Ulukhanyan, 2008), etc. These expansions are equally used to construct any thin body theory. In this case, the problem of mechanical interpretation of the expansion terms higher then the second order arises.

The third method, asymptotic integration, was proposed, for example, see the work of Gol'denveizer (1962, 1963, 1976). Mathematically, it leads to a uniform approximation of the solution for all elements of the theory (as kinematic as force elements), since terms of the same order are only considered.

The fourth method, also called the direct method (Naghdi, 1972; Zhilin, 1976; Filin, 1987), is rarely used, since it contradicts the traditional views on presenting the results of calculations in the form of stress fields. Such a representation for two-dimensional theories is very difficult, and sometimes it is an impracticable process. The possibilities of this method can be judged, for example, by the works Naghdi (1972); Zhilin (1976).

The fifth method is the method of sequential differentiation of the relations of the three-dimensional theory (Vekua, 1982). By this (little-known in the literature) method a consistent moment theory of shells is constructed. In particular, a system of equations of the 10th order has been obtained, which is consistent with five independently given physical or kinematic boundary conditions.

Finally, it should be noted that different variants of theories of thin bodies (theories of rods, plates, shells, and multilayered structures) have been developed. An analysis of the published works shows that the development of refined theories of shells and multilayered structures continues to be actively developed. At the same time nonlinear theories of thin bodies find more and more presentations in the literature. The used mathematical apparatus has been significantly expanded both for the realization of the posed problem and for the purpose of providing new statements of the problems. In parallel with the theoretical, the experimental investigations are realized.

In principle, any problem of the theory of shells can be considered (solved) in a three-dimensional formulation, which is more accurate in comparison with a two-dimensional formulation. However, it is not possible to realize this possibility in practice in the required amount due to the excessive complexity of solving three-dimensional problems and a large variety of practically necessary formulations of problems.

The behavior of thin bodies, obeying the general laws of the mechanics of a deformable solids, also depends on the specific proper laws (Pikul, 1992). Due to the relative smallness of the thickness, the strength of the shell in the transverse direction is substantially weaker than the strength in the tangential directions. The equations of the mechanics of a three-dimensional body, including Hooke's law, do not take this circumstance into account. Therefore, their direct use in the theory of shells leads to a significant error (Vekua, 1982). The specific laws of deformation of thin bodies are a physical prerequisite for the construction of new theories of thin bodies.

The widespread use of composite materials of a layered and fibre-reinforced structure in various branches of engineering has necessitated the development of new methods for calculating and designing thin bodies made from these materials. It turned out that the classical theory, which had previously dominated in the applied methods of calculating thin-walled structures, cannot satisfactorily describe the stress-strain state of composite thin bodies. In some cases, the need for multilayered thin bodies is caused by structural and operational considerations. This is very important with increased demands on the safety of structures, especially in aircraft and rocket production, taking into account that the progress of computing technology makes it possible to carry out more and more complex numerical calculations.

Due to the widespread use of thin bodies (one-, two-, three- and multilayered structures), there is a need to create new refined theories and improved methods for their calculation. Therefore, the construction of refined theories of thin bodies and the development of effective methods for their calculation are an important and actual task. Below an analytic method for constructing a theory of elastic thin bodies of various approximations is developed.

9.2 On Parametrization of a Thin Body Domain With one Small Size with an Arbitrary Base Surface

Let V be the domain of a three-dimensional Euclidean space occupied by a thin body.

Definition 9.1. A three-dimensional body, one or two sizes of which is smaller than the others, is called a thin body.

Definition 9.2. A two-dimensional domain one size of which is smaller than the other, is called a thin domain.

Definition 9.3. The assignment of a domain to any coordinate system is called the parametrization of this domain.

If the contrary is not specified, we will consider a three-dimensional body, one size of which is smaller than the others, i.e. three-dimensional thin body with one small size.

Let us consider some regular surface $S \in C_m$, $m \geq 3$, which has the following property: from any point of the region V onto the surface S we can drop a normal, which intersects with S only at one point. Generally speaking, a surface may not belong to V at all or only partially intersect with it. For definiteness, let us assume that $S \subset V$.

Let Σ be the set of lateral surfaces of the thin body. If we have a closed body, then Σ is absent. We assume that S and Σ intersect at a right angle. Moreover, we assume that Σ are oriented surfaces. Consequently, their generators are normals to S. As S we can also consider the median surface of the thin body, if such exists.

Thus, there is a wide choice of surface S, which is called the base surface (base) of the parametrization of a thin body domain. Therefore, this circumstance should be taken into account, and the surface of a relatively simple structure (for example, a plane, a sphere, a cylinder, etc.) should be chosen as the basis of parametrization, when calculating specific thin bodies. It should also be taken into account another circumstance. The problems of the theory of thin bodies with mechanical properties belong to the class of problems admitting rough approximations. Therefore, it is essential, without disturbing the mechanical and mathematical picture of the stress-strain state of a thin body, to express formulated problems with various approximations, allowing to obtain a relatively simple mathematical model to simplify the corresponding mathematical theory. In particular, a significant simplification can be achieved by a suitable choice of the basic parametrization surface. Replacing one base surface with another, which has simpler geometrical characteristics, can lead to simpler equations without great distorting the mechanical nature of the problem.

The outlines of the thin body are determined by its front surfaces. As face surfaces, the thin body can have piecewise smooth surfaces. Without significant distortion of the mechanical problem, we can assume that the smooth parts of the front surfaces are arbitrarily smooth. In particular, they can be considered as analytical surfaces. Let us denote the front surfaces of the thin body by $\overset{(+)}{S}$ and $\overset{(-)}{S}$ (Vekua, 1982).

The radius-vector of an arbitrary point of the domain of the thin body is represented as in Nikabadze (2016) (the normal section of the thin body is shown in Fig. 9.1)

$$\hat{\mathbf{r}}(x', x^3) = \mathbf{r}(x') + z\mathbf{n}(x'); \quad z = \bar{h}(x') + x^3 h(x'), \quad -1 \le x^3 \le 1, \quad (9.1)$$

where $\mathbf{r} = \mathbf{r}(x')$ is a vector parametric equation of the base surface S (in Fig. 9.1 the normal section of the body surface is shown), $x' = (x^1, x^2)$ is an arbitrary point[1] on S, i.e. x^1 and x^2 are the curvilinear (Gaussian) coordinates on the base surface S, $\bar{h}(x') = [\overset{(+)}{h}(x') - \overset{(-)}{h}(x')]/2$, $h(x') = [\overset{(+)}{h}(x') + \overset{(-)}{h}(x')]/2$, $\mathbf{n}(x')$ is the unit normal vector to S at the point x', $\overset{(-)}{h}(x')$ is the distance from the point x' to the corresponding point of the surface $\overset{(-)}{S}$, $\overset{(+)}{h}(x')$ is the distance from the same point x'

---

[1] Under x' we mean an arbitrary point of the base surface S, which has two coordinates x^1 and x^2, i.e. dependence of quantities on x' means their dependence on x^1 and x^2. In addition, the usual rules of tensor calculus (Lurie, 1990; Nikabadze, 2016, 2017a,b; Pobedrya, 1986; Sokol'nikov, 1971; Vekua, 1978) are applied. The upper and lower case Latin indices run through the values 1, 2 and 1, 2, 3, respectively.

Fig. 9.1 Normal section of thin body

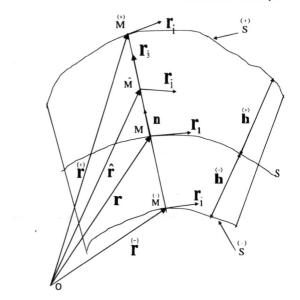

to the corresponding point of the surface $\overset{(+)}{S}$ (the normal sections of these surfaces are shown in Fig. 9.1), $2h(x') = \overset{(+)}{h}(x') + \overset{(-)}{h}(x')$ is the thickness of the thin body at x'. Note that the point O, in general, does not belong to the plane shown in Fig. 9.1.

It is easy to see that (9.1) with $\forall x'$ and $x^3 = -1$ define the front surface $\overset{(-)}{S}$, the vector parametric equation of which will have the form

$$\overset{(-)}{\mathbf{r}}(x') = \hat{\mathbf{r}}(x', x^3)\Big|_{x^3=-1} = \mathbf{r}(x') - \overset{(-)}{h}(x')\mathbf{n}(x'), \tag{9.2}$$

for $\forall x'$ and $x^3 = 1$ is the front surface $\overset{(+)}{S}$, the vector parametric equation of which will be represented in the form

$$\overset{(+)}{\mathbf{r}}(x') = \hat{\mathbf{r}}(x', x^3)\Big|_{x^3=1} = \mathbf{r}(x') + \overset{(+)}{h}(x')\mathbf{n}(x'), \tag{9.3}$$

and for $\forall x'$ and $x^3 = const$, where $x^3 \in (-1, 1)$ is an equidistant surface $\hat{S}$ from the base S. Obviously, the relation (9.1) for $x' \in S$ and $x^3 \in [-1, 1]$ is a vector parametric equation of a thin body domain.

Let us denote the point x' of the base surface S by M, and the points corresponding to the point M on the front surfaces $\overset{(-)}{S}$, $\overset{(+)}{S}$ and $\hat{S}$ by $\overset{(-)}{M}$, $\overset{(+)}{M}$ and $\hat{M}$, which are defined using (9.1) with $x^3 = -1$, $x^3 = 1$ and $x^3 = const \neq \pm 1$, respectively. For derivatives with respect to x^P from $\hat{\mathbf{r}}(x', x^3)$ and $\overset{(\sim)}{\mathbf{r}}$, $\sim \in \{-, \emptyset, +\}$ let us introduce[2] the notations $\mathbf{r}_{\hat{P}} = \partial_P \hat{\mathbf{r}}$ and $\mathbf{r}_{\tilde{P}} = \partial_P \overset{(\sim)}{\mathbf{r}}$, $\sim \in \{-, \emptyset, +\}$, respectively. Then, from (9.1)–(9.3) we get

$$\mathbf{r}_{\hat{P}} = \left(g_P^Q - zb_P^Q\right)\mathbf{r}_Q + \partial_P z\mathbf{n}, \quad \mathbf{r}_{\underset{P}{-}} = \left(g_P^Q + \overset{(-)}{h} b_P^Q\right)\mathbf{r}_Q - \partial_P \overset{(-)}{h}\,\mathbf{n},$$

$$\mathbf{r}_{\underset{P}{+}} = \left(g_P^Q + \overset{(+)}{h} b_P^Q\right)\mathbf{r}_Q - \partial_P \overset{(+)}{h}\,\mathbf{n} \quad \left(z = \bar{h}(x') + x^3 h(x')\right). \tag{9.4}$$

Writing (9.4), Weingarten's derivational formulas

$$\mathbf{n}_P = \partial_P \mathbf{n} = -b_P^Q \mathbf{r}_Q = -\underset{\sim}{\mathbf{b}} \cdot \mathbf{r}_P$$

were used, where $\underset{\sim}{\mathbf{b}}$ is the second surface tensor S.

Pairs of vectors $\mathbf{r}_{\underset{1}{*}}$, $\mathbf{r}_{\underset{2}{*}}$, $* \in \{-, \emptyset, \wedge, +\}$,, define at the points $\overset{(*)}{M} \in \overset{(*)}{S}$, $* \in \{-, \emptyset, \wedge, +\}$ and form two-dimensional covariant surface frames (bases) $\overset{(*)}{M}\mathbf{r}_{\underset{1}{*}}\mathbf{r}_{\underset{2}{*}}$, $* \in \{-, \emptyset, \wedge, +\}$. Using these frames (bases), we can construct contravariant frames $\overset{(*)}{M}\mathbf{r}^{\underset{*}{1}}\mathbf{r}^{\underset{*}{2}}$, $* \in \{-, \emptyset, \wedge, +\}$ (bases $\mathbf{r}^{\underset{*}{1}}\mathbf{r}^{\underset{*}{2}}$, $* \in \{-, \emptyset, \wedge, +\}$) by the appropriate formulas in the same way as with new parametrization (Nikabadze, 1988b, 2014a,b, 2016). Naturally, covariant and contravariant bases generate their typical geometric characteristics. In particular, the following matrices can be defined:

$$g_{\tilde{I}\tilde{J}} = \mathbf{r}_{\tilde{I}} \cdot \mathbf{r}_{\tilde{J}}, \quad g_{\tilde{I}}^{\breve{J}} = \mathbf{r}_{\tilde{I}} \cdot \mathbf{r}^{\breve{J}}, \quad g^{\tilde{I}\breve{J}} = \mathbf{r}^{\tilde{I}} \cdot \mathbf{r}^{\breve{J}}, \quad \sim, \smile \in * \in \{-, \emptyset, \wedge, +\}. \tag{9.5}$$

Differentiating (9.1) by x^3, we get

$$\mathbf{r}_{\hat{3}} = \partial_3 \hat{\mathbf{r}} = h(x')\mathbf{n}(x'), \quad \forall x^3 \in [-1, 1]. \tag{9.6}$$

According to (9.6) we can take

$$\mathbf{r}_{\underset{3}{-}} = \mathbf{r}_3 = \mathbf{r}_{\hat{3}} = \mathbf{r}_{\underset{3}{+}} = h(x')\mathbf{n}(x'), \quad \forall x^3 \in [-1, 1]. \tag{9.7}$$

Using the relation (9.4) and (9.7), one can define spatial covariant bases $\mathbf{r}_{\underset{p}{*}}$, $* \in \{-, \emptyset, +\}$ at points $\overset{(*)}{M} \in \overset{(*)}{S}$, $* \in \{-, \emptyset, +\}$. Thus, the third basis vector of spatial covariant bases at the point $\overset{(*)}{M} \in \overset{(*)}{S}$, $* \in \{-, \emptyset, \wedge, +\}$ is the same the vector $\mathbf{r}_3 = h(x')\mathbf{n}(x')$.

In view of (9.7), the first relation (9.4) and (9.6) can be represented as one formula

---

[2] In the following, short type records are used $\overset{(\sim)}{M} \in \overset{(\sim)}{S}$, $\sim \in \{-, \emptyset, \wedge, +\}$ or $\mathbf{r}_{\tilde{p}} = g_{\tilde{p}}^{\breve{q}}\mathbf{r}_{\breve{q}}$, $\sim, \smile \{-, \emptyset, \wedge, +\}$, where $\emptyset$ is an empty set. The first entry means: if $\sim = -$, then $\overset{(-)}{M} \in \overset{(-)}{S}$; if $\sim = \emptyset$, then $M \in S$; if $\sim = \wedge$, then $\hat{M} \in \hat{S}$; if $\sim = +$, then $\overset{(+)}{M} \in \overset{(+)}{S}$. The second entry means that if, for example, $\sim = \emptyset$, $\smile = -$, then $\mathbf{r}_p = g_p^{\bar{q}}\mathbf{r}_{\bar{q}}$; if $\sim = \wedge$, $\smile = \emptyset$, then $\mathbf{r}_{\hat{p}} = g_{\hat{p}}^q\mathbf{r}_q$, etc. Looking through all the values, we get all the relations.

$$\mathbf{r}_{\hat{p}} = \left(g_p^q - zb_p^q + h^{-1}\partial_I zg_p^I g_3^q\right)\mathbf{r}_q. \tag{9.8}$$

Similarly, using the last two formulas (9.4) and (9.7), we have

$$\mathbf{r}_{\underset{p}{-}} = \left(g_p^q + \overset{(-)}{h}\, b_p^q - h^{-1}\partial_I\, \overset{(-)}{h}\, g_p^I g_3^q\right)\mathbf{r}_q,$$
$$\mathbf{r}_{\underset{p}{+}} = \left(g_p^q - \overset{(+)}{h}\, b_p^q + h^{-1}\partial_I\, \overset{(+)}{h}\, g_p^I g_3^q\right)\mathbf{r}_q. \tag{9.9}$$

In Eqs. (9.8) and (9.9) $b_p^q = g_p^M g_N^q b_M^N$ are components of the extended second surface tensor S (Vekua, 1978, 1982). Note that the formulas (9.9) are obtained from (9.8) for $x^3 = -1$ and $x^3 = 1$, respectively.

The triples of vectors $\mathbf{r}_{\underset{1}{*}}, \mathbf{r}_{\underset{2}{*}}, \mathbf{r}_{\underset{3}{*}}$, $* \in \{-, \emptyset, \wedge, +\}$, defined at the points $M \in \overset{(*)}{S}$, $* \in \{-, \emptyset, \wedge, +\}$, form three-dimensional (spatial) covariant bases and $\overset{(*)}{M}\mathbf{r}_{\underset{1}{*}}\mathbf{r}_{\underset{2}{*}}\mathbf{r}_{\underset{3}{*}}$, $* \in \{-, \emptyset, \wedge, +\}$ are the three-dimensional space frames, which generate the corresponding parametrization. Using these frames (bases), we can construct the contravariant frames $\overset{(*)}{M}\mathbf{r}^1\mathbf{r}^2\mathbf{r}^3$, $* \in \{-, \emptyset, \wedge, +\}$ (bases $\mathbf{r}^1\mathbf{r}^2\mathbf{r}^3$, $* \in \{-, \emptyset, \wedge, +\}$) corresponding to them. Indeed, based on the definition of the contravariant basis (Vekua, 1978; Lurie, 1990; Pobedrya, 1986; Nikabadze, 2014a,b, 2016, 2017b), we have

$$\mathbf{r}^{\tilde{k}} = \frac{1}{2}C^{\tilde{k}\tilde{p}\tilde{q}}\mathbf{r}_{\tilde{p}} \times \mathbf{r}_{\tilde{q}}, \quad \sim \in \{-, \emptyset, \wedge, +\}. \tag{9.10}$$

where $C^{\tilde{k}\tilde{p}\tilde{q}} = \left(\mathbf{r}^{\tilde{k}} \times \mathbf{r}^{\tilde{p}}\right) \cdot \mathbf{r}^{\tilde{q}}, \sim \in \{-, \emptyset, \wedge, +\}$ are the contravariant components of the discriminant tensor (Vekua, 1978) at points $\overset{(*)}{M} \in \overset{(*)}{S}$, $* \in \{-, \emptyset, \wedge, +\}$, respectively.

Similarly to (9.5) let us introduce the next matrices:

$$g_{\tilde{p}\tilde{q}} = \mathbf{r}_{\tilde{p}} \cdot \mathbf{r}_{\tilde{q}}, \quad g_{\tilde{p}}^{\breve{q}} = \mathbf{r}_{\tilde{p}} \cdot \mathbf{r}^{\breve{q}}, \quad g^{\tilde{p}\breve{q}} = \mathbf{r}^{\tilde{p}} \cdot \mathbf{r}^{\breve{q}}, \quad \sim, \smile \in \{-, \emptyset, \wedge, +\}. \tag{9.11}$$

According to the first two relations (9.11) and (9.8) we have

$$g_{\hat{p}q} = g_{pq} - zb_{pq} + h^{-1}\partial_I zg_p^I g_{3q}, \quad g_{\hat{p}}^q = g_p^q - zb_p^q + h^{-1}\partial_I zg_p^I g_3^q. \tag{9.12}$$

Similarly with (9.9) and first two relations (9.11) we get

$$g_{\underset{pq}{-}} = g_{pq} + \overset{(-)}{h}\, b_{pq} - h^{-1}\partial_I\, \overset{(-)}{h}\, g_p^I g_{3q},$$
$$g_{\underset{p}{-}}^q = g_p^q + \overset{(-)}{h}\, b_p^q - h^{-1}\partial_I\, \overset{(-)}{h}\, g_p^I g_3^q,$$
$$g_{\underset{pq}{+}} = g_{pq} - \overset{(+)}{h}\, b_{pq} + h^{-1}\partial_I\, \overset{(+)}{h}\, g_p^I g_{3q},$$
$$g_{\underset{p}{+}}^q = g_p^q - \overset{(+)}{h}\, b_p^q + h^{-1}\partial_I\, \overset{(+)}{h}\, g_p^I g_3^q. \tag{9.13}$$

It is easy to see that from (9.12) and (9.13) we will have

$$g_{\hat{P}3}=h\partial_P z, \quad g_{\hat{P}}^3=h^{-1}\partial_P z, \quad \mathbf{r}^3=h^{-1}(x')\mathbf{n}(x'), \qquad (9.14)$$

$$g_{\underset{P3}{-}}=-h\partial_P \overset{(-)}{h}, \quad g_{\underset{P}{-}}^3=-h^{-1}\partial_P \overset{(-)}{h}, \quad g_{\underset{P3}{+}}=h\partial_P \overset{(+)}{h}, \quad g_{\underset{P}{+}}^3=h^{-1}\partial_P \overset{(+)}{h}. \qquad (9.15)$$

Note that according to second formula of (9.14) relations (9.12) can be represent in the form

$$g_{\hat{p}q}=g_{pq}-zb_{pq}+g_{\hat{M}}^3 g_p^M g_{3q}, \quad g_{\hat{p}}^q=g_p^q-zb_p^q+g_{\hat{M}}^3 g_p^M g_3^q, \qquad (9.16)$$

and with the second and fourth formulas (9.15) relations (9.13) can be written as

$$g_{\underset{pq}{-}}=g_{pq}+\overset{(-)}{h}b_{pq}+g_{\underset{M}{-}}^3 g_p^M g_{3q}, \quad g_{\underset{p}{-}}^q=g_p^q+\overset{(-)}{h}b_p^q+g_{\underset{M}{-}}^3 g_p^M g_3^q,$$
$$g_{\underset{pq}{+}}=g_{pq}-\overset{(+)}{h}b_{pq}+g_{\underset{M}{+}}^3 g_p^M g_{3q}, \quad g_{\underset{p}{+}}^q=g_p^q-\overset{(+)}{h}b_p^q+g_{\underset{M}{+}}^3 g_p^M g_3^q. \qquad (9.17)$$

Note also that according to (9.15) relations (9.14) will get the view

$$g_{\hat{P}3}=\frac{1}{2}\left[g_{\underset{P3}{+}}+g_{\underset{P3}{-}}+x^3(g_{\underset{P3}{+}}-g_{\underset{P3}{-}})\right], \quad g_{\hat{P}}^3=\frac{1}{2}\left[g_{\underset{P}{+}}^3+g_{\underset{P}{-}}^3+x^3(g_{\underset{P}{+}}^3-g_{\underset{P}{-}}^3)\right]. \qquad (9.18)$$

It is easy to see that with (9.11) there is the next connection between basis vectors:

$$\mathbf{r}_{\tilde{p}}=g_{\tilde{p}}^{\breve{q}}\mathbf{r}_{\breve{q}}, \quad \sim, \smile \in \{-,\emptyset,\wedge,+\}, \qquad (9.19)$$

which stay true when juggling with indices. Based on (9.19), we can prove the validity of the relation

$$g_{\tilde{p}}^{\breve{q}}=g_{\tilde{p}}^{\overset{*}{n}}g_{\overset{*}{n}}^{\breve{q}}, \quad \sim, \smile, * \in \{-,\emptyset,\wedge,+\}, \qquad (9.20)$$

which stay true when juggling with indices. It is easy to find expressions for $g_{\hat{p}\hat{q}}$. Indeed, using (9.12) from (9.20) we get

$$g_{\hat{p}\hat{q}}=g_{pq}-2zb_{pq}+z^2 b_p^I b_{Iq}+g_{\hat{S}}^3 g_p^S g_{3q}+g_{\hat{T}}^3 g_q^T g_{3q}+g_{33}g_{\hat{S}}^3 g_{\hat{T}}^3 g_p^S g_q^T. \qquad (9.21)$$

Now let us find the expression for $\sqrt{\hat{g}}=(\mathbf{r}_{\hat{1}}\times\mathbf{r}_{\hat{2}})\cdot\mathbf{r}_{\hat{3}}$. According to (9.19) when $\sim=\wedge$ and $\smile\in\{-,\emptyset,+\}$ we have

$$\sqrt{\hat{g}}=\frac{1}{2}\epsilon^{IJ}(\mathbf{r}_{\hat{I}}\times\mathbf{r}_{\hat{J}})\cdot\mathbf{r}_{\hat{3}}=\frac{1}{2}\epsilon^{IJ}g_{\hat{I}}^{\breve{K}}g_{\hat{J}}^{\breve{L}}(\mathbf{r}_{\breve{K}}\times\mathbf{r}_{\breve{L}})\cdot\mathbf{r}_{\hat{3}}=\sqrt{\overset{(\smile)}{g}}\det\big(g_{\hat{I}}^{\breve{K}}\big),$$

that is

$$\sqrt{\hat{g}}=\sqrt{\overset{(\smile)}{g}}\det\big(g_{\hat{I}}^{\breve{K}}\big), \quad \det\big(g_{\hat{I}}^{\breve{K}}\big)=\frac{1}{2}\epsilon^{IJ}\epsilon_{KL}g_{\hat{I}}^{\breve{K}}g_{\hat{J}}^{\breve{L}}, \quad \smile\in\{-,\emptyset,+\}. \qquad (9.22)$$

Here ϵ^{IJ}, ϵ_{KL} are two-dimensional Levi-Civita symbols, and

$$\sqrt{\overset{\smile}{g}} = (\mathbf{r}_{\check{1}} \times \mathbf{r}_{\check{2}}) \cdot \mathbf{r}_{\check{3}}, \quad \smile \in \{-, \emptyset, +\},$$

$$\overset{(-)}{g} = \hat{g}|_{x^3=-1}, \quad g = \hat{g}|_{z=0}, \quad \overset{(+)}{g} = \hat{g}|_{x^3=1}. \tag{9.23}$$

From (9.22) when $\smile = \emptyset$ we get

$$\hat{\vartheta} \equiv \sqrt{\hat{g}g^{-1}} = \det\left(g_{\hat{I}}^{K}\right) = \frac{1}{2}\,\epsilon^{IJ}\epsilon_{KL}g_{\hat{I}}^{K}g_{\hat{J}}^{L}. \tag{9.24}$$

Note that there is a more general relation than (9.22). Namely,

$$\sqrt{\overset{(\widetilde{\sim})}{g}} = \frac{1}{2}\sqrt{\overset{(\smile)}{g}}\,\epsilon^{IJ}\epsilon_{KL}g_{\check{I}}^{\check{K}}g_{\check{J}}^{\check{L}} = \sqrt{\overset{(\smile)}{g}}\,\det\left(g_{\check{P}}^{\check{Q}}\right), \quad \sim, \smile \in \{-, \emptyset, \wedge, +\}. \tag{9.25}$$

From (9.25) we find

$$\det\left(g_{\check{P}}^{\check{Q}}\right) = \sqrt{\overset{(\widetilde{\sim})}{g}\overset{(\smile)}{g}{}^{-1}} = \frac{1}{2}\,\epsilon^{IJ}\epsilon_{KL}g_{\check{I}}^{\check{K}}g_{\check{J}}^{\check{L}}, \quad \sim, \smile \in \{-, \emptyset, \wedge, +\}.$$

Now let us find the expression for $\mathbf{r}^{\check{k}}$. Due to (9.19), after simple calculations from (9.10) we will have

$$\mathbf{r}^{\check{k}} = \frac{1}{2}\sqrt{\overset{(\smile)}{g}{}^{-1}\overset{(\widetilde{\sim})}{g}}\,\epsilon^{kpq}\epsilon_{lmn}g_{\check{p}}^{\widetilde{m}}g_{\check{q}}^{\widetilde{n}}\mathbf{r}^{\check{l}} = \frac{1}{2}\sqrt{\overset{(\smile)}{g}{}^{-1}g}\,\epsilon^{kpq}\epsilon_{lmn}g_{\check{p}}^{m}g_{\check{q}}^{n}\mathbf{r}^{l},$$

$$\sim, \smile \in \{-, \emptyset, \wedge, +\}. \tag{9.26}$$

From here when $\smile = \wedge$ we will get

$$\mathbf{r}^{\check{k}} = \frac{1}{2}\sqrt{\hat{g}^{-1}\overset{(\widetilde{\sim})}{g}}\,\epsilon^{kpq}\epsilon_{lmn}g_{\hat{p}}^{\widetilde{m}}g_{\hat{q}}^{\widetilde{n}}\mathbf{r}^{\check{l}} = \frac{1}{2}\,\hat{\vartheta}^{-1}\epsilon^{kpq}\epsilon_{lmn}g_{\hat{p}}^{m}g_{\hat{q}}^{n}\mathbf{r}^{l},$$

$$\sim \in \{-, \emptyset, \wedge, +\}. \tag{9.27}$$

Here ϵ^{kpq}, ϵ_{lmn} are Levi-Civita symbols. Based on (9.26), we have

$$g_{\check{I}}^{\check{k}} = \mathbf{r}^{\check{k}} \cdot \mathbf{r}_{\check{I}} = \frac{1}{2}\sqrt{\overset{(\smile)}{g}{}^{-1}\overset{(\widetilde{\sim})}{g}}\,\epsilon^{kpq}\epsilon_{lmn}g_{\check{p}}^{\widetilde{m}}g_{\check{q}}^{\widetilde{n}},$$

$$g^{\check{k}\check{l}} = \mathbf{r}^{\check{k}} \cdot \mathbf{r}^{\check{l}} = \frac{1}{2}\sqrt{\overset{(\smile)}{g}{}^{-1}\overset{(\widetilde{\sim})}{g}}\,\epsilon^{kpq}\epsilon_{smn}g_{\check{p}}^{\widetilde{m}}g_{\check{q}}^{\widetilde{n}}g^{\widetilde{s}\check{l}}, \quad \sim, \smile \in \{-, \emptyset, \wedge, +\}.$$

From here or using (9.27) we can find

$$g_{\hat{I}}^{\hat{k}} = \frac{1}{2}\,\hat{\vartheta}^{-1}\epsilon^{kpq}\epsilon_{lmn}g_{\hat{p}}^{m}g_{\hat{q}}^{n}, \qquad g^{\hat{k}l} = \frac{1}{2}\,\hat{\vartheta}^{-1}\epsilon^{kpq}\epsilon_{smn}g_{\hat{p}}^{m}g_{\hat{q}}^{n}g^{sl}.$$

Let us find the expressions for $\mathbf{r}^{\hat{P}}$ and $\mathbf{r}^{\hat{3}}$. Obviously, expressions for these basis vectors can be found using (9.27), however, we will find them in a different way. Due to (9.19), we have

$$\mathbf{r}^{\hat{I}} = C^{\hat{I}\hat{J}\hat{3}}\mathbf{r}_{\hat{J}} \times \mathbf{r}_{\hat{3}} = \sqrt{\hat{g}^{-1}}\, g_{\hat{J}}^{K}\,\mathbf{r}_K \times \mathbf{r}_3 = \sqrt{\hat{g}^{-1}g}\,\epsilon^{IJ}\epsilon_{LK}g_{\hat{J}}^{K}\mathbf{r}^L,$$

that is

$$\mathbf{r}^{\hat{P}} = g_M^{\hat{P}}\mathbf{r}^M = \hat{\vartheta}^{-1}A_M^{\hat{P}}\mathbf{r}^M, \quad g_M^{\hat{P}} = \hat{\vartheta}^{-1}A_M^{\hat{P}}, \quad A_M^{\hat{P}} = \epsilon^{PK}\epsilon_{ML}g_{\hat{K}}^{L}. \quad (9.28)$$

According to (9.19) we have

$$\mathbf{r}^3 = g_{\hat{P}}^3\mathbf{r}^{\hat{P}} = \mathbf{r}^{\hat{3}} + g_{\hat{P}}^3\mathbf{r}^{\hat{P}} = \mathbf{r}^{\hat{3}} + g_{\hat{P}}^3 g_M^{\hat{P}}\mathbf{r}^M.$$

From here

$$\mathbf{r}^{\hat{3}} = \mathbf{r}^3 - g_{\hat{P}}^3 g_M^{\hat{P}}\mathbf{r}^M, \quad g_M^{\hat{3}} = -g_{\hat{P}}^3 g_M^{\hat{P}}, \quad (9.29)$$

and from (9.29) we get

$$\mathbf{r}^{\bar{3}} = \mathbf{r}^{\hat{3}}|_{x^3=-1} = \mathbf{r}^3 - g_{\underset{P}{-}}^3 g_M^{\bar{P}}\mathbf{r}^M, \quad g_M^{\bar{3}} = -g_{\underset{P}{-}}^3 g_M^{\bar{P}},$$

$$\mathbf{r}^{\overset{+}{3}} = \mathbf{r}^{\hat{3}}|_{x^3=1} = \mathbf{r}^3 - g_{\underset{P}{+}}^3 g_M^{\overset{+}{P}}\mathbf{r}^M, \quad g_M^{\overset{+}{3}} = -g_{\underset{P}{+}}^3 g_M^{\overset{+}{P}}. \quad (9.30)$$

9.2.1 Representation of the Second Rank Unit Tensor and Representation of its Components in the Form of Power Series

Using the usual representation of the second rank unit tensor (SRUT) (Vekua, 1978; Lurie, 1990; Pobedrya, 1986; Nikabadze, 2014a,b, 2017b), according to (9.19) and (9.20) we get the expression

$$\underset{\sim}{\mathbf{E}} = g_{\breve{m}}^{\breve{n}}\mathbf{r}^{\breve{m}}\mathbf{r}_{\breve{n}}, \quad \sim, \smile \in \{-, \emptyset, \wedge, +\}, \quad (9.31)$$

which stay true when juggling with indices. From (9.31) it can be seen that the elements of the matrices (9.11) introduced above represent the components of the SRUT. Let us introduce the following definition.

Definition 9.4. The components of $g_{\breve{p}}^{\breve{q}}$, $g_{\breve{q}}^{\breve{P}}$, $g^{\breve{p}\breve{q}}$, $g_{\breve{p}\breve{q}}$, $\sim, \smile \in \{-, \emptyset, \wedge, +\}$, $\sim \neq \smile$ are called the transfer components of the SRUT under the considered parametrization of the thin body domain.

Further, when finding the moments of various quantities and expressions with respect to systems of orthogonal polynomials, the expressions for the transfer components $g_M^{\hat{M}}$, $g_M^{\hat{3}}$ and the components $g^{\hat{P}\hat{Q}}$, $g^{\hat{P}\hat{3}}$, $g^{\hat{3}\hat{3}}$ of SRUT in the form of power series with respect to x^3 are of interest. Similar representations under the new parametrization are given in Nikabadze (2014a,b, 2017b). Under the classical parametrization, the representation of basis vectors and components of the SRUT, similar to $\mathbf{r}^{\hat{P}}$ and $g^{\hat{P}\hat{Q}}$ are given in Vekua (1978). The classical parametrization differs from the one considered in that in the classical case z is chosen as the transverse coordinate, but in this case $z = \bar{h} + x^3 h$, therefore we can obtain the representations for $\mathbf{r}^{\hat{P}}$ if in the representation of a similar object from z is replaced by $\bar{h} + x^3 h$ (Vekua, 1978). As a result, we get

$$
\begin{aligned}
\mathbf{r}^{\hat{P}} &= \mathbf{r}^P \cdot \left(\mathbf{\underset{\sim}{E}} - z\mathbf{\underline{b}}\right)^{-1} = \mathbf{r}^P \cdot \left(\mathbf{\underset{\sim}{E}} + z\mathbf{\underline{b}} + z^2 \mathbf{\underline{b}}^2 + \dots\right) \\
&= \left(g_M^P + z b_M^P + z^2 b_N^P b_M^N + \dots\right)\mathbf{r}^M.
\end{aligned}
\tag{9.32}
$$

Note that

$$
\left(\mathbf{\underset{\sim}{E}} - z\mathbf{\underline{b}}\right)^{-1} = \sum_{s=0}^{\infty} z^s \mathbf{\underline{b}}^s, \quad \left(\mathbf{\underset{\sim}{E}} - z\mathbf{\underline{b}}\right)^{-2} = \sum_{s=0}^{\infty} (1+s) z^s \mathbf{\underline{b}}^s.
\tag{9.33}
$$

Of course, the relations (9.33) take place if the norm $\| z\mathbf{\underline{b}} \| = \| (\bar{h} + x^3 h)\mathbf{\underline{b}} \| < 1$. According to (9.32) we have

$$
\begin{aligned}
g_M^{\hat{P}} &= \mathbf{r}^P \cdot \left(\mathbf{\underset{\sim}{E}} - z\mathbf{\underline{b}}\right)^{-1} \cdot \mathbf{r}_M = \sum_{s=0}^{\infty} \underset{(s)M}{A^P} z^s, \quad \underset{(0)M}{A^P} = g_M^P, \\
\underset{(1)M}{A^P} &= b_M^P, \quad \underset{(2)M}{A^P} = b_N^P b_M^N, \dots, \underset{(s)M}{A^P} = b_{N_1}^P b_{N_2}^{N_1} \dots b_{N_{s-1}}^{N_{s-2}} b_M^{N_{s-1}}.
\end{aligned}
\tag{9.34}
$$

Knowing (9.33) and (9.34), it is easy to find the presentations of other components. We will get

$$
\begin{aligned}
g^{\hat{P}\hat{Q}} &= g_M^{\hat{P}} g^{M\hat{Q}} = \mathbf{r}^P \cdot \left(\mathbf{\underset{\sim}{E}} - z\mathbf{\underline{b}}\right)^{-2} \cdot \mathbf{r}^Q = g^{QM} \sum_{s=0}^{\infty} (1+s) \underset{(s)}{A}_M^P z^s, \\
g_M^{\hat{3}} &= -g_{\hat{P}}^3 \sum_{s=0}^{\infty} \underset{(s)}{A}_M^P z^s, \quad g^{\hat{P}\hat{3}} = -g_{\hat{Q}}^3 g^{\hat{P}\hat{Q}} = -g_{\hat{Q}}^3 g^{QM} \sum_{s=0}^{\infty} (1+s) \underset{(s)M}{A}^P z^s, \\
g^{\hat{3}\hat{3}} &= g^{33} + g_{\hat{P}}^3 g_{\hat{Q}}^3 g^{\hat{P}\hat{Q}} = g^{33} + g_{\hat{P}}^3 g_{\hat{Q}}^3 g^{QM} \sum_{s=0}^{\infty} (1+s) \underset{(s)}{A}_M^P z^s.
\end{aligned}
\tag{9.35}
$$

Using the multiplication rule for series in the Cauchy form, by (9.34) we find

$$
g_M^{\hat{P}} g_N^{\hat{Q}} = \sum_{s=0}^{\infty} \underset{(s)}{B}_{MN}^{PQ} z^s, \quad \underset{(s)}{B}_{MN}^{PQ} = \sum_{r=0}^{s} \underset{(s-r)M}{A}^P \underset{(r)N}{A}^Q.
\tag{9.36}
$$

It should be noted that since the components of the SRUT participate in the representations of the equations and the constitutive relations of the mechanics of deformable bodies under the considered parametrization, the number of terms held in the right-hand sides of (9.34) – (9.36) depends on the problem and required accuracy of approximation. The relations (9.34) – (9.36) play an important role in the construction of various variants of mathematical theories of thin bodies with the use of expansion in systems of orthogonal polynomials.

9.2.2 Representations of Gradient, Divergence, Repeated Gradient, and Laplacian

In this case, using (9.28) and (9.29) for the gradient of some tensor $\hat{\mathbb{F}}(x', x^3)$, we have

$$\hat{\nabla}\mathbb{F} = \mathbf{r}^{\hat{P}} N_P \mathbb{F} + \mathbf{r}^3 \nabla_3 \mathbb{F} = \mathbf{r}^M g_M^{\hat{P}} N_P \mathbb{F} + \mathbf{r}^3 \partial_3 \mathbb{F}, \tag{9.37}$$

where we introduced the following differential operator

$$N_P = \partial_P - g_{\hat{P}}^3 \partial_3, \quad N = \mathbf{r}^{\hat{P}}(\partial_P - g_{\hat{P}}^3 \partial_3) = \mathbf{r}^M g_M^{\hat{P}}(\partial_P - g_{\hat{P}}^3 \partial_3). \tag{9.38}$$

Here the components $g_{\hat{P}}^3$ of SRUT (9.14) characterize the thickness change.

It is easy to see that divergence, for example, of the second rank tensor $\underset{\sim}{\mathbf{P}}$ according to (9.37) can be represent as

$$\hat{\nabla} \cdot \underset{\sim}{\mathbf{P}} = \nabla_p \mathbf{P}^{\hat{p}} = g_M^{\hat{P}} N_P \mathbf{P}^M + \partial_3 \mathbf{P}^3. \tag{9.39}$$

Other representations of divergence of the second rank tensor have the view

$$\hat{\nabla} \cdot \underset{\sim}{\mathbf{P}} = \frac{1}{\sqrt{g}\,\hat{\vartheta}} \partial_P (\sqrt{g}\,\hat{\vartheta}\mathbf{P}^{\hat{P}}) + \frac{1}{\hat{\vartheta}} \partial_3(\hat{\vartheta}\mathbf{P}^{\hat{3}}),$$
$$\hat{\nabla} \cdot \underset{\sim}{\mathbf{P}} = \mathbf{r}^{\hat{P}} \cdot \partial_P \underset{\sim}{\mathbf{P}} + \mathbf{r}^3 \cdot \partial_3 \underset{\sim}{\mathbf{P}} = g_M^{\hat{P}} \partial_P \mathbf{P}^M + (\mathbf{r}^3 - g_{\hat{P}}^3 g_M^{\hat{P}} \mathbf{r}^M) \cdot \partial_3 \underset{\sim}{\mathbf{P}}. \tag{9.40}$$

Writing the second ratio (9.40) we took into account (9.28) and (9.29).

By (9.37) representation of the repeated gradient of some tensor $\hat{\mathbb{F}}(x', x^3)$ under this parametrization has the form

$$\hat{\nabla}\hat{\nabla}\underset{\sim}{\mathbb{F}} = \mathbf{r}^M \mathbf{r}^N g_M^{\hat{P}} N_P (g_N^{\hat{Q}} N_Q \mathbb{F}) + \mathbf{r}^M \mathbf{r}^3 g_M^{\hat{P}} N_P \partial_3 \mathbb{F}$$
$$+ \mathbf{r}^3 \mathbf{r}^N \partial_3 (g_N^{\hat{Q}} N_Q \mathbb{F}) + \mathbf{r}^3 \mathbf{r}^3 \partial_3^2 \mathbb{F} = \mathbf{r}^M \mathbf{r}^N g_M^{\hat{P}} g_N^{\hat{Q}} N_P N_Q \mathbb{F} \tag{9.41}$$
$$+ \mathbf{r}^M \mathbf{r}^3 g_M^{\hat{P}} N_P \partial_3 \mathbb{F} + \mathbf{r}^3 \mathbf{r}^N g_N^{\hat{Q}} \nabla_3 N_Q \mathbb{F} + \mathbf{r}^3 \mathbf{r}^3 \partial_3^2 \mathbb{F}.$$

Here

$$N_P N_Q = \nabla_P \nabla_Q - (g_{\hat{P}}^3 \nabla_3 \nabla_Q + g_{\hat{Q}}^3 \nabla_P \nabla_3) + g_{\hat{P}}^3 g_{\hat{Q}}^3 \nabla_3^2.$$

We can see that according to (9.41) for Laplacian we will get the expression

$$\hat{\Delta}\mathbb{F} = g^{MN} g_M^{\hat{P}} N_P (g_N^{\hat{Q}} N_Q \mathbb{F}) + g^{33} \partial_3^2 \mathbb{F} = g^{MN} g_M^{\hat{P}} g_N^{\hat{Q}} N_P N_Q \mathbb{F} + g^{33} \partial_3^2 \mathbb{F}.$$

We can note that having the representations of gradient (9.37) and repeated gradient (9.41), it is not difficult to get the expressions for other differential operators (rotor, repeated rotor, repeated divergence, a gradient of divergence). In order to reduce this contribution, we will not dwell on them..

9.3 Presentations of the Equations of Motion, Heat Influx and Constitutive Relations of Micropolar Theory

9.3.1 Presentations of the Equations of Micropolar Theory

Using (9.39) and (9.40) the motion equations of the micropolar theory can be written in the form

$$
\begin{aligned}
& g_M^{\hat{P}} N_P \mathbf{P}^M + \partial_3 \mathbf{P}^3 + \rho \mathbf{F} = \rho \partial_t^2 \mathbf{u}, \\
& g_M^{\hat{P}} N_P \boldsymbol{\mu}^M + \partial_3 \boldsymbol{\mu}^3 + \underset{\sim}{\mathbf{C}} \overset{2}{\otimes} \underset{\sim}{\mathbf{P}} + \rho \mathbf{H} = \underset{\sim}{\mathbf{J}} \cdot \partial_t^2 \boldsymbol{\varphi},
\end{aligned}
\tag{9.42}
$$

$$
\begin{aligned}
& \frac{1}{\sqrt{g}} \partial_P \left(\sqrt{g} \hat{\vartheta} \mathbf{P}^{\hat{P}} \right) + \partial_3 \left(\hat{\vartheta} \mathbf{P}^3 \right) + \rho \hat{\vartheta} \mathbf{F} = \rho \hat{\vartheta} \partial_t^2 \mathbf{u}, \\
& \frac{1}{\sqrt{g}} \partial_P \left(\sqrt{g} \hat{\vartheta} \boldsymbol{\mu}^{\hat{P}} \right) + \partial_3 \left(\hat{\vartheta} \boldsymbol{\mu}^3 \right) + \underset{\sim}{\mathbf{C}} \overset{2}{\otimes} \hat{\vartheta} \underset{\sim}{\mathbf{P}} + \rho \hat{\vartheta} \mathbf{H} = \underset{\sim}{\mathbf{J}} \cdot \hat{\vartheta} \partial_t^2 \boldsymbol{\varphi},
\end{aligned}
\tag{9.43}
$$

$$
\begin{aligned}
& \mathbf{r}^{\hat{P}} \cdot \partial_P \underset{\sim}{\mathbf{P}} + \mathbf{r}^{\hat{3}} \cdot \partial_3 \underset{\sim}{\mathbf{P}} + \hat{\rho} \mathbf{F} = \rho \partial_t^2 \mathbf{u}, \\
& \mathbf{r}^{\hat{P}} \cdot \partial_P \underset{\sim}{\boldsymbol{\mu}} + \mathbf{r}^{\hat{3}} \cdot \partial_3 \underset{\sim}{\boldsymbol{\mu}} + \underset{\sim}{\mathbf{C}} \overset{2}{\otimes} \underset{\sim}{\mathbf{P}} + \rho \mathbf{H} = \underset{\sim}{\mathbf{J}} \cdot \partial_t^2 \boldsymbol{\varphi}.
\end{aligned}
\tag{9.44}
$$

Equations (9.42)–(9.44) denote the system of motion equation of the micropolar mechanics of deformable solids (MMDS) under the considered parametrization of the thin body domain. Therefore, is it appropriate to call them different representations of the system of motion equations of micropolar mechanics of deformable thin solids (MMDTS).

Let us introduce the next definition.

Definition 9.5. The equations of motion, constitutive equations (CR), etc., which are obtained from the corresponding representations under the considered parametrization of the thin body domain, are called relations of approximations of order r, if the first $r + 1$ terms are preserved in the expansion of $g_M^{\hat{P}}$.

Introducing the notation

$$
g_{(s)}^{\hat{P}} M = \sum_{s=0}^{r} A_{(s)}^{P} M z^s,
\tag{9.45}
$$

for example, from (9.42), if change $g_M^{\hat{P}}$ to $g_{(r)}^{\hat{P}} M$, then we will get the system of equations of MMDTS of approximations of order r

$$
g_{(r)}^{\hat{P}} M N_P \mathbf{P}^M + \partial_3 \mathbf{P}^3 + \rho \mathbf{F} = \rho \partial_t^2 \mathbf{u},
$$

$$
g_{(r)}^{\hat{P}} M N_P \boldsymbol{\mu}^M + \partial_3 \boldsymbol{\mu}^3 + \underset{\sim}{\mathbf{C}} \overset{2}{\otimes} \underset{\sim}{\mathbf{P}} + \rho \mathbf{H} = \underset{\sim}{\mathbf{J}} \cdot \partial_t^2 \boldsymbol{\varphi}.
$$

Therefore, for $r = 0$, we obtain the relations of the zero approximation, and for $r = \infty$ we assume that $\underset{(\infty)}{g}\overset{\hat{P}}{M} = g_M^{\hat{P}}$. In order to reduce this cotribution, we will not dwell on them.

9.3.2 Representation of the Equation of Heat Influx in Micropolar Mechanics of a Deformable Thin Solids

In general, the equation of heat influx in MMDTS can be present in the form (Pobedrya, 2006)

$$-\nabla \cdot \mathbf{q} + \rho q - T \frac{d}{dt}(\mathbf{a} \overset{2}{\otimes} \mathbf{P} + \mathbf{d} \overset{2}{\otimes} \boldsymbol{\mu}) + W^* = \rho c_p \partial_t T, \qquad (9.46)$$

where $\mathbf{q}$ is the external heat flux vector, q is the mass heat influx, T is the temperature, $\mathbf{a}$, $\mathbf{d}$ are the thermal expansion tensors, $\mathbf{P} \neq \mathbf{P}^T$ is the stress tensor, $\boldsymbol{\mu} \neq \boldsymbol{\mu}^T$ is the couple-stress tensor, W^* is the dissipation function, ρ is the density of the medium, c_p is the heat capacity at constant pressure. If a physically linear medium is considered, then non-linearity in (9.46) is revealed in the third term of the left side. A similar picture takes place in the particular case of this equation, which is obtained from (9.46) when $\mathbf{d} = 0$ (Pobedrya, 1995). In the latter case, since both heat capacities c_p and c_v (heat capacity with constant volume) cannot be simultaneously constant (be independent of the temperature), very often accept the assumption that the temperature T is replaced with the temperature $T_0 = \text{const}$. Given this assumption, the desired expression of the heat influx equation, similarly to (9.42), will have the form

$$-g_M^{\hat{P}} N_P q^M - \partial_3 q^3 + \rho q - T_0 \frac{d}{dt}(\mathbf{a} \overset{2}{\otimes} \mathbf{P} + \mathbf{d} \overset{2}{\otimes} \boldsymbol{\mu}) + W^* = \rho c_p \partial_t T. \qquad (9.47)$$

Note that based on the first relation (9.47), by the definition 9.5 and the notation (9.45) the heat influx equation of approximation of order r is represented as follows

$$-\underset{(r)}{g}\overset{\hat{P}}{M} N_P q^M - \partial_3 q^3 + \rho q - T_0 \frac{d}{dt}(\mathbf{a} \overset{2}{\otimes} \mathbf{P} + \mathbf{d} \overset{2}{\otimes} \boldsymbol{\mu}) + W^* = \rho c_p \partial_t T. \qquad (9.48)$$

9.3.3 Representations of Hooke's Law and Fourier's Heat Conduction Law

In the linear micropolar theory of elasticity, the Hooke law in non-isothermal processes, according to the generalized Duhamel-Neumann principle (Pobedrya, 2006, 1995), can be represented as

$$\underset{\sim}{\mathbf{P}} = \underset{\approx}{\mathbf{A}} \overset{2}{\otimes} (\boldsymbol{\gamma} - \underset{\sim}{\mathbf{a}}\vartheta) + \underset{\approx}{\mathbf{B}} \overset{2}{\otimes} (\boldsymbol{\varkappa} - \underset{\sim}{\mathbf{d}}\vartheta),$$

$$\boldsymbol{\mu} = \underset{\approx}{\mathbf{C}} \overset{2}{\otimes} (\boldsymbol{\gamma} - \underset{\sim}{\mathbf{a}}\vartheta) + \underset{\approx}{\mathbf{D}} \overset{2}{\otimes} (\boldsymbol{\varkappa} - \underset{\sim}{\mathbf{d}}\vartheta), \tag{9.49}$$

where $\boldsymbol{\gamma} = \nabla \mathbf{u} - \underset{\sim}{\mathbf{C}} \cdot \boldsymbol{\varphi}$ is the strain tensor of the micropolar theory (Kupradze,

1979), $\boldsymbol{\varkappa} = \nabla\boldsymbol{\varphi}$ is the torsion-bend tensor, $\underset{\approx}{\mathbf{A}}$, $\underset{\approx}{\mathbf{B}}^T = \underset{\approx}{\mathbf{C}}$, $\underset{\approx}{\mathbf{D}}$ are the fourth rank material tensors, ϑ is the temperature drop. Taking into account the expression for $\boldsymbol{\gamma}$, we can write (9.49) in form

$$\underset{\sim}{\mathbf{P}} = \underset{\approx}{\mathbf{A}} \overset{2}{\otimes} \nabla\mathbf{u} + \underset{\approx}{\mathbf{B}} \overset{2}{\otimes} \nabla\boldsymbol{\varphi} - \underset{\approx}{\mathbf{A}} \overset{2}{\otimes} \underset{\sim}{\mathbf{C}} \cdot \boldsymbol{\varphi} - \underset{\sim}{\mathbf{b}}\vartheta,$$

$$\boldsymbol{\mu} = \underset{\approx}{\mathbf{C}} \overset{2}{\otimes} \nabla\mathbf{u} + \underset{\approx}{\mathbf{D}} \overset{2}{\otimes} \nabla\boldsymbol{\varphi} - \underset{\approx}{\mathbf{C}} \overset{2}{\otimes} \underset{\sim}{\mathbf{C}} \cdot \boldsymbol{\varphi} - \boldsymbol{\beta}\vartheta, \tag{9.50}$$

where the tensors

$$\underset{\sim}{\mathbf{b}} = \underset{\approx}{\mathbf{A}} \overset{2}{\otimes} \underset{\sim}{\mathbf{a}} + \underset{\approx}{\mathbf{B}} \overset{2}{\otimes} \underset{\sim}{\mathbf{d}}, \quad \boldsymbol{\beta} = \underset{\approx}{\mathbf{C}} \overset{2}{\otimes} \underset{\sim}{\mathbf{a}} + \underset{\approx}{\mathbf{D}} \overset{2}{\otimes} \underset{\sim}{\mathbf{d}}$$

are called tensors of the thermomechanical behavior, $\underset{\sim}{\mathbf{a}}$ and $\underset{\sim}{\mathbf{d}}$ are the second-rank tensors of heat expansion. Note that the particular case of the law (9.50) is considered in Kupradze (1979); Nowacki (1975), more common relations are given in Pobedrya (2003, 2006).

Now we can find the desired representations of the Hooke law (9.50) under the new parametrization of the thin body domain. Indeed, by operator (9.38) after simple transformations from (9.50) we will have Nikabadze (2014a,b)

$$\underset{\sim}{\mathbf{P}} = \underset{\approx}{\mathbf{A}}^{M\cdot} \cdot g_M^{\hat{P}} N_P \mathbf{u} + \underset{\approx}{\mathbf{A}}^{3\cdot\cdot} \partial_3 \mathbf{u} + \underset{\approx}{\mathbf{B}}^{M\cdot} \cdot g_M^{\hat{P}} N_P \boldsymbol{\varphi} + \underset{\approx}{\mathbf{B}}^{3\cdot\cdot} \partial_3 \boldsymbol{\varphi} - \underset{\approx}{\mathbf{A}} \overset{2}{\otimes} \underset{\sim}{\mathbf{C}} \cdot \boldsymbol{\varphi} - \underset{\sim}{\mathbf{b}}\vartheta,$$

$$\boldsymbol{\mu} = \underset{\approx}{\mathbf{C}}^{M\cdot} \cdot g_M^{\hat{P}} N_P \mathbf{u} + \underset{\approx}{\mathbf{C}}^{3\cdot\cdot} \partial_3 \mathbf{u} + \underset{\approx}{\mathbf{D}}^{M\cdot} \cdot g_M^{\hat{P}} N_P \boldsymbol{\varphi} + \underset{\approx}{\mathbf{D}}^{3\cdot\cdot} \partial_3 \boldsymbol{\varphi} - \underset{\approx}{\mathbf{C}} \overset{2}{\otimes} \underset{\sim}{\mathbf{C}} \cdot \boldsymbol{\varphi} - \boldsymbol{\beta}\vartheta, \tag{9.51}$$

where we introduced the notations

$$\underset{\approx}{\mathbf{A}}^{m\cdot} = \underset{\approx}{\mathbf{A}} \overset{2}{\otimes} \mathbf{r}^m \underset{\sim}{\mathbf{E}}, \quad \underset{\approx}{\mathbf{B}}^{m\cdot} = \underset{\approx}{\mathbf{B}} \overset{2}{\otimes} \mathbf{r}^m \underset{\sim}{\mathbf{E}}, \quad \underset{\approx}{\mathbf{C}}^{m\cdot} = \underset{\approx}{\mathbf{C}} \overset{2}{\otimes} \mathbf{r}^m \underset{\sim}{\mathbf{E}}, \quad \underset{\approx}{\mathbf{D}}^{m\cdot} = \underset{\approx}{\mathbf{D}} \overset{2}{\otimes} \mathbf{r}^m \underset{\sim}{\mathbf{E}}.$$

It should be noted that relations (9.50) similarly to (9.44) can be represent as

$$\underset{\sim}{\mathbf{P}} = \underset{\approx}{\mathbf{A}} \overset{2}{\otimes} \mathbf{r}^{\hat{P}} \partial_P \mathbf{u} + \underset{\approx}{\mathbf{A}} \overset{2}{\otimes} \mathbf{r}^3 \partial_3 \mathbf{u} + \underset{\approx}{\mathbf{B}} \overset{2}{\otimes} \mathbf{r}^{\hat{P}} \partial_P \boldsymbol{\varphi} + \underset{\approx}{\mathbf{B}} \overset{2}{\otimes} \mathbf{r}^3 \partial_3 \boldsymbol{\varphi} - \underset{\approx}{\mathbf{A}} \overset{2}{\otimes} \underset{\sim}{\mathbf{C}} \cdot \boldsymbol{\varphi} - \underset{\sim}{\mathbf{b}}\vartheta,$$

$$\boldsymbol{\mu} = \underset{\approx}{\mathbf{C}} \overset{2}{\otimes} \mathbf{r}^{\hat{P}} \partial_P \mathbf{u} + \underset{\approx}{\mathbf{C}} \overset{2}{\otimes} \mathbf{r}^3 \partial_3 \mathbf{u} + \underset{\approx}{\mathbf{D}} \overset{2}{\otimes} \mathbf{r}^{\hat{P}} \partial_P \boldsymbol{\varphi} + \underset{\approx}{\mathbf{D}} \overset{2}{\otimes} \mathbf{r}^3 \partial_3 \boldsymbol{\varphi} - \underset{\approx}{\mathbf{C}} \overset{2}{\otimes} \underset{\sim}{\mathbf{C}} \cdot \boldsymbol{\varphi} - \boldsymbol{\beta}\vartheta. \tag{9.52}$$

Assuming (9.45), it is easy to see that the relations (9.51) contain an infinite set of terms. Therefore, in this form, they cannot be used. In the appendix the approximate constitutive relationships are represented using a finite number of terms.

It is easy to see that according to definition 9.5 and denotation (9.45) the CR of approximation of order s similarly to (9.48) are represented in the form

$$\mathop{P}_{\sim(s)} = \mathop{A}_{\approx}^{M\cdot} \cdot g^{\hat{P}}_{(s)M} N_P \mathbf{u} + \mathop{A}_{\approx}^{3\cdot} \cdot \partial_3 \mathbf{u} + \mathop{B}_{\approx}^{M\cdot} \cdot g^{\hat{P}}_{(s)M} N_P \boldsymbol{\varphi} + \mathop{B}_{\approx}^{3\cdot} \cdot \partial_3 \boldsymbol{\varphi} - \mathop{A}_{\approx}^{2} \otimes \mathop{C}_{\approx} \cdot \boldsymbol{\varphi} - \mathop{b}_{\sim} \vartheta,$$

$$\boldsymbol{\mu}_{\sim(s)} = \mathop{C}_{\approx}^{M\cdot} \cdot g^{\hat{P}}_{(s)M} N_P \mathbf{u} + \mathop{C}_{\approx}^{3\cdot} \cdot \partial_3 \mathbf{u} + \mathop{D}_{\approx}^{M\cdot} \cdot g^{\hat{P}}_{(s)M} N_P \boldsymbol{\varphi} + \mathop{D}_{\approx}^{3\cdot} \cdot \partial_3 \boldsymbol{\varphi} - \mathop{C}_{\approx}^{2} \otimes \mathop{C}_{\approx} \cdot \boldsymbol{\varphi} - \mathop{\beta}_{\sim} \vartheta.$$

$$(9.53)$$

Obviously, CR of zero approximation have the view

$$\mathop{P}_{\sim(0)} = \mathop{A}_{\approx}^{M\cdot} \cdot N_M \mathbf{u} + \mathop{A}_{\approx}^{3\cdot} \cdot \partial_3 \mathbf{u} + \mathop{B}_{\approx}^{M\cdot} \cdot N_M \boldsymbol{\varphi} + \mathop{B}_{\approx}^{3\cdot} \cdot \partial_3 \boldsymbol{\varphi} - \mathop{A}_{\approx}^{2} \otimes \mathop{C}_{\approx} \cdot \boldsymbol{\varphi} - \mathop{b}_{\sim} \vartheta,$$

$$\boldsymbol{\mu}_{\sim(0)} = \mathop{C}_{\approx}^{M\cdot} \cdot N_M \mathbf{u} + \mathop{C}_{\approx}^{3\cdot} \cdot \partial_3 \mathbf{u} + \mathop{D}_{\approx}^{M\cdot} \cdot N_M \boldsymbol{\varphi} + \mathop{D}_{\approx}^{3\cdot} \cdot \partial_3 \boldsymbol{\varphi} - \mathop{C}_{\approx}^{2} \otimes \mathop{C}_{\approx} \cdot \boldsymbol{\varphi} - \mathop{\beta}_{\sim} \vartheta.$$

$$(9.54)$$

Let us write the Fourier heat conduction law under the new parametrization of the thin body domain. Since the Fourier heat conduction law (Pobedrya, 1995; Nowacki, 1975) has the form $\mathbf{q} = -\mathop{\Lambda}_{\sim} \cdot \nabla T$, where the positive definite second-rank tensor $\mathop{\Lambda}_{\sim}$ is called the thermal conductivity tensor, therefore, by the definition 9.5 and (9.45), the Fourier heat conduction law of the zero approximation and approximation of order s represent in the form (Nikabadze, 2014a,b)

$$\mathbf{q}_{(0)} = -\boldsymbol{\Lambda}^M N_M T - \boldsymbol{\Lambda}^3 \partial_3 T, \quad \mathbf{q}_{(s)} = -\boldsymbol{\Lambda}^M g^{\hat{P}}_{(s)M} N_P T - \boldsymbol{\Lambda}^3 \partial_3 T, \quad \boldsymbol{\Lambda}^m = \mathop{\Lambda}_{\sim} \cdot \mathbf{r}^m.$$

$$(9.55)$$

Obviously, the Fourier heat conduction law, similarly to (9.52), can also be written in the form

$$\mathbf{q} = -\mathop{\Lambda}_{\sim} \cdot \mathbf{r}^{\hat{P}} \partial_P T - \mathop{\Lambda}_{\sim} \cdot \mathbf{r}^{\overset{3}{3}} \partial_3 T. \tag{9.56}$$

9.4 Some Recurrence Relations of the System of Legendre Polynomials on the Segment $[-1, 1]$

9.4.1 Main Recurrence Relations

In this case, the basic recurrence relations for the Legendre polynomial system are (Vekua, 1982; Sansone, 1959; Suyetin, 1976)

$$(2n + 1) x P_n(x) = n P_{n-1}(x) + (n + 1) P_{n+1}(x), \quad n \geq 1, \tag{9.57}$$

$$x P_n'(x) = n P_n(x) + P_{n-1}'(x), \quad n \geq 1, \tag{9.58}$$

$$P_n'(x) = (2n - 1) P_{n-1}(x) + P_{n-2}'(x), \quad n \geq 2, \quad -1 \leq x \leq 1. \tag{9.59}$$

9.4.2 Additional Recurrence Relations

Some of the desired relations, which can be obtained from (9.57)–(9.59), we write out below without proof. We have

$$
\begin{aligned}
P'_n(x) &= \sum_{p=0}^{[(n-1)/2]} [2n-(4p+1)]P_{n-(2p+1)}(x) \\
&= \frac{1}{2}\sum_{p=a}^{n-1}(2p+1)[1-(-1)^{n+p}]P_p(x),
\end{aligned}
\tag{9.60}
$$

where $a = \dfrac{1}{2}[1+(-1)^n]$, $[x]$ is the integer part of number x;

$$
xP'_n(x) = nP_n(x) + (2n-3)P_{n-2}(x) + (2n-7)P_{n-4}(x) + \ldots,
$$

$$
\begin{aligned}
x^2 P_n(x) &= \frac{n(n-1)}{(2n-1)(2n+1)}P_{n-2}(x) + \frac{2n^2+2n-1}{(2n-1)(2n+3)}P_n(x) \\
&\quad + \frac{(n+1)(n+2)}{(2n+1)(2n+3)}P_{n+2}(x),
\end{aligned}
$$

$$
\begin{aligned}
x^2 P'_n(x) &= \frac{3n^2-n-1}{2n+1}P_{n-1}(x) + \frac{n(n+1)}{2n+1}P_{n+1}(x) + (2n-5)P_{n-3}(x) \\
&\quad + (2n-9)P_{n-5}(x) + (2n-13)P_{n-7}(x) + (2n-17)P_{n-9}(x) + \ldots
\end{aligned}
$$

$$
\begin{aligned}
P''_n(x) &= (2n-1)(2n-3)P_{n-2}(x) + 2(2n-3)(2n-7)P_{n-4}(x) \\
&\quad + 3(2n-5)(2n-11)P_{n-6}(x) + 4(2n-7)(2n-15)P_{n-8}(x) + \ldots
\end{aligned}
$$

$$
\begin{aligned}
xP''_n(x) &= (n-1)(2n-1)P_{n-1}(x) + (2n-5)(3n-4)P_{n-3}(x) \\
&\quad + (2n-9)(5n-11)P_{n-5}(x) + (2n-13)(7n-22)P_{n-7}(x) + \ldots,
\end{aligned}
$$

$$
\begin{aligned}
x^2 P''_n(x) &= (n-1)nP_n(x) + (2n-3)(2n-3)P_{n-2}(x) \\
&\quad + (2n-7)(4n-8)P_{n-4}(x) + (2n-11)(6n-17)P_{n-6}(x) + \ldots.
\end{aligned}
$$

Note that the these recurrence relations play an important role in the construction of various variants of theories of thin bodies. Note also that in more detail the recurrence relations and their proofs presented in Nikabadze (2008b, 2014a,b).

9.5 Moments of Some Expressions Regarding the Legendre Polynomial System

Let $\{u_k(x^3)\}_{k=0}^{\infty}$ be some orthogonal system of polynomials on the segment $[a,b]$, and $\mathbb{F}(x',x^3)$ is some tensor field.

Definition 9.6. Integral, denoted by $\mathbb{M}_{u_k}(\mathbb{F})$,

$$\mathbb{M}_{u_k}(\mathbb{F}) = ||u_k||^{-2} \int\limits_a^b \mathbb{F}(x', x^3) u_k(x^3) h(x^3) dx^3 \tag{9.61}$$

is called the moment of the kth order of $\mathbb{F}(x', x^3)$ with respect to the system of polynomials $\{u_k(x^3)\}_{k=0}^\infty$.

Here $||u_k||$ is the norm of the polynomial $u_k(x^3)$, $h(x^3)$ is the weight function. Note that instead of $\mathbb{F}$ we can, for example, consider $[f(x') + x^3 g(x')]^s \mathbb{F}(x', x^3)$, where $f(x')$ and $g(x')$ are some functions, and s is a non-negative integer.

In the case of the Legendre polynomial system $\{P_k\}_{k=0}^\infty$ the norm is $||P_k|| = \sqrt{2/(2k+1)}$, and $h = 1$. Therefore, introducing the denotation $\overset{(k)}{\mathbb{M}}(\mathbb{F})$ for a moment of kth order with respect to the system of Legendre polynomials, instead of (9.61) we will have

$$\overset{(k)}{\mathbb{M}}(\mathbb{F}) = \frac{2k+1}{2} \int\limits_{-1}^1 \mathbb{F}(x', x^3) P_k(x^3) dx^3. \tag{9.62}$$

Here we assume that the tensor field $\mathbb{F}(x', x^3) \in C_m$, $m \geq 1$. Then, for a fixed value of x', it can be expanded into a Legendre-Fourier series

$$\mathbb{F}(x', x^3) = \sum_{k=0}^\infty \overset{(k)}{\mathbb{M}}(\mathbb{F}) P_k(x^3) = \sum_{k=0}^\infty \overset{(k)}{\mathbb{F}} P_k(x^3), \quad \overset{(k)}{\mathbb{M}}(\mathbb{F}) = \overset{(k)}{\mathbb{F}}. \tag{9.63}$$

Taking into account the values of the Legendre polynomials at the ends of the segment $[-1, 1]$ ($P_k(-1) = (-1)^k$, $P_k(1) = 1$), from (9.63) we have

$$\overset{(-)}{\mathbb{F}}(x') = \mathbb{F}|_{x^3=-1} = \sum_{p=0}^\infty (-1)^p \overset{(p)}{\mathbb{F}}, \quad \overset{(+)}{\mathbb{F}}(x') = \mathbb{F}|_{x^3=1} = \sum_{p=0}^\infty \overset{(p)}{\mathbb{F}}. \tag{9.64}$$

9.5.1 Moments of Some Expressions Regarding the Legendre Polynomial System

The moments of the partial derivatives of $\partial_i \mathbb{F}$ with respect to the Legendre polynomial system (Vekua, 1982; Nikabadze, 2014a,b) in the considered case will take the form

$$\overset{(k)}{\mathbb{M}}(\partial_i \mathbb{F}) = g_i^J \partial_J \overset{(k)}{\mathbb{F}} + g_i^3 \overset{(k)}{\mathbb{F}}', \tag{9.65}$$

where

$$\overset{(k)}{\mathbb{F}}' = \overset{(k)}{\mathbb{M}}(\partial_3 \mathbb{F}) = (2k+1) \sum_{p=0}^\infty \overset{(k+2p+1)}{\mathbb{F}} = \frac{2k+1}{2} \sum_{p=k}^\infty \left[1 - (-1)^{k+p}\right] \overset{(p)}{\mathbb{F}}. \tag{9.66}$$

Note that (9.66) can be represented in another form. To this end, we transform the right side. Taking into account (9.64), we have

$$\sum_{p=k}^{\infty} \left[1 - (-1)^{k+p}\right] \overset{(p)}{\mathbb{F}} = \sum_{p=0}^{\infty} \left[1 - (-1)^{k+p}\right] \overset{(p)}{\mathbb{F}} - \sum_{p=0}^{k-1} \left[1 - (-1)^{k+p}\right] \overset{(p)}{\mathbb{F}}$$

$$= \overset{(+)}{\mathbb{F}} - (-1)^k \overset{(-)}{\mathbb{F}} - \sum_{p=0}^{k-1} \left[1 - (-1)^{k+p}\right] \overset{(p)}{\mathbb{F}}.$$

Taking into account the latter, the ratio (9.66) can be represented in the form

$$\overset{(k)}{\mathbb{F}}' = \frac{2k+1}{2}\left[\overset{(+)}{\mathbb{F}} - (-1)^k \overset{(-)}{\mathbb{F}}\right] - \frac{2k+1}{2} \sum_{p=0}^{k-1} \left[1 - (-1)^{k+p}\right] \overset{(p)}{\mathbb{F}}. \tag{9.67}$$

The relation (9.67) is convenient to use when $\overset{(-)}{\mathbb{F}}$ and $\overset{(+)}{\mathbb{F}}$ are given on the front surfaces. We give another representation (9.66), namely

$$\overset{(k)}{\mathbb{F}}' = \frac{2k+1}{2} \sum_{p=k}^{N} \left[1 - (-1)^{k+p}\right] \overset{(p)}{\mathbb{F}} + \frac{2k+1}{2}\left[\overset{(+)}{\mathbb{F}}' - (-1)^k \overset{(-)}{\mathbb{F}}'\right], \tag{9.68}$$

where we introduced

$$\overset{(+)}{\mathbb{F}}' = \sum_{p=N+1}^{\infty} \overset{(p)}{\mathbb{F}}, \quad \overset{(-)}{\mathbb{F}}' = \sum_{p=N+1}^{\infty} (-1)^p \overset{(p)}{\mathbb{F}}, \quad N \geq k. \tag{9.69}$$

Introducing the definition of the operator "prime" by (9.66)–(9.68), it is easy to prove that the statement holds.

Proposition 9.1. *The moment operator (9.62) and the "prime" operator (9.66)–(9.68) with respect to the system of Legendre polynomials have the property of generalized linearity, i.e. for any tensor fields* $\mathbb{F}(x', x^3)$, $\mathbb{G}(x', x^3)$ *and any functions* $\alpha(x')$ *and* $\beta(x')$ *the relations*

$$\overset{(k)}{\mathbb{M}}[\alpha(x')\mathbb{F} + \beta(x')\mathbb{G}] = \alpha(x') \overset{(k)}{\mathbb{F}} + \beta(x') \overset{(k)}{\mathbb{G}},$$
$$\overset{(k)}{\mathbb{M}}'[\alpha(x')\mathbb{F} + \beta(x')\mathbb{G}] = \alpha(x') \overset{(k)}{\mathbb{F}}' + \beta(x') \overset{(k)}{\mathbb{G}}', \quad k \in \mathbb{N}_0 \tag{9.70}$$

are valid. Therefore, the operators of moment and "prime" are linear operators.

Using the second relation (9.60) and (9.62), it is easy to prove the following formulas

$$\overset{(k)}{\mathbb{M}}(x^3 \partial_3 \mathbb{F}) = k \overset{(k)}{\mathbb{F}} + (2k+1) \sum_{p=1}^{\infty} \overset{(k+2p)}{\mathbb{F}}$$

$$= k \overset{(k)}{\mathbb{F}} + \frac{2k+1}{2} \sum_{p=k+1}^{\infty} \left[1 + (-1)^{k+p}\right] \overset{(p)}{\mathbb{F}} \tag{9.71}$$

$$= -(k+1) \overset{(k)}{\mathbb{F}} + \frac{2k+1}{2} \sum_{p=k}^{\infty} \left[1 + (-1)^{k+p}\right] \overset{(p)}{\mathbb{F}}.$$

We can see that (9.71) similarly to (9.68) can be represent as

$$\overset{(k)}{\mathbb{M}}(x^3\partial_3\mathbb{F}) = -(k+1)\overset{(k)}{\mathbb{F}} + \frac{2k+1}{2}\sum_{p=k}^{N}\left[1+(-1)^{k+p}\right]\overset{(p)}{\mathbb{F}}$$
$$+\frac{2k+1}{2}\left[\overset{(+)}{\mathbb{F}'}+(-1)^k\overset{(-)}{\mathbb{F}'}\right]. \tag{9.72}$$

Let us prove that it is valid the next formula

$$\overset{(k)}{\mathbb{M}}(x^3\partial_3\mathbb{F}) = [\overset{(k)}{\mathbb{M}}(x^3\mathbb{F})]'. \tag{9.73}$$

Indeed, by (9.57), we have

$$\overset{(k)}{\mathbb{M}}(x^3\mathbb{F}) = \frac{k}{2k-1}\overset{(k-1)}{\mathbb{F}} + \frac{k+1}{2k+3}\overset{(k+1)}{\mathbb{F}}. \tag{9.74}$$

Applying the operator "prime" to (9.74) and taking into account its linearity and (9.66), we get

$$\overset{(k)}{\mathbb{M}'}(x^3\mathbb{F}) = \frac{k}{2k-1}\overset{(k-1)}{\mathbb{F}}' + \frac{k+1}{2k+3}\overset{(k+1)}{\mathbb{F}}' = \frac{k}{2k-1}\left[(2k-1)\sum_{p=0}^{\infty}\overset{(k+2p)}{\mathbb{F}}\right]$$
$$+\frac{k+1}{2k+3}\left[(2k+3)\sum_{p=0}^{\infty}\overset{(k+2p+2)}{\mathbb{F}}\right] = k\sum_{p=0}^{\infty}\overset{(k+2p)}{\mathbb{F}} + (k+1)\sum_{p=0}^{\infty}\overset{(k+2p+2)}{\mathbb{F}}$$
$$= k\overset{(k)}{\mathbb{F}} + k\sum_{p=1}^{\infty}\overset{(k+2p)}{\mathbb{F}} + (k+1)\sum_{p=1}^{\infty}\overset{(k+2p)}{\mathbb{F}} = k\overset{(k)}{\mathbb{F}} + (2k+1)\sum_{p=1}^{\infty}\overset{(k+2p)}{\mathbb{F}},$$

that is

$$\overset{(k)}{\mathbb{M}'}(x^3\mathbb{F}) = k\overset{(k)}{\mathbb{F}} + (2k+1)\sum_{p=1}^{\infty}\overset{(k+2p)}{\mathbb{F}}.$$

Comparing the last relation with (9.71), we come to (9.73).

Further, using the method of mathematical induction, it is easy to prove the validity of the relation

$$\overset{(k)}{\mathbb{M}}\left[(x^3)^s\partial_3\mathbb{F}\right] = \{\overset{(k)}{\mathbb{M}}\left[(x^3)^s\mathbb{F}\right]\}'. \tag{9.75}$$

It should be noted that, based on (9.75) and the generalized linearity of the operator "prime" (9.70), the following formula can be proved:

$$\overset{(k)}{\mathbb{M}}\{[f(x') + x^3g(x')]^s\partial_3\mathbb{F}\} = \{\overset{(k)}{\mathbb{M}}\{[f(x') + x^3g(x')]^s\mathbb{F}\}\}'. \tag{9.76}$$

Note also that for covariant derivatives, for example, of the components of the vector, the following relations are valid:

$$\overset{(k)}{\mathbb{M}}(\nabla_p u_{\tilde{q}}) = g_p^S\nabla_S\overset{(k)}{\mathbb{M}}(u_{\tilde{q}}) + g_p^3\overset{(k)}{\mathbb{M}'}(u_{\tilde{q}}),$$
$$\overset{(k)}{\mathbb{M}}(\nabla_p u^{\tilde{q}}) = g_p^S\nabla_S\overset{(k)}{\mathbb{M}}(u^{\tilde{q}}) + g_p^3\overset{(k)}{\mathbb{M}'}(u^{\tilde{q}}), \quad \sim \in \{-, \emptyset, \wedge, +\}. \tag{9.77}$$

9.6 Different Representations of the System of Motion Equations in Moments

These representations of the system of equations of motion in moments can be obtained by applying the moment operator of the system of Legendre polynomials to (9.42)–(9.44) and taking into account its property of generalized linearity. Here we get them based on (9.43) and (9.44). Applying the operators of moments of kth order to (9.44) and taking into account its linearity, as well as (9.76) and (9.77), we get

$$
\begin{aligned}
\nabla_P \overset{(k)}{\mathbf{M}}\big(\mathbf{P}^{\hat{P}}\big) + \overset{(k)}{\mathbf{M}'}\big(\mathbf{P}^{\hat{3}}\big) + \rho \overset{(k)}{\mathbf{F}} &= \rho \partial_t^2 \overset{(k)}{\mathbf{u}}, \\
\nabla_P \overset{(k)}{\mathbf{M}}\big(\boldsymbol{\mu}^{\hat{P}}\big) + \overset{(k)}{\mathbf{M}'}\big(\boldsymbol{\mu}^{\hat{3}}\big) + \underset{\sim}{\mathbf{C}} \overset{2}{\otimes} \overset{(k)}{\mathbf{P}} + \rho \overset{(k)}{\mathbf{H}} &= \mathbf{J} \cdot \partial_t^2 \overset{(k)}{\boldsymbol{\varphi}}, \quad k \in \mathbb{N}_0.
\end{aligned}
\tag{9.78}
$$

It should be noted that to obtain (9.78) it was assumed that the material was homogeneous with respect to x^3. In addition, we note that the system (9.78) in its present form holds for both the Legendre and Chebyshev polynomials, and for any polynomial system.

Similarly to (9.78), based on (9.43), we will have

$$
\begin{aligned}
\frac{1}{\sqrt{g}} \partial_P \big[\sqrt{g}\,\overset{(k)}{\mathbf{M}}(\hat{\vartheta}\mathbf{P}^{\hat{P}})\big] + \overset{(k)}{\mathbf{M}'}(\hat{\vartheta}\mathbf{P}^{\hat{3}}) + \rho \overset{(k)}{\mathbf{M}}(\hat{\vartheta}\mathbf{F}) &= \rho \overset{(k)}{\mathbf{M}}(\hat{\vartheta}\partial_t^2 \mathbf{u}), \\
\frac{1}{\sqrt{g}} \partial_P \big[\sqrt{g}\,\overset{(k)}{\mathbf{M}}(\hat{\vartheta}\boldsymbol{\mu}^{\hat{P}})\big] + \overset{(k)}{\mathbf{M}'}(\hat{\vartheta}\boldsymbol{\mu}^{\hat{3}}) + \underset{\sim}{\mathbf{C}} \overset{2}{\otimes} \underset{\sim}{\overset{(k)}{\mathbf{M}}}(\hat{\vartheta}\underset{\sim}{\mathbf{P}}) + \rho \overset{(k)}{\mathbf{M}}(\hat{\vartheta}\mathbf{H}) &= \mathbf{J} \cdot \overset{(k)}{\mathbf{M}}(\hat{\vartheta}\partial_t^2 \boldsymbol{\varphi}),
\end{aligned}
$$
$$k \in \mathbb{N}_0.$$

Obviously, this system can be represented as

$$
\begin{aligned}
\nabla_P \overset{(k)}{\mathbf{M}}(\hat{\vartheta}\mathbf{P}^{\hat{P}}) + \overset{(k)}{\mathbf{M}'}(\hat{\vartheta}\mathbf{P}^{\hat{3}}) + \rho \overset{(k)}{\mathbf{M}}(\hat{\vartheta}\mathbf{F}) &= \rho \overset{(k)}{\mathbf{M}}(\hat{\vartheta}\partial_t^2 \mathbf{u}), \\
\nabla_P \overset{(k)}{\mathbf{M}}(\hat{\vartheta}\boldsymbol{\mu}^{\hat{P}}) + \overset{(k)}{\mathbf{M}'}(\hat{\vartheta}\boldsymbol{\mu}^{\hat{3}}) + \underset{\sim}{\mathbf{C}} \overset{2}{\otimes} \underset{\sim}{\overset{(k)}{\mathbf{M}}}(\hat{\vartheta}\underset{\sim}{\mathbf{P}}) + \rho \overset{(k)}{\mathbf{M}}(\hat{\vartheta}\mathbf{H}) &= \mathbf{J} \cdot \overset{(k)}{\mathbf{M}}(\hat{\vartheta}\partial_t^2 \boldsymbol{\varphi}),
\end{aligned}
\tag{9.79}
$$
$$k \in \mathbb{N}_0.$$

Note that in (9.78) and (9.79) ∇_P is the operator of a covariant derivative with respect to the coordinate system associated with the base surface.

In order to obtain the required representations of the equations (9.78) and (9.79), it is necessary to find various expressions for $\overset{(k)}{\mathbf{M}'}(\mathbf{P}^{\hat{3}})$, $\overset{(k)}{\mathbf{M}'}(\hat{\vartheta}\mathbf{P}^{\hat{3}})$, $\overset{(k)}{\mathbf{M}'}(\boldsymbol{\mu}^{\hat{3}})$ and $\overset{(k)}{\mathbf{M}'}(\vartheta\boldsymbol{\mu}^{\hat{3}})$. It is not difficult to notice that the representations for $\overset{(k)}{\mathbf{M}'}(\hat{\vartheta}\mathbf{P}^{\hat{3}})$, $\overset{(k)}{\mathbf{M}'}(\boldsymbol{\mu}^{\hat{3}})$ and $\overset{(k)}{\mathbf{M}'}(\vartheta\boldsymbol{\mu}^{\hat{3}})$ can be obtained from the corresponding expressions for $\overset{(k)}{\mathbf{M}'}(\mathbf{P}^{\hat{3}})$ by replacing the letter $\mathbf{P}$ with $\hat{\vartheta}\mathbf{P}$, $\boldsymbol{\mu}$ and $\hat{\vartheta}\boldsymbol{\mu}$, respectively. In this regard, below we obtain the required representations only for $\overset{(k)}{\mathbf{M}'}(\mathbf{P}^{\hat{3}})$.

By the second relation (9.1) and the first formulas (9.18) and (9.29), we have

$$\mathbf{P}^{\hat{3}} = \mathbf{r}^{\hat{3}} \cdot \underset{\sim}{\mathbf{P}} = \left(\mathbf{r}^3 - g_{\hat{P}}^3 \mathbf{r}^{\hat{P}}\right) \cdot \underset{\sim}{\mathbf{P}} = \mathbf{P}^3 - h^{-1}\partial_P(\bar{h} + x^3 h)\mathbf{P}^{\hat{P}}. \tag{9.80}$$

Based on (9.80) and the properties of generalized linearity of the operator of moments of kth order and operator "prime" , we will have

$$\overset{(k)}{\mathbf{M}'}(\mathbf{P}^{\hat{3}}) = \overset{(k)}{\mathbf{M}'}(\mathbf{P}^3) - h^{-1}\partial_P \bar{h}\overset{(k)}{\mathbf{M}'}(\mathbf{P}^{\hat{P}}) - h^{-1}\partial_P h\overset{(k)}{\mathbf{M}'}(x^3\mathbf{P}^{\hat{P}}). \tag{9.81}$$

Note that (9.81) holds for both the Legendre polynomial system and the Chebyshev polynomial system. Taking into account

$$\bar{h} = (\overset{(+)}{h} - \overset{(-)}{h})/2, \quad h = (\overset{(+)}{h} + \overset{(-)}{h})/2,$$

and also the second and third formulas (9.15), the equality (9.81) can be written as

$$\overset{(k)}{\mathbf{M}'}(\mathbf{P}^{\hat{3}}) = \overset{(k)}{\mathbf{M}'}(\mathbf{P}^3) + \frac{1}{2}g_{\underset{P}{-}}^3 \left[\overset{(k)}{\mathbf{M}'}(x^3\mathbf{P}^{\hat{P}}) - \overset{(k)}{\mathbf{M}'}(\mathbf{P}^{\hat{P}})\right]$$

$$- \frac{1}{2}g_{\underset{P}{+}}^3 \left[\overset{(k)}{\mathbf{M}'}(x^3\mathbf{P}^{\hat{P}}) + \overset{(k)}{\mathbf{M}'}(\mathbf{P}^{\hat{P}})\right]. \tag{9.82}$$

9.6.1 Presentations of the System of Motion Equations in Moments with Respect to Systems of Legendre Polynomials

Based on (9.66), (9.71) and (9.73), it is not difficult to prove the following relations:

$$\overset{(k)}{\mathbf{M}'}(x^3\mathbf{P}^{\hat{P}}) - \overset{(k)}{\mathbf{M}'}(\mathbf{P}^{\hat{P}}) = -(k+1)\overset{(k)}{\mathbf{M}}(\mathbf{P}^{\hat{P}}) + 2(k+1)\sum_{p=k}^{\infty}(-1)^{k+p}\overset{(p)}{\mathbf{M}}(\mathbf{P}^{\hat{P}}),$$

$$\overset{(k)}{\mathbf{M}'}(x^3\mathbf{P}^{\hat{P}}) + \overset{(k)}{\mathbf{M}'}(\mathbf{P}^{\hat{P}}) = -(k+1)\overset{(k)}{\mathbf{M}}(\mathbf{P}^{\hat{P}}) + (2k+1)\sum_{p=k}^{\infty}\overset{(p)}{\mathbf{M}}(\mathbf{P}^{\hat{P}}).$$

$$\tag{9.83}$$

By virtue of (9.64), the formulas (9.83) can also be represented as

$$\overset{(k)}{\mathbf{M}'}\big(x^3\mathbf{P}^{\hat{P}}\big) - \overset{(k)}{\mathbf{M}'}\big(\mathbf{P}^{\hat{P}}\big)$$

$$= k\overset{(k)}{\mathbf{M}}\big(\mathbf{P}^{\hat{P}}\big) + (2k+1)(-1)^k \overset{(-)}{\mathbf{P}}{}^{\bar{P}} - (2k+1)\sum_{p=0}^{k}(-1)^{k+p}\overset{(p)}{\mathbf{M}}\big(\mathbf{P}^{\hat{P}}\big),$$

$$\overset{(k)}{\mathbf{M}'}\big(x^3\mathbf{P}^{\hat{P}}\big) + \overset{(k)}{\mathbf{M}'}\big(\mathbf{P}^{\hat{P}}\big) = k\overset{(k)}{\mathbf{M}}\big(\mathbf{P}^{\hat{P}}\big) + (2k+1)\overset{(+)}{\mathbf{P}}{}^{\bar{P}} - (2k+1)\sum_{p=0}^{k}\overset{(p)}{\mathbf{M}}\big(\mathbf{P}^{\hat{P}}\big).$$

$$(9.84)$$

Taking into account (9.66) and (9.83), the relation (9.82) can be written in the following form:

$$\overset{(k)}{\mathbf{M}'}\big(\mathbf{P}^{\hat{3}}\big) = \frac{2k+1}{2}\sum_{p=k}^{\infty}\big[1-(-1)^{k+p}\big]\overset{(p)}{\mathbf{M}}\big(\mathbf{P}^3\big) + \frac{k+1}{2}\big(g_{\underset{P}{+}}^3 - g_{\underset{P}{-}}^3\big)\overset{(k)}{\mathbf{M}}\big(\mathbf{P}^{\hat{P}}\big)$$

$$-\frac{2k+1}{2}\sum_{p=k}^{\infty}\big[g_{\underset{P}{+}}^3 - (-1)^{k+p}g_{\underset{P}{-}}^3\big]\overset{(p)}{\mathbf{M}}\big(\mathbf{P}^{\hat{P}}\big), \qquad (9.85)$$

$$\mathbf{P} \to \boldsymbol{\mu}, \quad \mathbf{P} \to \hat{\vartheta}\mathbf{P}, \quad \mathbf{P} \to \hat{\vartheta}\boldsymbol{\mu}.$$

Here, the last line means that instead of this line we should write out the relations obtained from the one written by replacing $\mathbf{P}$ with $\boldsymbol{\mu}$, $\vartheta\mathbf{P}$ and $\vartheta\boldsymbol{\mu}$, respectively, i.e. in this case, three relations should be written. Similar notes will be used further.

Next, using (9.67) and (9.84), from (9.82) we get

$$\overset{(k)}{\mathbf{M}'}\big(\mathbf{P}^{\hat{3}}\big) = -\frac{2k+1}{2}\sum_{p=0}^{k}\big[1-(-1)^{k+p}\big]\overset{(p)}{\mathbf{M}}\big(\mathbf{P}^3\big) - \frac{k}{2}\big(g_{\underset{P}{+}}^3 - g_{\underset{P}{-}}^3\big)\overset{(k)}{\mathbf{M}}\big(\mathbf{P}^{\hat{P}}\big)$$

$$+\frac{2k+1}{2}\sum_{p=0}^{k}\big[g_{\underset{P}{+}}^3 - (-1)^{k+p}g_{\underset{P}{-}}^3\big]\overset{(p)}{\mathbf{M}}\big(\mathbf{P}^{\hat{P}}\big)$$

$$+\frac{2k+1}{2}\Big[\big(\overset{(+)}{\mathbf{P}}{}^3 - g_{\underset{P}{+}}^3\overset{(+)}{\mathbf{P}}{}^{\bar{P}}\big) - (-1)^k\big(\overset{(-)}{\mathbf{P}}{}^3 - g_{\underset{P}{-}}^3\overset{(-)}{\mathbf{P}}{}^{\bar{P}}\big)\Big].$$

Hence, taking into account the formulas

$$\overset{(+)}{\mathbf{P}}{}^3 = \mathbf{P}^3\big|_{x^3=1} = \overset{(+)}{\mathbf{P}}{}^3 - g_{\underset{P}{+}}^3\overset{(+)}{\mathbf{P}}{}^{\bar{P}}, \quad \overset{(-)}{\mathbf{P}}{}^3 = \mathbf{P}^3\big|_{x^3=-1} = \overset{(-)}{\mathbf{P}}{}^3 - g_{\underset{P}{-}}^3\overset{(-)}{\mathbf{P}}{}^{\bar{P}}, \quad (9.86)$$

that are obtained from (9.80) with $x^3 = -1$ and $x^3 = 1$, we get

$$\overset{(k)}{\mathbf{M}'}\big(\mathbf{P}^{\hat{3}}\big) = -\frac{2k+1}{2}\sum_{p=0}^{k}\big[1-(-1)^{k+p}\big]\overset{(p)}{\mathbf{M}}\big(\mathbf{P}^3\big) - \frac{k}{2}\big(g_{\underset{P}{+}}^3 - g_{\underset{P}{-}}^3\big)\overset{(k)}{\mathbf{M}}\big(\mathbf{P}^{\hat{P}}\big)$$

$$+\frac{2k+1}{2}\sum_{p=0}^{k}\big[g_{\underset{P}{+}}^3 - (-1)^{k+p}g_{\underset{P}{-}}^3\big]\overset{(p)}{\mathbf{M}}\big(\mathbf{P}^{\hat{P}}\big) + \frac{2k+1}{2}\Big[\overset{(+)}{\mathbf{P}}{}^3 - (-1)^k\overset{(-)}{\mathbf{P}}{}^3\Big], \quad k \in \mathbb{N}_0.$$

$$(9.87)$$

It is not difficult to see that the relation, similar to (9.87), for $\overset{(k)}{\mathbf{M}'}\big(\hat{\vartheta}\mathbf{P}^{\hat{3}}\big)$ will be

$$
\overset{(k)}{\mathbf{M}'}(\hat{\vartheta}\mathbf{P}^{\hat{3}}) = -\frac{2k+1}{2}\sum_{p=0}^{k}\big[1-(-1)^{k+p}\big]\overset{(p)}{\mathbf{M}}(\hat{\vartheta}\mathbf{P}^{\hat{3}}) - \frac{k}{2}\big(g^{3}_{\underset{P}{+}}-g^{3}_{\underset{P}{-}}\big)\overset{(k)}{\mathbf{M}}(\hat{\vartheta}\mathbf{P}^{\hat{P}})
$$

$$
+\frac{2k+1}{2}\sum_{p=0}^{k}\big[g^{3}_{\underset{P}{+}}-(-1)^{k+p}g^{3}_{\underset{P}{-}}\big]\overset{(p)}{\mathbf{M}}(\hat{\vartheta}\mathbf{P}^{\hat{P}}) + \frac{2k+1}{2}\Big[\overset{(+)}{\vartheta}\overset{(+)\overset{+}{}}{\mathbf{P}^{3}} - (-1)^{k}\overset{(-)}{\vartheta}\overset{(-)\overset{-}{}}{\mathbf{P}^{3}}\Big].
$$

$$(9.88)$$

It should be noted that in (9.87) and (9.88), as well as in the relations obtained from (9.87) and (9.88) by replacing $\mathbf{P}$ with $\boldsymbol{\mu}$, the values $\overset{(-)\overset{-}{}}{\mathbf{P}^{3}}$, $\overset{(+)\overset{+}{}}{\mathbf{P}^{3}}$ and $\overset{(-)\overset{-}{}}{\boldsymbol{\mu}^{3}}$, $\overset{(+)\overset{+}{}}{\boldsymbol{\mu}^{3}}$ are determined using the boundary conditions on the front surfaces. Indeed, let $\overset{(-)}{\mathbf{P}}$ and $\overset{(+)}{\mathbf{P}}$ be the given stress vectors, and $\overset{(-)}{\boldsymbol{\mu}}$ and $\overset{(+)}{\boldsymbol{\mu}}$ are given vectors of couple stress on the front surfaces $\overset{(-)}{S}$ and $\overset{(+)}{S}$, respectively. Then the boundary conditions of physical content on the front surfaces will be presented as

$$
\overset{(-)\overset{-}{}}{\mathbf{P}^{3}} = \mathbf{r}^{\bar{3}}\cdot\overset{(-)}{\mathbf{P}} = -\sqrt{\overset{--}{g^{33}}}\,\overset{(-)}{\mathbf{P}},\quad \overset{(+)\overset{+}{}}{\mathbf{P}^{3}} = \mathbf{r}^{\overset{+}{3}}\cdot\overset{(+)}{\mathbf{P}} = \sqrt{\overset{++}{g^{33}}}\,\overset{(+)}{\mathbf{P}},
$$

$$
\overset{(-)\overset{-}{}}{\boldsymbol{\mu}^{3}} = \mathbf{r}^{\bar{3}}\cdot\overset{(-)}{\boldsymbol{\mu}} = -\sqrt{\overset{--}{g^{33}}}\,\overset{(-)}{\boldsymbol{\mu}},\quad \overset{(+)\overset{+}{}}{\boldsymbol{\mu}^{3}} = \mathbf{r}^{\overset{+}{3}}\cdot\overset{(+)}{\boldsymbol{\mu}} = \sqrt{\overset{++}{g^{33}}}\,\overset{(+)}{\boldsymbol{\mu}}.
$$

$$(9.89)$$

Here, analogically to (9.86), we have

$$
\overset{(+)\overset{+}{}}{\boldsymbol{\mu}^{3}} = \hat{\boldsymbol{\mu}}^{3}\big|_{x^{3}=1} = \overset{(+)3}{\boldsymbol{\mu}} - g^{3}_{\underset{P}{+}}\overset{(+)\overset{+}{}}{\boldsymbol{\mu}^{P}},\quad \overset{(-)\overset{-}{}}{\boldsymbol{\mu}^{3}} = \hat{\boldsymbol{\mu}}^{3}\big|_{x^{3}=-1} = \overset{(-)3}{\boldsymbol{\mu}} - g^{3}_{\underset{P}{-}}\overset{(-)\overset{-}{}}{\boldsymbol{\mu}^{P}}.
$$

Substituting (9.89) into (9.87) and (9.88), as well as in the relations obtained from (9.87) and (9.88) by replacing the letter $\mathbf{P}$ with $\boldsymbol{\mu}$, we obtain them taking into account the boundary conditions of physical content on the front surfaces. The advantage of such a record of these relations in comparison with (9.85) is that they contain a finite number of unknowns (moments of the quantities values). In order to reduce the letters we will not write them out.

Taking into account (9.85), from (9.78) and (9.79) we find the following representations of the system of motion equations in moments:

$$
\nabla_{P}\overset{(k)}{\mathbf{M}}(\mathbf{P}^{\hat{P}}) + \frac{k+1}{2}\big(g^{3}_{\underset{P}{+}}-g^{3}_{\underset{P}{-}}\big)\overset{(k)}{\mathbf{M}}(\mathbf{P}^{\hat{P}}) - \frac{2k+1}{2}\sum_{p=k}^{\infty}\big[g^{3}_{\underset{P}{+}}-(-1)^{k+p}g^{3}_{\underset{P}{-}}\big]\overset{(p)}{\mathbf{M}}(\mathbf{P}^{\hat{P}})
$$

$$
+\frac{2k+1}{2}\sum_{p=k}^{\infty}\big[1-(-1)^{k+p}\big]\overset{(p)}{\mathbf{M}}(\mathbf{P}^{3}) + \rho\overset{(k)}{\mathbf{M}}(\mathbf{F}) = \rho\overset{(k)}{\mathbf{M}}(\partial_{t}^{2}\mathbf{u}),
$$

$$
\nabla_{P}\overset{(k)}{\mathbf{M}}(\boldsymbol{\mu}^{\hat{P}}) + \frac{k+1}{2}\big(g^{3}_{\underset{P}{+}}-g^{3}_{\underset{P}{-}}\big)\overset{(k)}{\mathbf{M}}(\boldsymbol{\mu}^{\hat{P}}) - \frac{2k+1}{2}\sum_{p=k}^{\infty}\big[g^{3}_{\underset{P}{+}}-(-1)^{k+p}g^{3}_{\underset{P}{-}}\big]\overset{(p)}{\mathbf{M}}(\boldsymbol{\mu}^{\hat{P}})
$$

$$
+\frac{2k+1}{2}\sum_{p=k}^{\infty}\big[1-(-1)^{k+p}\big]\overset{(p)}{\mathbf{M}}(\boldsymbol{\mu}^{3}) + \underset{\approx}{\mathbf{C}}\overset{2}{\otimes}\overset{(k)}{\underset{\sim}{\mathbf{M}}}(\underset{\sim}{\mathbf{P}}) + \rho\overset{(k)}{\mathbf{M}}(\mathbf{H}) = \mathbf{J}\cdot\overset{(k)}{\mathbf{M}}(\partial_{t}^{2}\boldsymbol{\varphi});
$$

$$(9.90)$$

$$\nabla_P \overset{(k)}{\mathbf{M}}\big(\hat{\vartheta}\mathbf{P}^{\hat{P}}\big) - \frac{k}{2}\big(g^3_{\underset{P}{+}} - g^3_{\underset{P}{-}}\big)\overset{(k)}{\mathbf{M}}\big(\hat{\vartheta}\mathbf{P}^{\hat{P}}\big) - \frac{2k+1}{2}\sum_{p=k+1}^{\infty}\big[g^3_{\underset{P}{+}} - (-1)^{k+p}g^3_{\underset{P}{-}}\big]\overset{(p)}{\mathbf{M}}\big(\hat{\vartheta}\mathbf{P}^{\hat{P}}\big)$$

$$+ \frac{2k+1}{2}\sum_{p=k}^{\infty}\big[1-(-1)^{k+p}\big]\overset{(p)}{\mathbf{M}}(\hat{\vartheta}\mathbf{P}^3) + \rho\overset{(k)}{\mathbf{M}}(\hat{\vartheta}\mathbf{F}) = \rho\overset{(k)}{\mathbf{M}}(\hat{\vartheta}\partial_t^2\mathbf{u}),$$

$$\nabla_P \overset{(k)}{\mathbf{M}}\big(\hat{\vartheta}\boldsymbol{\mu}^{\hat{P}}\big) - \frac{k}{2}\big(g^3_{\underset{P}{+}} - g^3_{\underset{P}{-}}\big)\overset{(k)}{\mathbf{M}}\big(\hat{\vartheta}\boldsymbol{\mu}^{\hat{P}}\big) - \frac{2k+1}{2}\sum_{p=k+1}^{\infty}\big[g^3_{\underset{P}{+}} - (-1)^{k+p}g^3_{\underset{P}{-}}\big]\overset{(p)}{\mathbf{M}}\big(\hat{\vartheta}\boldsymbol{\mu}^{\hat{P}}\big)$$

$$+ \frac{2k+1}{2}\sum_{p=k}^{\infty}\big[1-(-1)^{k+p}\big]\overset{(p)}{\mathbf{M}}(\hat{\vartheta}\boldsymbol{\mu}^3) + \underset{\approx}{\mathbf{C}}\overset{2}{\otimes}\underset{\sim}{\overset{(k)}{\mathbf{M}}}(\hat{\vartheta}\underset{\sim}{\mathbf{P}}) + \rho\overset{(k)}{\mathbf{M}}(\hat{\vartheta}\mathbf{H}) = \underset{\sim}{\mathbf{J}}\cdot\overset{(k)}{\mathbf{M}}(\hat{\vartheta}\partial_t^2\boldsymbol{\varphi}).$$

$$\tag{9.91}$$

Using (9.87) and (9.88) and similar relations for $\overset{(k)}{\mathbf{M}}'(\boldsymbol{\mu}^{\hat{3}})$ and $\overset{(k)}{\mathbf{M}}'(\hat{\vartheta}\boldsymbol{\mu}^{\hat{3}})$ from (9.78) and (9.79) we will get other representations of the system of motion equations

$$\nabla_P \overset{(k)}{\mathbf{M}}\big(\mathbf{P}^{\hat{P}}\big) - \frac{k}{2}\big(g^3_{\underset{P}{+}} - g^3_{\underset{P}{-}}\big)\overset{(k)}{\mathbf{M}}\big(\mathbf{P}^{\hat{P}}\big) + \frac{2k+1}{2}\sum_{p=0}^{k}\big[g^3_{\underset{P}{+}} - (-1)^{k+p}g^3_{\underset{P}{-}}\big]\overset{(p)}{\mathbf{M}}\big(\mathbf{P}^{\hat{P}}\big)$$

$$- \frac{2k+1}{2}\sum_{p=0}^{k}\big[1-(-1)^{k+p}\big]\overset{(p)}{\mathbf{M}}(\mathbf{P}^3) + \underline{\overset{(k)}{\boldsymbol{\Phi}}} = \rho\overset{(k)}{\mathbf{M}}(\partial_t^2\mathbf{u}),$$

$$\nabla_P \overset{(k)}{\mathbf{M}}\big(\boldsymbol{\mu}^{\hat{P}}\big) - \frac{k}{2}\big(g^3_{\underset{P}{+}} - g^3_{\underset{P}{-}}\big)\overset{(k)}{\mathbf{M}}\big(\boldsymbol{\mu}^{\hat{P}}\big) + \frac{2k+1}{2}\sum_{p=0}^{k}\big[g^3_{\underset{P}{+}} - (-1)^{k+p}g^3_{\underset{P}{-}}\big]\overset{(p)}{\mathbf{M}}\big(\boldsymbol{\mu}^{\hat{P}}\big)$$

$$- \frac{2k+1}{2}\sum_{p=0}^{k}\big[1-(-1)^{k+p}\big]\overset{(p)}{\mathbf{M}}(\boldsymbol{\mu}^3) + \underset{\approx}{\mathbf{C}}\overset{2}{\otimes}\underset{\sim}{\overset{(k)}{\mathbf{M}}}(\underset{\sim}{\mathbf{P}}) + \underline{\overset{(k)}{\mathbf{M}}} = \underset{\sim}{\mathbf{J}}\cdot\overset{(k)}{\mathbf{M}}(\partial_t^2\boldsymbol{\varphi}), \quad k\in\mathbb{N}_0,$$

$$\tag{9.92}$$

$$\nabla_P \overset{(k)}{\mathbf{M}}\big(\hat{\vartheta}\mathbf{P}^{\hat{P}}\big) - \frac{k}{2}\big(g^3_{\underset{P}{+}} - g^3_{\underset{P}{-}}\big)\overset{(k)}{\mathbf{M}}\big(\hat{\vartheta}\mathbf{P}^{\hat{P}}\big) + \frac{2k+1}{2}\sum_{p=0}^{k}\big[g^3_{\underset{P}{+}} - (-1)^{k+p}g^3_{\underset{P}{-}}\big]\overset{(p)}{\mathbf{M}}\big(\hat{\vartheta}\mathbf{P}^{\hat{P}}\big)$$

$$- \frac{2k+1}{2}\sum_{p=0}^{k}\big[1-(-1)^{k+p}\big]\overset{(p)}{\mathbf{M}}(\hat{\vartheta}\mathbf{P}^3) + \underline{\overset{(k)}{\boldsymbol{\Phi}}} = \rho\overset{(k)}{\mathbf{M}}(\hat{\vartheta}\partial_t^2\mathbf{u}),$$

$$\nabla_P \overset{(k)}{\mathbf{M}}\big(\hat{\vartheta}\boldsymbol{\mu}^{\hat{P}}\big) - \frac{k}{2}\big(g^3_{\underset{P}{+}} - g^3_{\underset{P}{-}}\big)\overset{(k)}{\mathbf{M}}\big(\hat{\vartheta}\boldsymbol{\mu}^{\hat{P}}\big) + \frac{2k+1}{2}\sum_{p=0}^{k}\big[g^3_{\underset{P}{+}} - (-1)^{k+p}g^3_{\underset{P}{-}}\big]\overset{(p)}{\mathbf{M}}\big(\hat{\vartheta}\boldsymbol{\mu}^{\hat{P}}\big)$$

$$- \frac{2k+1}{2}\sum_{p=0}^{k}\big[1-(-1)^{k+p}\big]\overset{(p)}{\mathbf{M}}(\hat{\vartheta}\boldsymbol{\mu}^3) + \underset{\approx}{\mathbf{C}}\overset{2}{\otimes}\underset{\sim}{\overset{(k)}{\mathbf{M}}}(\hat{\vartheta}\underset{\sim}{\mathbf{P}}) + \underline{\overset{(k)}{\mathbf{M}}} = \underset{\sim}{\mathbf{J}}\cdot\overset{(k)}{\mathbf{M}}(\hat{\vartheta}\partial_t^2\boldsymbol{\varphi}), \quad k\in\mathbb{N}_0.$$

$$\tag{9.93}$$

Here we introduce

$$\underline{\overset{(k)}{\boldsymbol{\Phi}}} = \frac{2k+1}{2}\big[\overset{(+)+}{\mathbf{P}^3} - (-1)^k\overset{(-)-}{\mathbf{P}^3}\big] + \rho\overset{(k)}{\mathbf{M}}(\mathbf{F})$$

$$= \frac{2k+1}{2}\big[\sqrt{g^{++}_{33}}\overset{(+)}{\mathbf{P}} + (-1)^k\sqrt{g^{--}_{33}}\overset{(-)}{\mathbf{P}}\big] + \rho\overset{(k)}{\mathbf{M}}(\mathbf{F}),$$

$$\underline{\overset{(k)}{\mathbf{M}}} = \frac{2k+1}{2}\big[\overset{(+)+}{\boldsymbol{\mu}^3} - (-1)^k\overset{(-)-}{\boldsymbol{\mu}^3}\big] + \rho\overset{(k)}{\mathbf{M}}(\mathbf{H})$$

$$= \frac{2k+1}{2}\big[\sqrt{\overset{++}{g^{33}}}\overset{(+)}{\boldsymbol{\mu}} + (-1)^k \sqrt{\overset{--}{g^{33}}}\overset{(-)}{\boldsymbol{\mu}}\big] + \rho\overset{(k)}{\mathbf{M}}(\mathbf{H}),$$

$$\overset{(k)}{\boldsymbol{\Phi}} = \frac{2k+1}{2}\big[\overset{(+)(+)+}{\vartheta\,\mathbf{P}^{3}} - (-1)^k \overset{(-)(+)-}{\vartheta\,\mathbf{P}^{3}}\big] + \rho\overset{(k)}{\mathbf{M}}(\hat{\vartheta}\mathbf{F}) \tag{9.94}$$

$$= \frac{2k+1}{2}\big[\overset{(+)}{\vartheta}\sqrt{\overset{++}{g^{33}}}\overset{(+)}{\mathbf{P}} + (-1)^k \overset{(-)}{\vartheta}\sqrt{\overset{--}{g^{33}}}\overset{(-)}{\mathbf{P}}\big] + \rho\overset{(k)}{\mathbf{M}}(\hat{\vartheta}\mathbf{F}),$$

$$\overset{(k)}{\mathbf{M}} = \frac{2k+1}{2}\big[\overset{(+)(+)+}{\vartheta\,\boldsymbol{\mu}^{3}} - (-1)^k \overset{(-)(-)-}{\vartheta\,\boldsymbol{\mu}^{3}}\big] + \rho\overset{(k)}{\mathbf{M}}(\hat{\vartheta}\mathbf{H})$$

$$= \frac{2k+1}{2}\big[\overset{(+)}{\vartheta}\sqrt{\overset{++}{g^{33}}}\overset{(+)}{\boldsymbol{\mu}} + (-1)^k \overset{(-)}{\vartheta}\sqrt{\overset{--}{g^{33}}}\overset{(-)}{\boldsymbol{\mu}}\big] + \rho\overset{(k)}{\mathbf{M}}(\hat{\vartheta}\mathbf{H}).$$

It should be noted that (9.94) and, therefore, (9.92) and (9.93) are written taking into account the boundary conditions of physical content on the front surfaces (9.89). Furthermore, the (9.92) and (9.93) representations compared to (9.90) and (9.91) have an advantage in the sense that each equation of the systems of equations (9.92) and (9.93) contains a finite number of unknowns.

The Eqs. (9.90)–(9.93) are the required representations of the motion equations in moments with respect to the Legendre polynomial system. Let us call them different representations of the system of equations of motion in moments with respect to the system of Legendre polynomials of MMDTB (micropolar mechanics of a deformable thin body). It should be noted that similarly to (9.90)–(9.93), it is possible to obtain various representations of the system of equations of motion in moments with respect to the system of Chebyshev polynomials of the first and second kind, but we will not dwell on this.

9.7 Representations of Constitutive Relations in Moments

Different representations of the required CR can be obtained from (9.51)–(9.54). However, we only get these relations, based only on (9.52). With respect to (9.76), (9.77) and (9.85) we will have

$$\overset{(k)}{\mathbf{M}}\big(r^3\partial_3\mathbf{u}\big) = \overset{(k)}{\mathbf{M}'}\big(r^3\mathbf{u}\big) = \frac{2k+1}{2}r^3 \sum_{p=k}^{\infty} \big[1 - (-1)^{k+p}\big]\overset{(p)}{\mathbf{M}}(\mathbf{u})$$

$$+\frac{k+1}{2}\big(g^3_{\underset{P}{+}} - g^3_{\underset{P}{-}}\big)\overset{(k)}{\mathbf{M}}\big(r^{\hat{P}}\mathbf{u}\big) - \frac{2k+1}{2} \sum_{p=k}^{\infty} \big[g^3_{\underset{P}{+}} - (-1)^{k+p}g^3_{\underset{P}{-}}\big]\overset{(p)}{\mathbf{M}}\big(r^{\hat{P}}\mathbf{u}\big), \tag{9.95}$$

$$\overset{(k)}{\mathbf{M}}\big(r^{\hat{P}}\partial_P\mathbf{u}\big) = \nabla_P\overset{(k)}{\mathbf{M}}\big(r^{\hat{P}}\mathbf{u}\big) = \mathbf{r}^M\nabla_P\overset{(k)}{\mathbf{M}}\big(g^{\hat{P}}_{M}\mathbf{u}\big), \quad \mathbf{u} \to \boldsymbol{\varphi}.$$

Note that the corresponding (9.95) relations of the zeroth approximation and approximations of order r are represented in the forms

$$\overset{(k)}{\underset{\sim}{M}}(\mathbf{r}^3\partial_3\mathbf{u}) = \overset{(k)}{\underset{\sim}{M}}'(\mathbf{r}^3\mathbf{u}) \approx \mathbf{r}^3 \frac{2k+1}{2} \sum_{p=k}^{\infty} \left[1-(-1)^{k+p}\right]\overset{(p)}{\underset{\sim}{M}}(\mathbf{u})$$

$$+\mathbf{r}^P \frac{k+1}{2}\left(g_{\underset{P}{+}}^3 - g_{\underset{P}{-}}^3\right)\overset{(k)}{\underset{\sim}{M}}(\mathbf{u}) - \mathbf{r}^P \frac{2k+1}{2}\sum_{p=k}^{\infty}\left[g_{\underset{P}{+}}^3 - (-1)^{k+p}g_{\underset{P}{-}}^3\right]\overset{(p)}{\underset{\sim}{M}}(\mathbf{u}),\quad(9.96)$$

$$\overset{(k)}{\underset{\sim}{M}}(\mathbf{r}^{\hat P}\partial_P\mathbf{u}) = \nabla_P\overset{(k)}{\underset{\sim}{M}}(\mathbf{r}^{\hat P}\mathbf{u}) \approx \mathbf{r}^M \nabla_M \overset{(k)}{\underset{\sim}{M}}(\mathbf{u}),\quad \mathbf{u}\to\boldsymbol{\varphi};$$

$$\overset{(k)}{\underset{\sim}{M}}(\mathbf{r}^3\partial_3\mathbf{u}) = \overset{(k)}{\underset{\sim}{M}}'(\mathbf{r}^3\mathbf{u}) = \frac{2k+1}{2}\mathbf{r}^3 \sum_{p=k}^{\infty} \left[1-(-1)^{k+p}\right]\overset{(p)}{\underset{\sim}{M}}(\mathbf{u})$$

$$+\mathbf{r}^M \frac{k+1}{2}\left(g_{\underset{P}{+}}^3 - g_{\underset{P}{-}}^3\right)\overset{(k)}{\underset{\sim}{M}}(g_{\underset{(r)M}{\hat P}}\mathbf{u}) - \mathbf{r}^M \frac{2k+1}{2}\sum_{p=k}^{\infty}\left[g_{\underset{P}{+}}^3 - (-1)^{k+p}g_{\underset{P}{-}}^3\right]\overset{(p)}{\underset{\sim}{M}}(g_{\underset{(r)M}{\hat P}}\mathbf{u}),$$

$$\overset{(k)}{\underset{\sim}{M}}(\mathbf{r}^{\hat P}\partial_P\mathbf{u}) = \nabla_P\overset{(k)}{\underset{\sim}{M}}(\mathbf{r}^{\hat P}\mathbf{u}) \approx \mathbf{r}^M \nabla_P \overset{(k)}{\underset{\sim}{M}}(g_{\underset{(r)M}{\hat P}}\mathbf{u}),\quad \mathbf{u}\to\boldsymbol{\varphi}.$$

$$(9.97)$$

It should be noted that in the simplified reduction method of an infinite system to a finite (Nikabadze, 2014a) (see also below), fixing some non-negative integer N, we neglect the moments of displacement and rotation vectors in CR, the order of which is greater than N. Taking this into account, from (9.96) and (9.97) we obtain the relations of the approximations $(0, N)$ and (r, N), respectively, if the limit of the sums ∞ is replaced by N. We will have

$$\overset{(k)}{\underset{\sim}{M}}(\mathbf{r}^3\partial_3\mathbf{u}) = \overset{(k)}{\underset{\sim}{M}}'(\mathbf{r}^3\mathbf{u}) \approx \mathbf{r}^3 \frac{2k+1}{2} \sum_{p=k}^{N} \left[1-(-1)^{k+p}\right]\overset{(p)}{\underset{\sim}{M}}(\mathbf{u})$$

$$+\mathbf{r}^P \frac{k+1}{2}\left(g_{\underset{P}{+}}^3 - g_{\underset{P}{-}}^3\right)\overset{(k)}{\underset{\sim}{M}}(\mathbf{u}) - \mathbf{r}^P \frac{2k+1}{2}\sum_{p=k}^{N}\left[g_{\underset{P}{+}}^3 - (-1)^{k+p}g_{\underset{P}{-}}^3\right]\overset{(p)}{\underset{\sim}{M}}(\mathbf{u}),\quad(9.98)$$

$$\overset{(k)}{\underset{\sim}{M}}(\mathbf{r}^{\hat P}\partial_P\mathbf{u}) = \nabla_P\overset{(k)}{\underset{\sim}{M}}(\mathbf{r}^{\hat P}\mathbf{u}) \approx \mathbf{r}^M \nabla_M \overset{(k)}{\underset{\sim}{M}}(\mathbf{u}),\quad \mathbf{u}\to\boldsymbol{\varphi};$$

$$\overset{(k)}{\underset{\sim}{M}}(\mathbf{r}^3\partial_3\mathbf{u}) = \overset{(k)}{\underset{\sim}{M}}'(\mathbf{r}^3\mathbf{u}) = \frac{2k+1}{2}\mathbf{r}^3 \sum_{p=k}^{N} \left[1-(-1)^{k+p}\right]\overset{(p)}{\underset{\sim}{M}}(\mathbf{u})$$

$$+\mathbf{r}^M \frac{k+1}{2}\left(g_{\underset{P}{+}}^3 - g_{\underset{P}{-}}^3\right)\overset{(k)}{\underset{\sim}{M}}(g_{\underset{(r)M}{\hat P}}\mathbf{u}) - \mathbf{r}^M \frac{2k+1}{2}\sum_{p=k}^{N}\left[g_{\underset{P}{+}}^3 - (-1)^{k+p}g_{\underset{P}{-}}^3\right]\overset{(p)}{\underset{\sim}{M}}(g_{\underset{(r)M}{\hat P}}\mathbf{u}),$$

$$\overset{(k)}{\underset{\sim}{M}}(\mathbf{r}^{\hat P}\partial_P\mathbf{u}) = \nabla_P\overset{(k)}{\underset{\sim}{M}}(\mathbf{r}^{\hat P}\mathbf{u}) \approx \mathbf{r}^M \nabla_P \overset{(k)}{\underset{\sim}{M}}(g_{\underset{(r)M}{\hat P}}\mathbf{u}),\quad \mathbf{u}\to\boldsymbol{\varphi}.$$

$$(9.99)$$

Note also that it is not difficult to get similar to (9.95)–(9.99) relations with respect to the Chebyshev polynomials of the first and second kind (Nikabadze, 2014a), but we will not dwell on them.

Applying the kth order moment operator to (9.52) and (9.56) we get

$$\underset{\sim}{\mathbf{P}} = \underset{\approx}{\overset{2}{\mathbf{A}}} \otimes \overset{(k)}{\underset{\sim}{\mathbf{M}}}(\mathbf{r}^{\hat{P}}\partial_P \mathbf{u}) + \underset{\approx}{\overset{2}{\mathbf{A}}} \otimes \overset{(k)}{\underset{\sim}{\mathbf{M}}}(\mathbf{r}^3 \partial_3 \mathbf{u}) + \underset{\approx}{\overset{2}{\mathbf{B}}} \otimes \overset{(k)}{\underset{\sim}{\mathbf{M}}}(\mathbf{r}^{\hat{P}}\partial_P \boldsymbol{\varphi})$$

$$+ \underset{\approx}{\overset{2}{\mathbf{B}}} \otimes \overset{(k)}{\underset{\sim}{\mathbf{M}}}(\mathbf{r}^3 \partial_3 \boldsymbol{\varphi}) - \underset{\approx}{\overset{2}{\mathbf{A}}} \otimes \underset{\sim}{\mathbf{C}} \cdot \overset{(k)}{\boldsymbol{\varphi}} - \underset{\sim}{\mathbf{b}} \overset{(k)}{\vartheta},$$

$$\underset{\sim}{\boldsymbol{\mu}} = \underset{\approx}{\overset{2}{\mathbf{C}}} \otimes \overset{(k)}{\underset{\sim}{\mathbf{M}}}(\mathbf{r}^{\hat{P}}\partial_P \mathbf{u}) + \underset{\approx}{\overset{2}{\mathbf{C}}} \otimes \overset{(k)}{\underset{\sim}{\mathbf{M}}}(\mathbf{r}^3 \partial_3 \mathbf{u}) + \underset{\approx}{\overset{2}{\mathbf{D}}} \otimes \overset{(k)}{\underset{\sim}{\mathbf{M}}}(\mathbf{r}^{\hat{P}}\partial_P \boldsymbol{\varphi})$$

$$+ \underset{\approx}{\overset{2}{\mathbf{D}}} \otimes \overset{(k)}{\underset{\sim}{\mathbf{M}}}(\mathbf{r}^3 \partial_3 \boldsymbol{\varphi}) - \underset{\approx}{\overset{2}{\mathbf{C}}} \otimes \underset{\sim}{\mathbf{C}} \cdot \overset{(k)}{\boldsymbol{\varphi}} - \underset{\sim}{\boldsymbol{\beta}} \overset{(k)}{\vartheta};$$

(9.100)

$$\mathbf{q} = -\underset{\sim}{\boldsymbol{\varLambda}} \cdot \overset{(k)}{\underset{\sim}{\mathbf{M}}}(\mathbf{r}^{\hat{P}}\partial_P T) - \underset{\sim}{\boldsymbol{\varLambda}} \cdot \overset{(k)}{\underset{\sim}{\mathbf{M}}}(\mathbf{r}^3 \partial_3 T). \qquad (9.101)$$

Taking into account (9.95)–(9.99) separately, from (9.100) and (9.101) we obtain the CR of physical and thermal content in moments of the corresponding approximations with respect to the Legendre polynomials, respectively. In order to reduce this contribution, we will not dwell on them. However, below we will consider them known.

9.8 On Boundary and Initial Conditions in Micropolar Mechanics of a Deformable Thin Body

9.8.1 The Boundary Conditions on the Front Surface

We consider the boundary conditions of physical content on the front surfaces and present them under the new parametrization of the thin body domain. Let $\overset{(+)}{\mathbf{P}}$ and $\overset{(-)}{\mathbf{P}}$ be the given stress vectors, and $\overset{(+)}{\boldsymbol{\mu}}$ and $\overset{(-)}{\boldsymbol{\mu}}$ are the given couple stress vectors on the front surfaces $\overset{(+)}{S}$ and $\overset{(-)}{S}$, respectively. Let us denote by $\overset{(+)}{\mathbf{n}}$ and $\overset{(-)}{\mathbf{n}}$ the unit vectors of outer normals to $\overset{(+)}{S}$ and $\overset{(-)}{S}$, respectively. It is easy to see that for $\overset{(-)}{\mathbf{n}}$ and $\overset{(+)}{\mathbf{n}}$ under the new parametrization we will have expressions

$$\overset{(-)}{\mathbf{n}} = -\frac{\mathbf{r}^{\bar{3}}}{\sqrt{g^{\bar{3}\bar{3}}}} = -\frac{1}{\sqrt{g^{\bar{3}\bar{3}}}}(\mathbf{r}^3 - g^3_{\bar{P}} g^{\bar{P}}_M \mathbf{r}^M),$$

$$\overset{(+)}{\mathbf{n}} = \frac{\mathbf{r}^{\overset{+}{3}}}{\sqrt{g^{\overset{++}{3}\overset{+}{3}}}} = \frac{1}{\sqrt{g^{\overset{++}{3}\overset{+}{3}}}}(\mathbf{r}^3 - g^3_{\overset{+}{P}} g^{\overset{+}{P}}_M \mathbf{r}^M).$$

(9.102)

Then, using (9.102), the boundary conditions of physical content on the front surfaces of the thin body (9.89) can be represented as

$$\mathbf{r}^{\bar{3}} \cdot \overset{(-)}{\underset{\sim}{\mathbf{P}}} = (\mathbf{r}^3 - g^3_{\underset{P}{-}} g^{\bar{P}}_M \mathbf{r}^M) \cdot \overset{(-)}{\underset{\sim}{\mathbf{P}}} = -\sqrt{g^{\bar{3}\bar{3}}} \overset{(-)}{\mathbf{P}},$$

$$\mathbf{r}^{\overset{+}{3}} \cdot \overset{(+)}{\underset{\sim}{\mathbf{P}}} = (\mathbf{r}^3 - g^3_{\underset{P}{+}} g^{\overset{+}{P}}_M \mathbf{r}^{\bar{M}}) \cdot \overset{(+)}{\underset{\sim}{\mathbf{P}}} = \sqrt{g^{\overset{++}{33}}} \overset{(+)}{\mathbf{P}}, \qquad (9.103)$$

$$\overset{(-)}{\underset{\sim}{\mathbf{P}}} = \underset{\sim}{\mathbf{P}}\Big|_{x^3=0}, \quad \overset{(+)}{\underset{\sim}{\mathbf{P}}} = \underset{\sim}{\mathbf{P}}\Big|_{x^3=1}; \quad \mathbf{P} \to \boldsymbol{\mu}, \quad x' \in S.$$

In non-isothermal processes, the normal components $\overset{(+)}{q}$ and $\overset{(-)}{q}$ of the heat flux vector $\mathbf{q}$ on the face surfaces of $\overset{(+)}{S}$ and $\overset{(-)}{S}$ can be given, respectively. The boundary conditions (second-kind conditions or Neumann-type conditions) (Pobedrya, 1995) on the front surfaces, similarly to (9.103), are represented in the form

$$\mathbf{r}^{\bar{3}} \cdot \overset{(-)}{\mathbf{q}} = (\mathbf{r}^3 - g^3_{\underset{P}{-}} g^{\bar{P}}_M \mathbf{r}^M) \cdot \overset{(-)}{\mathbf{q}} = -\sqrt{g^{\bar{3}\bar{3}}} \overset{(-)}{q},$$

$$\mathbf{r}^{\overset{+}{3}} \cdot \overset{(+)}{\mathbf{q}} = (\mathbf{r}^3 - g^3_{\underset{P}{+}} g^{\overset{+}{P}}_M \mathbf{r}^M) \cdot \overset{(+)}{\mathbf{q}} = \sqrt{g^{\overset{++}{33}}} \overset{(+)}{q}, \qquad (9.104)$$

$$\overset{(-)}{\mathbf{q}} = \mathbf{q}\Big|_{x^3=0}, \quad \overset{(+)}{\mathbf{q}} = \mathbf{q}\Big|_{x^3=1}, \quad x' \in S.$$

Boundary conditions can also be given that correspond to heat exchange with the environment according to Newton's law (third kind boundary conditions) (Pobedrya, 1995). In this case, the boundary conditions on $\overset{(+)}{S}$ and $\overset{(-)}{S}$ have the form

$$\mathbf{r}^{\bar{3}} \cdot \overset{(-)}{\mathbf{q}} = (\mathbf{r}^3 - g^3_{\underset{P}{-}} g^{\bar{P}}_M \mathbf{r}^M) \cdot \overset{(-)}{\mathbf{q}} = -\sqrt{g^{\bar{3}\bar{3}}} \overset{(-)}{\beta} (\overset{(-)}{T_c} - \overset{(-)}{T}),$$

$$\mathbf{r}^{\overset{+}{3}} \cdot \overset{(+)}{\mathbf{q}} = (\mathbf{r}^3 - g^3_{\underset{P}{+}} g^{\overset{+}{P}}_M \mathbf{r}^M) \cdot \overset{(+)}{\mathbf{q}} = \sqrt{g^{\overset{++}{33}}} \overset{(+)}{\beta} (\overset{(+)}{T_c} - \overset{(+)}{T}), \quad x' \in S, \qquad (9.105)$$

where T_c is the environment temperature, β is the heat transfer coefficient in kal/cm^2 °C, and

$$\overset{(-)}{\beta} = \beta\Big|_{x^3=0}, \quad \overset{(+)}{\beta} = \beta\Big|_{x^3=1}, \quad \overset{(-)}{T} = T\Big|_{x^3=0},$$

$$\overset{(+)}{T} = T\Big|_{x^3=1}, \quad \overset{(-)}{T_c} = T_c\Big|_{x^3=0}, \quad \overset{(+)}{T_c} = T_c\Big|_{x^3=1}. \qquad (9.106)$$

Below we will consider the boundary conditions of the first kind or the boundary conditions of Dirichlet type.

Further, we note that in a simplified reduction method for each approximate solution of a boundary-value problem, in the same way as it is done in Vekua (1982) for the classical version of the theory when applying the Legendre polynomials, a correction term is constructed to ensure the fulfillment of boundary conditions on the front surfaces. In this work, we will not stop at the construction of corrective terms under different boundary conditions on the front surfaces. At the new parametrization of the thin body domain these problems are considered in Nikabadze (2008b,

2014a) and also similar issues in the case of a thin body with two small sizes can be viewed in Nikabadze (2008a, 2014a).

From the above, it can be seen that the three-dimensional laws of Hooke and Fourier's thermal conductivity in the theory of thin bodies are replaced by the corresponding infinite systems of laws in moments. In addition, each law contains an infinite number of terms. Therefore, similarly to the systems of equations of motion and heat influx in moments, they should be reduced to finite systems of laws in moments, each law of which will contain a finite number of terms. The reduction is as follows: we fix some non-negative integers r and N, and then from the infinite system of laws in the normalized moments of the stress and couple stress tensors of approximations of order r choose the set of the first $N + 1$ laws. In the simplified reduction method, from the infinite system of laws in moments approximation of order r, we choose the set of the first $N + 1$ laws, in each law of which we neglect the moments of the unknown quantities whose order is greater than N. In this regard, it is advisable to introduce definitions.

Definition 9.7. The set of laws of Hooke (thermal conductivity of Fourier) in moments, which consists of the first $N + 1$ laws of the corresponding infinite laws of Hooke (thermal conductivity of Fourier) in normalized moments of the stress and moment stress tensors of the order r, is called the system of laws of Hooke (thermal conductivity of Fourier) in normalized moments of the stress and moment stress tensors of the approximation (r, N).

Definition 9.8. The set of laws of Hooke (thermal conductivity of Fourier) in moments, which consists of the first $N + 1$ laws of the corresponding infinite laws of Hooke (thermal conductivity of Fourier) in moments of the order r and each law of which does not contain moments of the unknown quantities, the order of which is greater than N, is called the system of laws of Hooke (thermal conductivity of Fourier) in moments of the approximation (r, N).

9.8.2 Boundary Conditions in Moments in the Theory of Thin Bodies

For the correct formulation of problems in the theory of thin bodies, to any system of equations that are consistent or inconsistent (with a simplified reduction scheme to a system of finite order) with boundary conditions on the front surfaces, the boundary conditions on the contour ∂S of the main base surface S should be attached. On the lateral surface of Σ, kinematic conditions (displacement and rotation vectors) or static (stress and moment stress vectors) can be given. On one part Σ_1 can be given kinematic conditions, and on the other part Σ_2 can be given static ones $(\Sigma_1 \cup \Sigma_2 = \Sigma, \Sigma_1 \cap \Sigma_2 = \emptyset)$. In non-isothermal processes on some part of the lateral face, the boundary conditions of the thermal content of the first kind (Dirichlet type) or the second kind (Neumann type), or the third kind (heat exchange with the environment according to Newton's law) are given, too. Below, we consider the kinematic, physical, and thermal boundary conditions on the lateral face, and from here we obtained the corresponding boundary conditions in moments on the boundary contour of the main base surface.

In the following we assume that the lateral surface of Σ consists of ruled surfaces and some non-negative integers r and N are fixed, which means that we consider the systems of equations in moments of the approximation (r, N), systems of laws of Hooke and Fourier thermal conductivity in normalized moments or in moments of approximation (r, N), as well as the corresponding boundary conditions on the boundary contour of the main base surface and the initial conditions in moments. Then, obviously, the moments of $\overset{(m)}{\mathbf{P}}$, $\overset{(m)}{\boldsymbol{\mu}}$, $\overset{(m)}{\mathbf{u}}$, $\overset{(m)}{\boldsymbol{\varphi}}$, $\overset{(m)}{T}$; $m = \overline{0, N}$ will be unknown. For example, for the micropolar theory of thin bodies in non-isothermal processes, the problem will be correctly posed if on the boundary contour ∂S of the main base surface S there are given $2N + 2$ vector kinematic boundary conditions, and on the part $\partial S_q \subseteq \partial S$ there are given $N + 1$ boundary condition of thermal content (in case of the first boundary-value problem) or on ∂S there are given $2N + 2$ vector static boundary conditions, and on $\partial S_q \subseteq \partial S$ there are given $N + 1$ boundary condition of the thermal content (in case of the second boundary-value problem), or on one part ∂S_1 there are given $2N + 2$ vector kinematic boundary conditions, on the others ∂S_2 ($\partial S_1 \cup \partial S_2 = \partial S$, $\partial S_1 \cap \partial S_2 = \emptyset$) there are given $2N + 2$ vector static boundary conditions, and on $\partial S_q \subseteq \partial S$ there are given $N + 1$ boundary condition of thermal content (in case of a mixed boundary value problem). We note that in case of dynamic problems, the initial conditions in moments, which will be discussed, should be attached to the boundary conditions in moments.

9.8.3 Kinematic Boundary Conditions in Moments

Let the vectors of displacement $\mathbf{u}$ and rotation $\boldsymbol{\varphi}$ are given on the lateral surfaces of Σ, i.e.

$$\mathbf{u}(x', x^3, t)\Big|_\Sigma = \mathbf{f}(x', x^3, t), \quad \boldsymbol{\varphi}(x', x^3, t)\Big|_\Sigma = \mathbf{g}(x', x^3, t)$$

Then the kinematic boundary conditions in moments approximation of N with respect to some system of polynomials are represented as

$$\overset{(k)}{\mathbf{u}}(x', t) = \overset{(k)}{\mathbf{f}}(x', t), \quad \overset{(k)}{\boldsymbol{\varphi}}(x', t) = \overset{(k)}{\mathbf{g}}(x', t), \quad k = \overline{0, N}, \quad x' \in \partial S \qquad (9.107)$$

Here $\overset{(k)}{\mathbf{f}}(x', t)$ and $\overset{(k)}{\mathbf{g}}(x', t)$, $k = \overline{0, N}$, are known vector fields on ∂S as moments of known vector fields $\mathbf{f}(x', x^3, t)$ and $\mathbf{g}(x', x^3, t)$, respectively.

9.8.4 Physical Boundary Conditions in Moments

Suppose that on the lateral surface Σ the vectors of stress $\mathbf{P}(x', x^3, t)$ and moment stress $\boldsymbol{\mu}(x', x^3, t)$ are given. Then the boundary conditions according to the Cauchy formulas on the lateral surface Σ can be written in the form

$$\hat{\mathbf{m}} \cdot \underset{\sim}{\mathbf{P}}(x', x^3, t) = \mathbf{P}(x', x^3, t), \quad \hat{\mathbf{m}} \cdot \underset{\sim}{\boldsymbol{\mu}}(x', x^3, t) = \boldsymbol{\mu}(x', x^3, t), \quad x' \in \partial S. \quad (9.108)$$

Here $\hat{\mathbf{m}}$ is the unit normal vector in an arbitrary point on the lateral surface. Before we obtain the boundary conditions in moments, let us derive some geometrical relations on the lateral surface under the considered parametrization of the thin body domain. Denoting by $d\hat{\Sigma}$ an elementary area with one vertex at the point with coordinates (x^1, x^2, x^3) and with sides $d\hat{\mathbf{r}} = \mathbf{r}_{\hat{I}} dx^I$ and $\mathbf{r}_3 dx^3$, we find

$$d\hat{\Sigma} = d\hat{\Sigma}\hat{\mathbf{m}} = \mathbf{r}_{\hat{I}} \times \mathbf{r}_3 dx^I dx^3 = \sqrt{\hat{g}} \epsilon_{IJ} \mathbf{r}^{\hat{I}} dx^J dx^3 = \sqrt{\hat{g}} g_{\hat{I}}^K \epsilon_{JK} \mathbf{r}^J dx^I dx^3.$$

Thus, we have

$$d\hat{\Sigma} = \sqrt{\hat{g}} \epsilon_{IJ} \mathbf{r}^{\hat{I}} dx^J dx^3 = \sqrt{\hat{g}} g_{\hat{I}}^K \epsilon_{JK} \mathbf{r}^J dx^I dx^3,$$

$$\overset{(+)}{d\Sigma} = \overset{(+)}{d\Sigma} \overset{(+)}{\mathbf{m}} = \sqrt{\overset{(+)}{g}} \epsilon_{IJ} \overset{+}{\mathbf{r}}^{\hat{I}} dx^J dx^3 = \sqrt{\overset{+}{g}} \epsilon_{JK} g_{\overset{+}{I}}^K \mathbf{r}^J dx^I dx^3, \quad (9.109)$$

$$d\Sigma = d\Sigma \mathbf{m} = \sqrt{g} \epsilon_{IJ} \mathbf{r}^I dx^J dx^3, \quad \overset{(-)}{d\Sigma} = \overset{(-)}{d\Sigma} \overset{(-)}{\mathbf{m}} = \sqrt{\overset{-}{g}} \epsilon_{JK} g_{\overset{-}{I}}^K \mathbf{r}^J dx^I dx^3.$$

The last two relations (9.109) are obtained similarly to the first. Therefore, they can be obtained from the first relation for $x^3 = 1$ and $x^3 = 0$, respectively. Here $\overset{(+)}{d\Sigma}$ is the elementary area with one vertex at the point $\overset{(+)}{M}$ with coordinates $(x^1, x^2, 1)$, sides $\overset{(+)}{d\mathbf{r}} = \mathbf{r}_{\overset{+}{I}} dx^I$ and $\mathbf{r}_3 dx^3$; $d\Sigma$ is the elementary area with one vertex at M with coordinates $(x^1, x^2, 0)$ and with sides $d\mathbf{r} = \mathbf{r}_I dx^I$ and $\mathbf{r}_3 dx^3$; $\overset{(-)}{d\Sigma}$ is the elementary area with one vertex at $\overset{(-)}{M}$ with coordinates $(x^1, x^2, -1)$ and with sides $\overset{(-)}{d\mathbf{r}} = \mathbf{r}_{\overset{-}{I}} dx^I$ and $\mathbf{r}_3 dx^3$; $\overset{(+)}{\mathbf{m}}, \overset{(+)}{\mathbf{m}}$ and $\overset{(-)}{\mathbf{m}}$ are the unit normal vectors at the points $M, \overset{(+)}{M}$ and $\overset{(-)}{M}$, respectively.

Further, by the first and third relations (9.109) we obtain

$$d\hat{\Sigma} = \hat{\vartheta} \frac{\sqrt{g^{\hat{K}\hat{L}} \epsilon_{KI} \epsilon_{LJ} dx^I dx^J}}{\sqrt{g^{KL} \epsilon_{KI} \epsilon_{LJ} dx^I dx^J}} d\Sigma = \frac{\sqrt{g^{MN} \epsilon_{MK} \epsilon_{NL} g_{\hat{I}}^K g_{\hat{J}}^L dx^I dx^J}}{\sqrt{g^{KL} \epsilon_{KI} \epsilon_{LJ} dx^I dx^J}} d\Sigma. \quad (9.110)$$

Multiplying the first, second, third, and fourth relations (9.109) by $\mathbf{r}_{\hat{K}}, \mathbf{r}_{\overset{+}{K}}, \mathbf{r}_K$ and $\mathbf{r}_{\overset{-}{K}}$ respectively, we will have

$$d\hat{\Sigma}\hat{m}_{\hat{I}} = \sqrt{\hat{g}}\epsilon_{IJ}dx^J dx^3, \quad d\overset{(+)}{\Sigma}\overset{(+)}{m}_{\underset{I}{+}} = \sqrt{\overset{(+)}{g}}\epsilon_{IJ}dx^J dx^3,$$

$$d\Sigma m_I = \sqrt{g}\epsilon_{IJ}dx^J dx^3, \quad d\overset{(-)}{\Sigma}\overset{(-)}{m}_{\underset{I}{-}} = \sqrt{\overset{(-)}{g}}\epsilon_{IJ}dx^J dx^3.$$

From here

$$d\hat{\Sigma}\hat{m}_{\hat{I}} = \hat{\vartheta}d\Sigma m_I = \overset{(+)}{\vartheta}\,d\overset{(+)}{\Sigma}\overset{(+)}{m}_{\underset{I}{+}} = \overset{(-)}{\vartheta}\,d\overset{(-)}{\Sigma}\overset{(-)}{m}_{\underset{I}{-}}. \tag{9.111}$$

Now let us find the boundary conditions in moments. First, we note that (9.108) can be represented as

$$\hat{m}_{\hat{I}}\mathbf{P}^{\hat{I}} = \mathbf{P}(x', x^3, t), \quad \hat{m}_{\hat{I}}\boldsymbol{\mu}^{\hat{I}} = \boldsymbol{\mu}(x', x^3, t), \quad x' \in \partial S. \tag{9.112}$$

Multiplying each relation (9.112) by $d\hat{\Sigma}$ and taking into account (9.111), we find

$$m_I g_{\hat{J}}^{\hat{I}}\mathbf{P}^J = \mathbf{P}\frac{d\hat{\Sigma}}{d\Sigma}\hat{\vartheta}^{-1}, \quad m_I g_{\hat{J}}^{\hat{I}}\boldsymbol{\mu}^J = \boldsymbol{\mu}\frac{d\hat{\Sigma}}{d\Sigma}\hat{\vartheta}^{-1}, \quad x' \in \partial S. \tag{9.113}$$

Introducing the notation

$$a(x', x^3) = \frac{d\hat{\Sigma}}{d\Sigma}\hat{\vartheta}^{-1} = \frac{\sqrt{g^{MN}\epsilon_{MK}\epsilon_{NL}g_{\hat{i}}^{K}g_{\hat{j}}^{L}dx^I dx^J}}{\sqrt{g^{KL}\epsilon_{KI}\epsilon_{LJ}dx^I dx^J}}\hat{\vartheta}^{-1}, \tag{9.114}$$

Eq. (9.113) can be written in the form

$$m_I g_{\hat{J}}^{\hat{I}}\mathbf{P}^J = a(x', x^3)\mathbf{P}(x', x^3, t), \quad m_I g_{\hat{J}}^{\hat{I}}\boldsymbol{\mu}^J = a(x', x^3)\boldsymbol{\mu}(x', x^3, t), \quad x' \in \partial S. \tag{9.115}$$

Representing $a(x', x^3)$ as a series with respect to $z = \bar{h} + x^3 h$

$$a(x', x^3) = \sum_{s=0}^{\infty} A_{(s)}(x')z^s, \quad A_{(s)}(x') = \frac{1}{s!}\left(\frac{\partial^s a}{\partial(x^3)^s}\right)_{(z=0)} \tag{9.116}$$

and taking into account (9.45), from (9.115) we obtain the following boundary conditions of approximation of order r:

$$m_I \underset{(r)}{g}_{\hat{J}}^{\hat{I}}\mathbf{P}^J = a_{(r)}(x', x^3)\mathbf{P}, \quad m_I \underset{(r)}{g}_{\hat{J}}^{\hat{I}}\boldsymbol{\mu}^J = a_{(r)}(x', x^3)\boldsymbol{\mu}, \quad x' \in \partial S,$$

$$a_{(r)}(x', x^3) = \sum_{s=0}^{r} A_s(x')z^s, \quad r \in \mathbb{N}_0. \tag{9.117}$$

Taking into account (9.45) and the first formula (9.116) and equating the coefficients with the same powers of x^3 in the right and left sides of the first two relations (9.117), we get

$$m_I \underset{(s)}{A}_{J}^{I}\mathbf{P}^J = A_{(s)}(x')\mathbf{P}, \quad m_I \underset{(s)}{A}_{J}^{I}\boldsymbol{\mu}^J = A_{(s)}(x')\boldsymbol{\mu}, \quad s \in \mathbb{N}_0, \quad x' \in \partial S. \tag{9.118}$$

The relations (9.115) and (9.118) are equivalent and the relations (9.117) are equivalent to the first $r+1$ Eqs. (9.118). Applying the kth order moment operator of some polynomial system (Legendre, Chebyshev) to (9.117), we find

$$m_I \overset{(k)}{\mathbf{M}}(\underset{(r)}{g}{}^I_J \mathbf{P}^J) = \overset{(k)}{\mathbf{M}}(a_{(r)}\mathbf{P}), \quad m_I \overset{(k)}{\mathbf{M}}(\underset{(r)}{g}{}^I_J \boldsymbol{\mu}^J) = \overset{(k)}{\mathbf{M}}(a_{(r)}\boldsymbol{\mu}),$$

$$r = \overline{0,N}, \quad x' \in \partial \overset{(-)}{S}. \tag{9.119}$$

Taking into account (9.118), from (9.119) we arrive to the relations

$$m_I \underset{(s)}{A}{}^I_J \overset{(k)}{\mathbf{P}}{}^J = \underset{(s)}{A} \overset{(k)}{\mathbf{P}}, \quad m_I \underset{(s)}{A}{}^I_J \overset{(k)}{\boldsymbol{\mu}}{}^J = \underset{(s)}{A} \overset{(k)}{\boldsymbol{\mu}}, \quad s = \overline{0,r}, \quad k = \overline{0,N}, \quad x' \in \partial S, \tag{9.120}$$

that can be also obtained by applying the moment operator of the kth order to (9.118). Note that, based on (9.118) from (9.119), we can exclude the moments of the unknown and known quantities, the order of which more than N. Then we obtain relations that we call static boundary conditions in moments of the approximation (r, N). They are equivalent to (9.120), therefore, it is efficient to consider the relations (9.120) as static boundary conditions in moments of (r, N) approximation.

9.8.5 Boundary Conditions of Heat Content in Moments

The boundary conditions of the first (Dirichlet type), the second (Neumann type) and the third (heat exchange with the environment according to Newton's law) type Pobedrya (1995) are considered and the corresponding boundary conditions are obtained in moments.

9.8.5.1 Boundary Conditions of the First Kind in Moments

In this case, the temperature is posed at the part $\Sigma_q \subseteq \Sigma$ of the lateral surface Σ

$$T(x', x^3, t)\Big|_{\Sigma_q} = T_0(x', x^3, t).$$

From here, similarly to (9.107), the desired boundary conditions of the first kind in the moments will have the form

$$\overset{(k)}{T}(x', t) = \overset{(k)}{T}_0(x', t), \quad k = \overline{0,N}, \quad x' \in \partial S_q \subseteq \partial S, \tag{9.121}$$

where $\overset{(k)}{T}_0(x', t)$, $k = \overline{0,N}$ are the known moments of the known scalar field $T_0(x', x^3, t)$.

9.8.5.2 Boundary Conditions of the Second Kind in Moments

In this case, at the part $\Sigma_q \subseteq \Sigma$ of the lateral face the condition is posed

$$\mathbf{m} \cdot \mathbf{q}(x', x^3, t)\Big|_{\Sigma_q} = q_0(x', x^3, t) \quad (m_I q^I\Big|_{\Sigma_q} = q_0).$$

Hence, similarly to (9.120), we obtain the required conditions in the form

$$m_I \underset{(s)}{A} \overset{I \ (k) \ J}{_{J}} \overset{(k)}{q}{}^J(x', t) = A_{(s)} \overset{(k)}{q}_0(x', t), \quad s = \overline{0, r}, \quad k = \overline{0, N}, \quad x' \in \partial S_q \subseteq \partial S. \quad (9.122)$$

The relations (9.122) are called the boundary conditions of the thermal content of the second kind in moments of the approximation (r, N).

9.8.5.3 Boundary Conditions of the Third Kind in Moments

In this case, the boundary conditions are represented as

$$\mathbf{m} \cdot \mathbf{q}(x', x^3, t)\Big|_{\Sigma_q} = \beta\left(T_c - T\Big|_{\Sigma_q}\right) \quad \left(m_I q^I\Big|_{\Sigma_q} = \beta\left(T_c - T\Big|_{\Sigma_q}\right)\right). \quad (9.123)$$

Then, similarly to (9.122) from (9.123), for the required conditions we will have the expressions

$$m_I \underset{(s)}{A} \overset{I \ (k) \ J}{_{J}} \overset{(k)}{q}{}^J(x', t) = A_{(s)} \beta(\overset{(k)}{T}_c - \overset{(k)}{T}), \quad s = \overline{0, r}, \quad k = \overline{0, N}, \quad x' \in \partial S_q \subseteq \partial S. \quad (9.124)$$

The relations (9.124) are called the boundary conditions of the thermal content of the third kind in moments of the approximation (r, N).

Writing (9.124), it is assumed that the heat transfer coefficient β does not depend on x^3. We also note that it was possible to consider the boundary conditions of a more general form, than those given above (see, for example, Pobedrya (1995), where the boundary conditions of the general form are given for the classical case). Note that if the systems of equations of motion and heat influx in moments are obtained from other their representations, similar to (9.43), then we would have to present the systems of boundary conditions of physical and thermal (second and third kinds) contents in moments in the corresponding form. We will not dwell on them in this paper.

9.8.6 Initial Conditions in Moments

When considering non-stationary problems at some instant $t = t_0$, initial conditions must be given. Let for the non-stationary (dynamic) problem of the micropolar mechanics of a deformable solid initial conditions are represented as

$$
\mathbf{u}(x', x^3, t)\Big|_{t=t_0} = \mathbf{u}_0(x', x^3), \quad \partial_t \mathbf{u}(x', x^3, t)\Big|_{t=t_0} = \mathbf{v}_0(x', x^3),
$$
$$
\boldsymbol{\varphi}(x', x^3, t)\Big|_{t=t_0} = \boldsymbol{\varphi}_0(x', x^3), \quad \partial_t \boldsymbol{\varphi}(x', x^3, t)\Big|_{t=t_0} = \mathbf{w}_0(x', x^3), \tag{9.125}
$$

and for the non-stationary problem of heat conduction the initial condition represented in the form

$$
T(x', x^3, t)\Big|_{t=t_0} = T^0(x', x^3). \tag{9.126}
$$

From (9.125) for the sought initial conditions in moments we will have the expression

$$
\overset{(k)}{\mathbf{u}}(x', t)\Big|_{t=t_0} = \overset{(k)}{\mathbf{u}}_0(x'), \quad \partial_t \overset{(k)}{\mathbf{u}}(x', t)\Big|_{t=t_0} = \overset{(k)}{\mathbf{v}}_0(x'),
$$
$$
\overset{(k)}{\boldsymbol{\varphi}}(x', t)\Big|_{t=t_0} = \overset{(k)}{\boldsymbol{\varphi}}_0(x'), \quad \partial_t \overset{(k)}{\boldsymbol{\varphi}}(x', t)\Big|_{t=t_0} = \overset{(k)}{\mathbf{w}}_0(x'), \quad k = \overline{0, N}, \quad x' \in \overset{(-)}{S}. \tag{9.127}
$$

Similarly to the first relation (9.127) from (9.126) for the non-stationary problem of heat conduction, we obtain the following system of initial conditions in moments

$$
\overset{(k)}{T}(x', t)\Big|_{t=t_0} = \overset{(k)}{T^0}(x'), \quad k = \overline{0, N}, \quad x' \in \overset{(-)}{S}. \tag{9.128}
$$

Note that (9.127) and (9.128) represent a system of initial conditions in moments of approximation N of the dynamic problem of micropolar thermomechanics of a deformable thin body (TMDTB).

9.9 Problem Statements in Moments of Micropolar Thermomechanics of a Deformable Thin Body

9.9.1 Statement of the Coupled Dynamic Problem in Moments of (r, N) Approximation

This considered statement of the problem includes:

1) the system of equations of motion in moments of (r, N) approximation of the micropolar TMDTB;
2) the system of equations of heat influx in moments of (r, N) approximation of the micropolar TMDTB;
3) the system of CR in moments of (r, N) approximation of the micropolar TMDTB under the simplified reduction method;
4) the system of Fourier's laws of thermal conductivity in moments of (r, N) approximation of the micropolar TMDTB under the simplified reduction method;

5) depending on the type of boundary-value problems one of the following systems of boundary conditions in moments:

5a) the system of kinematic boundary conditions in moments of N approximation for the first boundary-value problem and any system of boundary conditions of thermal content in moments;

5b) the system of static boundary conditions in moments of (r, N) approximation of the micropolar TMDTB for the second boundary problem and any system of boundary conditions of thermal content in moments;

5c) the system of kinematic boundary conditions in moments of N approximation on one part of the boundary contour and the system of static boundary conditions in moments of (r, N) approximation of the micropolar TMDTB on the other part of the boundary contour for a mixed boundary-value problem and any systems of boundary conditions of thermal content in moments;

6) the system of initial conditions of kinematic and thermal contents in moments of N approximation.

9.9.2 Statement of a Non-stationary Temperature Problem in Moments of (r, N) Approximation

If the system of equations of heat influx in moments of approximation (r, N) does not include mechanical characteristics (moments of tensors of stress $\overset{(k)}{\underset{\sim}{P}}$ and moment stress $\overset{(k)}{\mu}$), then the non-stationary temperature problem in moments of (r, N) approximation is considered separately, which includes:

1) the system of equations of heat influx in moments of (r, N) approximation without mechanical characteristics;

2) the system of Fourier's laws of thermal conductivity in moments of (r, N) approximation of micropolar TMDTB under the simplified reduction method;

3) any system of boundary conditions of thermal content in moments;

4) the system of initial conditions of the thermal content in moments of N approximation.

In this case, the dynamic problem in moments of (r, N) approximation of the micropolar TMDTB is divided into two problems: the non-stationary temperature problem in moments of (r, N) approximation, the solution of which determines the temperature field, later considered known, and the dynamic problem in moments of (r, N) approximation of the micropolar TMDTB in non-isothermal processes with a known temperature field.

9.9.3 Statement of the Uncoupled Dynamic Problem in Moments of the (r, N) Approximation

The formulation of this problem includes:

1) the system of equations of motion in moments of (r, N) approximation of a micropolar TMDTB in non-isothermal processes with a known temperature field;
2) the system of CR in moments of the (r, N) approximation of the micropolar TMDTB with a known temperature field under the simplified reduction method;
3) depending on the type of boundary-value problems, one of the following systems of boundary conditions in moments:
3a) the system of kinematic boundary conditions in moments of N approximation for the first boundary-value problem;
3b) the system of static boundary conditions in moments of (r, N) approximation of the micropolar TMDTB for the second boundary-value problem;
3c) the system of kinematic boundary conditions in moments of the N approximation of the micropolar TMDTB on one part of the boundary contour and the system of static boundary conditions in moments of the (r, N) approximation of the micropolar TMDTB on the other part of the boundary contour for a mixed boundary-value problem;
4) the system of kinematic initial conditions in moments of N approximation.

It should be noted that when composing equations, CR, boundary and initial conditions included in the formulation of initial-boundary value problems, definition 9.8 should be used.

Thus, the statements of coupled and uncoupled dynamic problems in moments of the (r, N) approximation of a micropolar TMDTB, as well as of the non-stationary temperature problem in moments of the (r, N) approximation are given. From these statements, it is easy to obtain the formulations of the corresponding static and quasistatic problems, as well as by changing values of r and N, we can easily formulate the statements in moments of the required approximations. In addition, it is possible to obtain the statements of problem in isothermal processes. Finally, if in all the mentioned statements neglect the moments of moment stresses and the internal rotation vector, then we obtain the corresponding statements of problems in moments of (r, N) approximation of the classical TMDTB and MDTB.

It should be noted that the formulation of problems under the new parametrization of the thin body domain with one small size was considered in Nikabadze (2008b, 2014a), and in case of thin bodies with two small sizes the formulation of problems under the different parameterizations of the thin body domain can be found in Nikabadze (2008a, 2014a). Consequently, in case of thin bodies with two small sizes, we can consider similar statements of boundary-value problems.

In the foreseeable future, we are going the formulated initial-boundary value problems for micropolar theories of multilayered (single-layer) thin bodies of various approximations in moments via the Legendre polynomial systems to compare with different known formulations of the original problems of micropolar theory of multilayered (single-layer) shells and plates. It is also supposed to construct the corresponding theories of thin bodies based on the known 3D non-linear micropolar theories. In this regard, the following works are of some interest: Della Corte et al (2019); Eremeyey and Zubov (2008); Altenbach et al (2010); Javili et al (2011); Seppecher et al (2013); Eremeyey et al (2013).

Acknowledgements This work was supported by the Russian Foundation for Basic Research, grants 18–29–10085–mk and 19-01-00016-A.

References

Alekseev AE (1994) Derivation of equations for a layer of variable thickness based on expansions in terms of Legendre's polynomials. Journal of Applied Mechanics and Technical Physics 35(4):612–622

Alekseev AE (1995) Bending of a three-layer orthotropic beam. Journal of Applied Mechanics and Technical Physics 36(3):458–465

Alekseev AE (2000) Iterative method for solving problems of deformation of layered structures, taking into account the slippage of layers (in Russ.). Dinamika sploshnoy sredy: Sb nauch tr 116:170–174

Alekseev AE, Annin BD (2003) Equations of deformation of an elastic inhomogeneous laminated body of revolution. Journal of Applied Mechanics and Technical Physics 44(3):432–437

Alekseev AE, Demeshkin AG (2003) Detachment of a beam glued to a rigid plate. Journal of Applied Mechanics and Technical Physics 44(4):577–583

Alekseev AE, Alekhin VV, Annin BD (2001) Plane elastic problem for an inhomogeneous layered body. Journal of Applied Mechanics and Technical Physics 42(6):1038–1042

Altenbach H (1991) Modelling of viscoelastic behaviour of plates. In: Zyczkowski M (ed) Creep in Structures, Springer, Berlin Heidelberg, pp 531–537

Altenbach J, Altenbach H, Eremeyev V (2010) On generalized Cosserat-type theories of plates and shells: a short review and bibliography. Archive of Applied Mechanics 80(1):73–92

Ambartsumyan SA (1958) On the theory of bending of anisotropic plates and shallow shells. Izv AN SSSR OTN (5):69–77

Ambartsumyan SA (1970) A new refined theory of anisotropic shells. Polymer Mechanics 6(5):766–776

Ambartsumyan SA (1974) General Theory of Anisotropic Shells (in. Russ.). Nauka, Moscow

Ambartsumyan SA (1987) Theory of Anisotropic Plates (in Russ.). Nauka, Moscow

Chepiga VE (1976) To the improved theory of laminated shells (in Russ.). Appl Mech 12(11):45–49

Chepiga VE (1977) Construction of the theory of multilayer anisotropic shells with given conditional accuracy of order h^N (in Russ.). Mekh Tverdogo Tela (4):111–120

Chepiga VE (1986a) Asymptotic error of some hypotheses in the theory of laminated shells (in Russ.). Theory and calculation of elements of thin-walled structures pp 118–125

Chepiga VE (1986b) Numerical analysis of equations of the improved theory of laminated shells (in Russ.). 290-B1986, VINITI

Chepiga VE (1986c) The study of stability of multilayer shells by an improved theory (in Russ.). 289-B1986, VINITI

Chernykh KF (1986) Nonlinear Theory of Elasticity in Engineering Computations (in Russ.). Mashinostroenie, Leningrad

Chernykh KF (1988) Introduction into Anisotropic Elasticity (in Russ.). Nauka, Moscow

Della Corte A, Battista A, dell'Isola F, et al (2019) Large deformations of Timoshenko and Euler beams under distributed load. Math Phys 70(52)

Dergileva LA (1976) Solution method for a plane contact problem for an elastic layer (in Russ.). Continuum Dynamics 25:24–32

Egorova O, Zhavoronok S, Kurbatov A (2015) The variational equations of the extended Nth order shell theory and its application to some problems of dynamics (in Russ.). Perm National Polytechnic University Mechanics Bulletin (2):36–59

Eremeyey VA, Zubov LM (2008) Mechanics of Elastic Shells

Eremeyey VA, Lebedev LP, Altenbach H (2013) Foundations of Micropolar Mechanics. Springer-Verlag

Fellers J, Soler A (1970) Approximate solution of the finite cylinder problem using Legendre polynomials. AIAA 8(11)

Filin AP (1987) Elements of the Theory of Shells (in Russ.). Stroyizdat, Leningrad

Gol'denveizer AL (1976) Theory of Elastic Shells (in Russ.). Nauka, Moscow

Gol'denveizer AL (1962) Derivation of an approximate theory of bending of a plate by the method of asymptotic integration of the equations of the theory of elasticity. Journal of Applied Mathematics and Mechanics 26(4):1000–1025

Gol'denveizer AL (1963) Derivation of an approximate theory of shells by means of asymptotic integration of the equations of the theory of elasticity. Journal of Applied Mathematics and Mechanics 27(4):903–924

Grigolyuk EI, Selezov IT (1973) Nonclassic oscillation theories of rods, plates, and shells (in Russ.), vol 5. VINITI. Itogi nauki i tekniki, Moscow

Hencky H (1947) Über die berücksichtigung der schubverzerrung in ebenen platten. Ingenieur-Archiv 16:72–76

Hertelendy P (1968) An approximate theory governing symmetric motions of elastic rods of rectangular or square cross section. Trans ASME Journal of Applied Mechanics 35(2):333–341

Ivanov GV (1976) Solution of the plane mixed problem of the theory of elasticity in the form of a series in Legendre polynomials (in Russ.). Z Prikl Mekh Tekhn Fiz (6):126–137

Ivanov GV (1977) Solutions of plane mixed problems for the Poisson equation in the form of series over Legendre polynomials (in Russ.). Continuum Dynamics 28:43–54

Ivanov GV (1979) Reduction of a three-dimensional problem for an inhomogeneous shell to a two-dimensional problem (in Russ.). Dynamic Problems of Continuum Mechanics 39

Ivanov GV (1980) Theory of Plates and Shells (in Russ.). Novosib. State Univ., Novosibirsk

Javili A, dell'Isola F, Stemmann P (2011) Geometrically nonlinear higher-gradient elasticity with energetic boundaries. J Phys: Conf Ser

Kantor MM, Nikabadze MU, Ulukhanyan AR (2013) Equations of motion and boundary conditions of physical meaning of micropolar theory of thin bodies with two small cuts. Mechanics of Solids 48(3):317–328

Khoroshun LP (1978) On the construction of equations of layered plates and shells (in russ.). Prikladnaya Mekhanika (10):3–21

Khoroshun LP (1985) The concept of a mixture in the construction of the theory of layered plates and shells in russ.). Prikladnaya Mekhanika 21(4):110–118

Kienzler R (1982) Eine Erweiterung der klassischen Schalentheorie; der Einfluß von Dickenverzerrungen und Querschnittsverwölbungen. Ingenieur-Archiv 52(5):311–322

Kirchhoff G (1850) Über das gleichgewicht und die bewegung einer elastischen scheibe. Journal für die reine und angewandte Mathematik (Crelles Journal) (40):51–88

Kupradze VD (ed) (1979) Three-dimensional Problems of the Mathematical Theory of Elasticity and Thermoelasticity, North-Holland Series in Applied Mathematics and Mechanics, vol 25. North Holland

Kuznetsova E, Kuznetsova EL, Rabinskiy LN, Zhavoronok SI (2018) On the equations of the analytical dynamics of the quasi-3D plate theory of I. N. Vekua type and some their solutions. Journal of Vibroengineering 20(2):1108–1117

Levinson M (1980) An accurate, simple theory of the statics and dynamics of elastic plates. Mech Res Commun 7(6):343–350

Lewiński T (1987) On refined plate models based on kinematical assumptions. Ingenieur-Archiv 57(2):133–146

Lo KH, Christensen RM, Wu EM (1977a) A high-order theory of plate deformation. Part 1: Homogeneous plates. Trans ASME Journal of Applied Mechanics 44(4):663–668

Lo KH, Christensen RM, Wu EM (1977b) A high-order theory of plate deformation. Part 2: Laminated Plates. Trans ASME Journal of Applied Mechanics 44(4):669–676

Lurie AI (1990) Non-linear Theory of Elasticity, North-Holland Series in Applied Mathematics and Mechanics, vol 36. North Holland

Medick MA (1966) One-dimensional theories of wave propagation and vibrations in elastic bars of rectangular cross section. Trans ASME Journal of Applied Mechanics 33(3):489–495

Meunargiya TV (1987) Development of the method of I. N. Vekua for problems of the three-dimensional moment elasticity (in Russ.). Tbilisi State Univ., Tbilisi

Mindlin RD, Medick MA (1959) Extensional vibrations of elastic plates. Trans ASME J Appl Mech 26(4):561–569

Naghdi PM (1972) The theory of shells and plates. In: Flügge S (ed) Handbuch der Physik, vol VIa/2, Springer, Berlin, Heidelberg, pp 425–640

Nikabadze MU (1988a) On the theory of shells with two base surfaces (in Russ.). 8149-B88, VINITI

Nikabadze MU (1988b) Parameterization of shells with two base surfaces (in Russ.). 5588-B88, VINITI

Nikabadze MU (1989) Deformation of layered viscoelastic shells. In: Actual problems of strength in mechanical engineering (in Russ.), SVVMIU, Sevastopol, p 1

Nikabadze MU (1990a) Modeling of nonlinear deformation of elastic shells (in Russ.). PhD thesis, Lomonosov Moscow State University

Nikabadze MU (1990b) Plane curvilinear rods (in Russ.). 4509-B90, VINITI

Nikabadze MU (1990c) To the theory of shells with two base surfaces (in Russ.). 1859-B90, VINITI

Nikabadze MU (1990d) To the theory of shells with two base surfaces (in Russ.). 2676-B90, VINITI

Nikabadze MU (1991) New kinematic hypothesis and new equations of motion and equilibrium theories of shells and plane curvilinear rods (in Russ.). Vestn Mosk Univ, Matem Mekhan (6):54–61

Nikabadze MU (1998a) Constitutive relations of the new linear theory of thermoelastic shells (in Russ.). In: Actual problems of shell mechanics, UNIPRESS, Kazan, pp 158–162

Nikabadze MU (1998b) Different representations of the cauchy-green deformation tensor and the linear deformation tensor and their components in the new theory of shells (in Russ.). Mathematical modeling of systems and processes (6):59–65

Nikabadze MU (1999a) Constitutive relations of the new linear theory of thermoelastic shells of TS class (in Russ.). Mathematical modeling of systems and processes (7):52–56

Nikabadze MU (1999b) New rod space parametrization (in Russ.). 1663-B99, VINITI

Nikabadze MU (1999c) New rod theory (in Russ.). In: 16th inter-republican conference on numerical methods for solving problems of the theory of elasticity and plasticity, Novosibirsk

Nikabadze MU (1999d) Various forms of the equations of motion and boundary conditions of the new theory of shells (in Russ.). Mathematical modeling of systems and processes (7):49–51

Nikabadze MU (2000a) Some geometric relations of the theory of shells with two basic surfaces (in Russ.). Izv RAN MTT (4):129–139

Nikabadze MU (2000b) To the parametrization of the multilayer shell domain of 3d space (in Russ.). Mathematical modeling of systems and processes (8):63–68

Nikabadze MU (2001a) Dynamic equations of the theory of multilayer shell constructions under the new kinematic hypothesis (in Russ.). In: Elasticity and non-elasticity, 1, Izd. MGU, pp 389–395

Nikabadze MU (2001b) Equations of motion and boundary conditions of the theory of rods with several basic curves (in Russ.). Vestn Mosk Univ, Matem Mekhan (3):35–39

Nikabadze MU (2001c) Location gradients in the theory of shells with two basic surfaces (in Russ.). Mech Solids 36(4):64–69

Nikabadze MU (2001d) To the variant of the theory of multilayer structures (in Russ.). Izv RAN MTT (1):143–158

Nikabadze MU (2002a) Equations of motion and boundary conditions of a variant of the theory of multilayer plane curvilinear rods (in Russ.). Vestn Mosk Univ, Matem Mekhan (6):41–46

Nikabadze MU (2002b) Modern State of Multilayer Shell Structures (in Russ.). 2289–B2002, VINITI

Nikabadze MU (2003) Variant of the theory of shallow shells (in Russ.). In: Lomonosovskiye chteniya. Section mechanics., Izd. Moscov. Univ., Moscow

Nikabadze MU (2004a) Generalization of the Huygens-Steiner theorem and the Boer formulas and some of their applications (in Russ.). Izv RAN MTT (3):64–73

Nikabadze MU (2004b) Variants of the theory of shells with the use of expansions in Legendre polynomials (in Russ.). In: Lomonosovskiye chteniya. Section mechanics., Izd. Moscov. Univ., Moscow

Nikabadze MU (2005) To the variant of the theory of multilayer curvilinear rods (in Russ.). Izv RAN MTT (6):145–156

Nikabadze MU (2006) Application of Classic Orthogonal Polynomials to the Construction of the Theory of Thin Bodies (in Russ.). Elasticity and non-elasticity pp 218–228

Nikabadze MU (2007a) Application of Chebyshev Polynomials to the Theory of Thin Bodies. Moscow University Mechanics Bulletin 62(5):141–148

Nikabadze MU (2007b) Some issues concerning a version of the theory of thin solids based on expansions in a system of Chebyshev polynomials of the second kind. Mechanics of Solids 42(3):391–421

Nikabadze MU (2007c) To theories of thin bodies (in Russ.). In: Non-classical problems of mechanics, Proceedings of the international conference, Kutaisi, vol 1, pp 225–242

Nikabadze MU (2008a) Mathematical modeling of elastic thin bodies with two small dimensions with the use of systems of orthogonal polynomials (in Russ.). 722 – B2008, VINITI

Nikabadze MU (2008b) The application of systems of Legendre and Chebyshev polynomials at modeling of elastic thin bodies with a small size (in Russ.). 720-B2008, VINITI

Nikabadze MU (2014a) Development of the method of orthogonal polynomials in the classical and micropolar mechanics of elastic thin bodies (in Russ.). Moscow Univ. Press, Moscow

Nikabadze MU (2014b) Method of orthogonal polynomials in mechanics of micropolar and classical elasticity thin bodies (in Russ.). Doctoral dissertation. Moscow, MAI

Nikabadze MU (2016) Eigenvalue problems of a tensor and a tensor-block matrix (tmb) of any even rank with some applications in mechanics. In: Altenbach H, Forest S (eds) Generalized continua as models for classical and advanced materials, Advanced Structured Materials, vol 42, pp 279–317

Nikabadze MU (2017a) Eigenvalue problem for tensors of even rank and its applications in mechanics. Journal of Mathematical Sciences 221(2):174–204

Nikabadze MU (2017b) Topics on tensor calculus with applications to mechanics. Journal of Mathematical Sciences 225(1):1–194

Nikabadze MU, Ulukhanyan A (2005a) Formulation of the problem for thin deformable 3d body (in Russ.). Vestn Mosk Univ, Matem Mekhan (5):43–49

Nikabadze MU, Ulukhanyan A (2005b) Formulations of problems for a shell domain according to three-dimensional theories (in Russ.). 83–B2005, VINITI

Nikabadze MU, Ulukhanyan A (2008) Mathematical modeling of elastic thin bodies with one small dimension with the use of systems of orthogonal polynomials (in Russ.). 723 – B2008, VINITI

Nikabadze MU, Ulukhanyan AR (2016) Analytical solutions in the theory of thin bodies. In: Altenbach H, Forest S (eds) Generalized continua as models for classical and advanced materials, Advanced Structured Materials, vol 42, pp 319–361

Nowacki W (1975) Theory of Elasticity. Mir, Moscow, (Russian translation)

Pelekh BL (1973) Theory of shells with finite shear stiffness (in Russ.). Naukova Dumka, Kiev

Pelekh BL (1978) The Generalized Theory of Shells (in Russ.). Vischa shkola, Lvov

Pelekh BL, Sukhorolskii MA (1977) Construction of the generalized theory of transversal-isotropic shells in application to contact problems (in Russ.). Composites and New Structures pp 27–39

Pelekh BL, Sukhorolskii MA (1980) Contact problems of the theory of elastic anisotropic shells (in Russ.). Naukova Dumka, Kiev

Pelekh BL, Maksimuk AV, Korovaichuk IM (1988) Contact problems for laminated elements of constructions and bodies with coating (in Russ.). Naukova Dumka, Kiev

Pikul VV (1992) To the problem of constructing a physically correct theory of shells (in Russ.). Izv RAN MTT (3):18–25

Pobedrya BE (1986) Lectures on tensor analysis (in Russ.). M: Izd. Moscov. Univ.

Pobedrya BE (1995) Numerical methods in the theory of elasticity and plasticity (in Russ.). Izd. Moscov. Univ., Moscow

Pobedrya BE (2003) On the theory of constitutive relations in the mechanics of a deformable solid (in Russ.). In: Problemy mekhaniki, Fiszmatlit, Moscow, pp 635–657

Pobedrya BE (2006) Theory of thermomechanical processes (in Russ.). In: Elasticity and non-elasticity, Izd. MGU, pp 70–85

Preußer G (1984) Eine systematische Herleitung verbesserter Plattengleichungen. Ingenieur-Archiv 54(1):51–61

Reissner E (1985) Reflections on the theory of elastic plates. Applied Mechanics Reviews 38(11):1453–1464

Reissner E (1944) On the theory of bending of elastic plates. Journal of Mathematics and Physics 23(1-4):184–191

Sansone G (1959) Orthogonal Functions. Interscience Publishers Inc, New York

Seppecher P, Alibert J, dell'Isola F (2013) Linear elastic trusses leading to continua with exotic mechanical interactions. J of the Mech and Phys of Solids 61(12):2381–2401

Sokol'nikov IS (1971) Tensor analysis (in Russ.). Nauka, Moscow

Soler AI (1969) Higher-order theories for structural analysis using Legendre polynomial expansions. Trans ASME Journal of Applied Mechanics 36(4):757–762

Suyetin PK (1976) Classical orthogonal polynomials (in Russ.). Nauka, Moscow

Tvalchrelidze AK (1984) Theory of elastic shells using several base surfaces (in Russ.). In: Theory and numerical methods for calculating plates and shells, Tbilisi

Tvalchrelidze AK (1986) Basic equations of the theory of shells, taking into account large deformations and shears (in Russ.). Soobshch AN GruzSSR 121(1):53–56

Tvalchrelidze AK (1994) Shell theory using several base surfaces and some applications (in Russ.). PhD thesis, Kutaisi

Tvalchrelidze AK, Tvaltvadze DV, Nikabadze MU (1984) To the calculation of large axisymmetric deformations of the shells of rotation of elastomers (in Russ.). In: XXII scientific and technical. conf., Tbilisi

Ulukhanyan AR (2011) Dynamic equations of the theory of thin prismatic bodies with expansion in the system of Legendre polynomials. Mechanics of Solids 46(3):467–479

Vajeva DV, Volchkov YM (2005) The equations for determination of stress-deformed state of multilayer shells (in Russ.). In: In Proc. 9th Russian–Korean Symp. Sci. and Technol., Novosib. State Univ., Novosibirsk, pp 547–550

Vasiliev VV, Lurie SA (1990a) On the problem of constructing non-classical theories of plates (in Russ.). Izv RAN MTT (2):158–167

Vasiliev VV, Lurie SA (1990b) To the problem of clarifying the theory of shallow shells (in Russ.). Izv RAN MTT (6):139–146

Vekua IN (1955) On a method of calculating of prismatic shells (in Russ.). In: Tr. Tbilis. matem. ins-ta im. A.M.Razmadze, Izd-vo Metsniereba, Tbilisi, vol 21, pp 191–259

Vekua IN (1964) The theory of thin and shallow shells of variable thickness (in Russ.). Novosibirsk

Vekua IN (1965) Theory of thin shallow shells of variable thickness (in Russ.). In: Tr. Tbilis. matem. ins-ta im. A.M.Razmadze, Izd-vo Metsniyereba, Tbilisi, vol 30, pp 1–104

Vekua IN (1970) Variational principles for constructing the theory of shells (in Russ.). Izd-vo Tbil. Un-ta, Tbilisi

Vekua IN (1972) On one direction of constructing the theory of shells (in Russ.). In: Mechanics in the USSR for 50 years, vol 3, Nauka, Moscow, pp 267–290

Vekua IN (1978) Fundamentals of tensor analysis and the theory of covariant (in Russ.). Nauka

Vekua IN (1982) Some common methods for constructing various variants of the theory of shells (in Russ.). Nauka, Moscow

Volchkov YM (2000) Finite elements with adjustment conditions on their edges (in Russ.). Dinamika sploshnoy sredi 116:175–180

Volchkov YM, Dergileva LA (1977) Solution of elastic layer problems by approximate equations and comparison with solutions of the theory of elasticity (in Russ.). Dinamika sploshnoy sredy 28:43–54

Volchkov YM, Dergileva LA (1999) Edge effects in the stress state of a thin elastic interlayer (in Russ.). Journal of Applied Mechanics and Technical Physics 40(2):354–359

Volchkov YM, Dergileva LA (2004) Equations of an elastic anisotropic layer. Journal of Applied Mechanics and Technical Physics 45(2):301–309

Volchkov YM, Dergileva LA (2007) Reducing three-dimensional elasticity problems to two-dimensional problems by approximating stresses and displacements by Legendre polynomials. Journal of Applied Mechanics and Technical Physics 48(3):450–459

Volchkov YM, Dergileva LA, Ivanov GV (1994) Numerical modeling of stress states in two-dimensional problems of elasticity by the layers method (in Russ.). Journal of Applied Mechanics and Technical Physics 35(6):936–941

Wunderlich W (1973) Vergleich verschiedener Approximationen der Theorie dünner Schalen (mit numerischen Ergebnissen). Allgemeine Schalentheorien, Techn Wiss Mitteilungen (73):3.1–3.24

Zhavoronok S (2014) A Vekua type linear theory of thick elastic shells. ZAMM-Journal of Applied Mathematics and Mechanics/Zeitschrift für Angewandte Mathematik und Mechanik 94(1/2):164–184

Zhavoronok SI (2017) On Hamiltonian formulations and conservation laws for plate theories of Vekua–Amosov type. International Journal for Computational Civil and Structural Engineering 13(4):82–95

Zhavoronok SI (2018) On the use of extended plate theories of Vekua–Amosov type for wave dispersion problems. International Journal for Computational Civil and Structural Engineering 14(1):36–48

Zhilin PA (1976) Mechanics of deformable directed surfaces. Int J Solids Structures 12:635–648

Zozulya VV (2017a) Couple stress theory of curved rods. 2-D, high order, Timoshenko's and Euler-Bernoulli models. Curved and Layered Structures 4(1):119–133

Zozulya VV (2017b) Micropolar curved rods. 2-D, high order, Timoshenko's and Euler-Bernoulli models. Curved and Layered Structures 4(1):104–118

Zozulya VV (2017c) Nonlocal theory of curved rods. 2-D, high order, Timoshenko's and Euler-Bernoulli models. Curved and Layered Structures 4(1):221–236

Zozulya VV, Saez A (2014) High-order theory for arched structures and its application for the study of the electrostatically actuated MEMS devices. Archive of Applied Mechanics 84(7):1037–1055

Zozulya VV, Saez A (2016) A high-order theory of a thermoelastic beams and its application to the MEMS/NEMS analysis and simulations. Archive of Applied Mechanics 86(7):1255–1272

Chapter 10
Application of Eigenvalue Problems Under the Study of Wave Velocity in Some Media

Mikhail Nikabadze and Armine Ulukhanyan

Abstract The wave velocities in some media under different types of anisotropy are estimated applying the eigenvalue problem of material tensors. In this connection, the canonical representations of material tensors, as well as kinematic and dynamic conditions on the strong discontinuity surface are given. Using them the problem of finding the wave velocities is reduced to the eigenvalue problem for the corresponding dispersion tensor. In addition, dispersion equations for determining the wave velocities are obtained. In particular, classical and micropolar materials with different symbols of anisotropy (structure) are considered and the wave velocities through the eigenvalues of material tensors are found.

Keywords: Wave velocity · Eigenvalues of material tensors · Anisotropy symbol · Micropolar medium · Tensor-block matrix

10.1 Kinematic and Dynamic Conditions on the Strong Discontinuity Surface in Micropolar Mechanics

Let us consider a moving regular surface with respect to an unbounded space whose equation in a fixed Cartesian coordinate system is given by

$$\psi(x_1, x_2, x_3, t) = 0. \tag{10.1}$$

M. Nikabadze
Lomonosov Moscow State University
Bauman Moscow State Technical University, Moscow, Russia
e-mail: nikabadze@mail.ru

A. Ulukhanyan
Bauman Moscow State Technical University, Moscow, Russia
e-mail: armine_msu@mail.ru

© Springer Nature Switzerland AG 2019
H. Altenbach et al. (eds.), *Higher Gradient Materials and Related Generalized Continua*, Advanced Structured Materials 120,
https://doi.org/10.1007/978-3-030-30406-5_10

201

The regularity (10.1) means the existence of the unit normal vector $\mathbf{n}(x_1, x_2, x_3, t)$ at each point of the surface at any time

$$\mathbf{n}(x_1, x_2, x_3, t) = \frac{\nabla_x \psi}{|\nabla_x \psi|}, \quad \nabla_x \psi = \mathbf{k}_i \partial_i \psi,$$

as well as continuous differentiability the required number of times. Although, for our purpose, the existence of the unit normal is sufficient, that is $\psi \subset C^1$. Here $\mathbf{k}_i$ is an orthonormal basis of the Cartesian coordinate system, and C^1 is the set of continuously differentiable functions.

Definition 10.1. If the vectors $\mathbf{u}$ and φ are continuous while passing through the surface (10.1), and the first derivatives $\partial_k \mathbf{u}$, $\partial_t \mathbf{u}$, $\partial_k \varphi$, $\partial_t \varphi$ are subjected to a discontinuity so that on each side of this surface they take different finite values, then the surface (10.1) is called a strong discontinuity surface or the wave of stress and couple stress.

Definition 10.2. If the vectors $\mathbf{u}$, $\partial_k \mathbf{u}$, $\partial_t \mathbf{u}$, φ, $\partial_k \varphi$ and $\partial_t \varphi$ are continuous while passing through the surface (10.1), and the second derivatives of $\mathbf{u}$ and φ with respect to x_k and t are subjected to a discontinuity so that on each side of this surface they take different finite values, then the surface (10.1) is called a weak discontinuity surface or the wave of acceleration.

It should be noted that in this paper we use the notation $\partial_t = \partial/\partial t$, $\partial_i = \partial/\partial x_i$. We also note that we will mainly deal with the strong discontinuity surface.

As it is known (Petrashen, 1978, 1980; Sagomonyan, 1985; Poruchikov, 1986), the displacement velocity c of an arbitrary point of the surface (10.1) in the direction of the normal of this surface at this point is expressed by

$$c = -\frac{1}{|\nabla_x \psi|} \partial_t \psi, \quad \nabla \psi = \mathbf{k}_i \partial_i \psi. \tag{10.2}$$

If the wavefront (the strong discontinuity surface) (10.1) moves in a medium having a velocity field $\mathbf{v}(x_1, x_2, x_3, t)$, then the wavefront velocity with respect to the medium particles is determined by

$$\theta = c - v_n, \quad v_n = \mathbf{n} \cdot \mathbf{v}. \tag{10.3}$$

10.1.1 Kinematic Conditions on the Strong Discontinuity Surface

These conditions can be obtained in the same way as it was done, for example, in Petrashen (1978, 1980); Sagomonyan (1985) for the classical case. In this case, we are considering two independent vectors $\mathbf{u}(x_1, x_2, x_3, t)$ and $\varphi(x_1, x_2, x_3, t)$ in contrast to the classical case where only a single displacement vector is considered. Let us write out the required relations without derivation. They have the form

$$c[\partial_i\mathbf{u}] + n_i[\partial_t\mathbf{u}] = 0, \quad c[\partial_i\boldsymbol{\varphi}] + n_i[\partial_t\boldsymbol{\varphi}] = 0, \tag{10.4}$$

where c is determined by the first formula of (10.2), n_i are the components of the unit normal $\mathbf{n}$ to the wave front, and $[\mathbf{w}] = \mathbf{w}^+ - \mathbf{w}^-$, where $\mathbf{w} = \mathbf{u}$ or $\mathbf{w} = \boldsymbol{\varphi}$, means a jump in the value of $\mathbf{w}$ relative to the wave front, $\mathbf{w}^-$ ($\mathbf{w}^+$) is the limiting value of $\mathbf{w}$ when the arbitrarily chosen point ahead (behind) of the wave front tend to the considered point on the wave front. Subsequently, the square brackets « [] » is called the operator of a jump. Note, that some authors, for example, Altenbach et al (2010); Eringen (1999), use double square brackets for operator of a jump.

10.1.2 The Mass and the Tensor of the Moment of Inertia Conservation Laws on the Wave Front

Applying the mass and the tensor of the moments of inertia conservation laws

$$\frac{d}{dt}\int_V \rho dV = 0, \quad \frac{d}{dt}\int_V \underset{\sim}{\mathbf{J}} dV = 0,$$

where ρ is the density of the material, $\underset{\sim}{\mathbf{J}}$ is the density of the inertia tensor (the special dynamic characteristic of the medium) of the medium particle (Kupradze, 1979; Nowacki, 1970; Eringen, 1999), to the marked elementary cylinder, and producing simple calculations similar to the classical case (Petrashen, 1978, 1980; Sagomonyan, 1985), we find the required laws in the following form

$$[\theta\rho] = 0, \quad [\theta\underset{\sim}{\mathbf{J}}] = 0, \tag{10.5}$$

where θ is determined by (10.3).

10.1.3 Dynamic Conditions on the Wave Front

These conditions can be easily obtained using the law of variation of momentum and the theorem of variation of angular momentum of the internal rotational motions of the medium particles in integral form

$$\frac{d}{dt}\int_V \rho\mathbf{v}dV = \int_V \rho\mathbf{F}dV + \int_\Sigma \mathbf{P}_{(n)}d\Sigma,$$
$$\frac{d}{dt}\int_V \underset{\sim}{\mathbf{J}}\cdot\boldsymbol{\omega}dV = \int_V (\rho\mathbf{m} + \underset{\approx}{\mathbf{C}}\overset{2}{\otimes}\underset{\sim}{\mathbf{P}})dV + \int_\Sigma \boldsymbol{\mu}_{(n)}d\Sigma, \tag{10.6}$$

where $\mathbf{F}$ is the mass force vector, $\mathbf{m}$ is the mass moment vector, $\mathbf{P}_{(n)}$ and $\boldsymbol{\mu}_{(n)}$ are the stress and the couple stress vectors on the elementary area with the unit normal vector $\mathbf{n}$, $\mathbf{C}$ is the Levi-Civita third rank pseudo-tensor, $\overset{2}{\otimes}$ denotes the double inner product, V is the volume occupied by the body, Σ is the body boundary, $\mathbf{v} = \dot{\mathbf{u}}$, $\boldsymbol{\omega} = \dot{\boldsymbol{\varphi}}$, where the dot above the letter indicates the time derivative. In fact, applying (10.6) like in the classical case (Petrashen, 1978, 1980; Sagomonyan, 1985) to the cylinder mentioned above and taking into account (10.5), we obtain the required conditions

$$\rho\theta[\mathbf{v}] = -[\mathbf{P}_{(n)}] = -\mathbf{n} \cdot [\mathbf{P}], \quad \theta\mathbf{J} \cdot [\boldsymbol{\omega}] = -[\boldsymbol{\mu}_{(n)}] = -\mathbf{n} \cdot [\boldsymbol{\mu}], \tag{10.7}$$

where introduced

$$\rho\theta = \rho^-\theta^- = \rho^+\theta^+, \quad \theta\mathbf{J} = \theta^-\mathbf{J}^- = \theta^+\mathbf{J}^+. \tag{10.8}$$

It should be noted that in order to get the relations (10.7) we were taken into account the Cauchy formulas $\mathbf{P}_{(n)} = \mathbf{n} \cdot \mathbf{P}$ and $\boldsymbol{\mu}_{(n)} = \mathbf{n} \cdot \boldsymbol{\mu}$, and they are valid for any medium. Note also that the conditions (10.7) and analogous to them for different media are derived in Eringen (1999) (see also Baskakov, 1991; Konchakova, 1998).

10.2 Equations for Determining the Wave Velocities in an Infinite Micropolar Solid

Having the kinetic (10.4) and dynamical (10.7) conditions on the wave front, it is not difficult to find an equation for determining the wave velocities in any infinite micropolar medium, including in an infinite micropolar solid. Let us consider a micropolar rigid body with the constitutive relations (Nikabadze, 2007, 2014; Nikabadze and Ulukhanyan, 2016)

$$\mathbf{P} = \mathbf{A} \overset{2}{\otimes} \nabla\mathbf{u} + \mathbf{B} \overset{2}{\otimes} \nabla\boldsymbol{\varphi} - \mathbf{A} \overset{2}{\otimes} \mathbf{C} \cdot \boldsymbol{\varphi} - \mathbf{b}\vartheta,$$

$$\boldsymbol{\mu} = \mathbf{C} \overset{2}{\otimes} \nabla\mathbf{u} + \mathbf{D} \overset{2}{\otimes} \nabla\boldsymbol{\varphi} - \mathbf{C} \overset{2}{\otimes} \mathbf{C} \cdot \boldsymbol{\varphi} - \boldsymbol{\beta}\vartheta, \tag{10.9}$$

where

$$\mathbf{b} = \mathbf{A} \overset{2}{\otimes} \mathbf{a} + \mathbf{B} \overset{2}{\otimes} \mathbf{d},$$

$$\boldsymbol{\beta} = \mathbf{C} \overset{2}{\otimes} \mathbf{a} + \mathbf{D} \overset{2}{\otimes} \mathbf{d}. \tag{10.10}$$

Here $\mathbf{C} = \mathbf{B}^T$, $\mathbf{A}$, $\mathbf{D}$ are the fourth-rank material tensors (elastic moduli tensor), $\mathbf{a}$ and $\mathbf{d}$ are the second rank tensors of heat expansion, $\vartheta = T - T_0$ is the temperature drop, $\mathbf{b}$ and $\boldsymbol{\beta}$ are the tensors of thermomechanical properties, superscript "T" means transposition.

Note that some authors (see, for example, Eringen, 1999; Nowacki, 1970) believe that $\mathbf{B}$ is the asymmetric tensor, while other authors prove that $\mathbf{B}$ is the symmetric tensor (see, for example, Baskakov et al, 2001). In the latter case $\underset{\approx}{\mathbf{C}} = \underset{\approx}{\mathbf{B}}$, which according to the authors, is more justified. Anyway, we derive the relations in case $\underset{\approx}{\mathbf{C}} = \underset{\approx}{\mathbf{B}}^T$, since using them it is easy to obtain the corresponding relations for the case $\underset{\approx}{\mathbf{C}} = \underset{\approx}{\mathbf{B}}$.

Further, for the sake of a clear representation of the proposed methodology, we simplify the problem by assuming that the studied process is isothermal, i.e. we assume that $\vartheta = 0$. Then, assuming also that the material tensors are not subjected to the discontinuity when passing through the wave front and applying the operator of a jump to (10.9) we have

$$[\underline{\mathbf{P}}] = \underset{\approx}{\mathbf{A}} \overset{2}{\otimes} [\nabla \mathbf{u}] + \underset{\approx}{\mathbf{B}} \overset{2}{\otimes} [\nabla \varphi], \quad [\underline{\mu}] = \underset{\approx}{\mathbf{C}} \overset{2}{\otimes} [\nabla \mathbf{u}] + \underset{\approx}{\mathbf{D}} \overset{2}{\otimes} [\nabla \varphi]. \tag{10.11}$$

It is not difficult to see that multiplying (10.4) by $\mathbf{k}_i$ and summing over i, we obtain

$$[\nabla \mathbf{u}] = -\frac{1}{c}\mathbf{n}[\mathbf{v}], \quad [\nabla \varphi] = -\frac{1}{c}\mathbf{n}[\boldsymbol{\omega}]. \tag{10.12}$$

Taking into account (10.12), from (10.11) we get

$$[\underline{\mathbf{P}}] = -\frac{1}{c}(\underset{\approx}{\mathbf{A}} \overset{2}{\otimes} \mathbf{n}[\mathbf{v}] + \underset{\approx}{\mathbf{B}} \overset{2}{\otimes} \mathbf{n}[\boldsymbol{\omega}]), \quad [\underline{\mu}] = -\frac{1}{c}(\underset{\approx}{\mathbf{C}} \overset{2}{\otimes} \mathbf{n}[\mathbf{v}] + \underset{\approx}{\mathbf{D}} \overset{2}{\otimes} \mathbf{n}[\boldsymbol{\omega}]),$$

with the help of which from the dynamic conditions (10.7) we get the relations

$$(\mathbf{n} \cdot \underset{\approx}{\mathbf{A}} \overset{2}{\otimes} \mathbf{n}\underline{\mathbf{E}}) \cdot [\mathbf{v}] + (\mathbf{n} \cdot \underset{\approx}{\mathbf{B}} \overset{2}{\otimes} \mathbf{n}\underline{\mathbf{E}}) \cdot [\boldsymbol{\omega}] = c\rho\theta\underline{\mathbf{E}} \cdot [\mathbf{v}],$$
$$(\mathbf{n} \cdot \underset{\approx}{\mathbf{C}} \overset{2}{\otimes} \mathbf{n}\underline{\mathbf{E}}) \cdot [\mathbf{v}] + (\mathbf{n} \cdot \underset{\approx}{\mathbf{D}} \overset{2}{\otimes} \mathbf{n}\underline{\mathbf{E}}) \cdot [\boldsymbol{\omega}] = c\theta\underline{\mathbf{J}} \cdot [\boldsymbol{\omega}], \tag{10.13}$$

where $\underline{\mathbf{E}}$ is the second rank unit tensor. We note that, according to the mass conservation law and the law of conservation of the inertia tensor (10.5), $\rho\theta$ and $\underline{\mathbf{J}}\theta$ in (10.13) can be replaced by ρ^+ and $\underline{\mathbf{J}}^+$, respectively, since ahead of the wave front ρ^+, $\underline{\mathbf{J}}^+$, and v_n^+ may be considered known, but we will not dwell on this.

Introducing

$$\mathbf{A} = (1/\rho)\mathbf{n} \cdot \underset{\approx}{\mathbf{A}} \overset{2}{\otimes} \mathbf{n}\underline{\mathbf{E}}, \quad \mathbf{B} = (1/\rho)\mathbf{n} \cdot \underset{\approx}{\mathbf{B}} \overset{2}{\otimes} \mathbf{n}\underline{\mathbf{E}},$$
$$\underline{\mathbf{C}} = \mathbf{J}^{-1} \cdot (\mathbf{n} \cdot \underset{\approx}{\mathbf{C}} \overset{2}{\otimes} \mathbf{n}\underline{\mathbf{E}}), \quad \mathbf{D} = \mathbf{J}^{-1} \cdot (\mathbf{n} \cdot \underset{\approx}{\mathbf{D}} \overset{2}{\otimes} \mathbf{n}\underline{\mathbf{E}}), \tag{10.14}$$

with

$$\mathbb{M} = \begin{pmatrix} \mathbf{A} & \mathbf{B} \\ \underline{\mathbf{C}} & \mathbf{D} \end{pmatrix}, \quad [\mathbb{V}] = \begin{pmatrix} [\mathbf{v}] \\ [\boldsymbol{\omega}] \end{pmatrix}, \tag{10.15}$$

where $\mathbb{M}$ is called the tensor-block matrix (TBM), $\mathbb{V}$ is the vector column of the linear and angular velocities vectors, and $[\mathbb{V}]$ is the jump of this vector column, (10.13) can be represented as

$$\begin{pmatrix} \mathbf{A} & \mathbf{B} \\ \underline{\mathbf{C}} & \mathbf{D} \end{pmatrix} \cdot \begin{pmatrix} [\mathbf{v}] \\ [\boldsymbol{\omega}] \end{pmatrix} = \lambda \begin{pmatrix} [\mathbf{v}] \\ [\boldsymbol{\omega}] \end{pmatrix}, \quad \lambda = c\theta \tag{10.16}$$

or shortly

$$\underset{\sim}{\mathbb{M}} \cdot [\mathbb{V}] = \lambda[\mathbb{V}]. \tag{10.17}$$

The relations (10.17) (see also (10.16)) represent a homogeneous system of six algebraic equations with respect to six unknowns (two vectors $[\mathbf{v}]$ and $[\boldsymbol{\omega}]$)) having a nontrivial solution. In order to have a non-trivial solution it is necessary and sufficient that its determinant be zero. Since the system is sixth-order and due to $\theta^+ = c - v_n^+$, from the equality the determinant to zero we obtain the algebraic equation of the 6th degree with respect to c^2, which is the required dispersion equation for determination the wave velocities and their number in a given direction in an unbounded anisotropic micropolar body.

So, we got the eigenvalue problem for TBM in the form of (10.17). It is seen that $\lambda = c\theta$ is an eigenvalue, and $[\mathbb{V}]$ is the corresponding eigen-jump of the vector column. It is easy to see that, according to (10.17), the dispersion equation mentioned in the previous paragraph (the characteristic equation for $\underset{\sim}{\mathbb{M}}$) can be written in the form

$$\det(\underset{\sim}{\mathbb{M}} - \lambda\underset{\sim}{\mathbb{E}}) = 0, \tag{10.18}$$

where $\underset{\sim}{\mathbb{E}}$ is the second-rank unit TBM. It is easy to see that (10.18) can be written as

$$\lambda^6 - I_1(\underset{\sim}{\mathbb{M}})\lambda^5 + I_2(\underset{\sim}{\mathbb{M}})\lambda^4 - I_3(\underset{\sim}{\mathbb{M}})\lambda^3 + I_4(\underset{\sim}{\mathbb{M}})\lambda^2 - I_5(\underset{\sim}{\mathbb{M}})\lambda + I_6(\underset{\sim}{\mathbb{M}}) = 0,$$
$$I_6(\underset{\sim}{\mathbb{M}}) = \det(\underset{\sim}{\mathbb{M}}). \tag{10.19}$$

It is seen that the dispersion equation (10.19) is an algebraic equation of the sixth-degree and must have six roots (eigenvalues), counting each root as many times as its multiplicity. Each multiple root determines the square of the velocity of one wave. Hence, in an arbitrary anisotropic infinite micropolar medium, in general, no more than six waves can arise in each direction. Note that based on the dispersion equation (10.19), it is not difficult to find the number of waves arising in a micropolar elastic medium under various anisotropy. In this connection, it is sufficient to find the invariants of the TBM that appear in (10.19), and then solve the equation (10.19). The invariants are easily found through the degrees of the first invariants $\underset{\sim}{\mathbb{M}}$. In fact, we have (Nikabadze, 2014, 2017d; Nikabadze and Ulukhanyan, 2016)

$$S_k = I_k(\underset{\sim}{\mathbb{M}}) = \frac{1}{k!} \begin{vmatrix} s_1 & 1 & \cdots & 0 & 0 \\ s_2 & s_1 & \cdots & 0 & 0 \\ \cdots & \cdots & \cdots\cdots & \cdots \\ s_{k-1} & s_{k-2} & \cdots & s_1 & k-1 \\ s_k & s_{k-1} & \cdots & s_2 & s_1 \end{vmatrix}, \quad k = \overline{1,6}, \tag{10.20}$$

$$s_k = I_1(\underset{\sim}{\mathbb{M}}^k), \quad k = \overline{1,6}, \quad \underset{\sim}{\mathbb{M}}^k = \overbrace{\underset{\sim}{\mathbb{M}} \cdot \underset{\sim}{\mathbb{M}} \cdot \ldots \cdot \underset{\sim}{\mathbb{M}}}^{k}.$$

Here $I_k(\underset{\sim}{\mathbb{M}})$, $k = \overline{1,6}$, mean the invariants of TBM $\underset{\sim}{\mathbb{M}}$. In addition, the inverse relations to (10.20) are represented in the form (Nikabadze, 2014, 2017d; Nikabadze and Ulukhanyan, 2016)

$$s_k = I_1(\underset{\sim}{\mathbb{M}}^k) = \begin{vmatrix} S_1 & 1 & 0 & \cdots & 0 \\ 2S_2 & S_1 & 1 & \cdots & 0 \\ \cdots & \cdots & \cdots & \cdots & \cdots \\ kS_k & S_{k-1} & S_{k-2} & \cdots & S_1 \end{vmatrix}, \quad k = \overline{1,6}.$$

If the material has a center of symmetry, then $\underset{\approx}{\mathbf{B}} = \underset{\approx}{\mathbf{C}}^T = \underset{\approx}{\mathbf{0}}$. In this case, the TBM $\underset{\sim}{\mathbb{M}}$ becomes a tensor-block-diagonal matrix. For such a matrix, the characteristic equation and eigenvalues (wave velocities) are easily found. In fact, we will have

$$\det(\underset{\sim}{\mathbb{M}} - \lambda\underset{\sim}{\mathbb{E}}) = \det\begin{pmatrix} \underset{\approx}{\mathbf{A}} - \lambda\underset{\approx}{\mathbf{E}} & \underset{\approx}{\mathbf{0}} \\ \underset{\approx}{\mathbf{0}} & \underset{\approx}{\mathbf{D}} - \lambda\underset{\approx}{\mathbf{E}} \end{pmatrix} = $$
$$= \det(\underset{\approx}{\mathbf{A}} - \lambda\underset{\approx}{\mathbf{E}})\det(\underset{\approx}{\mathbf{D}} - \lambda\underset{\approx}{\mathbf{E}}) = 0, \tag{10.21}$$

where $\underset{\sim}{\mathbf{0}}$ is the second-rank zero tensor. We can see that (10.21) is equivalent to the following two equations

$$\lambda^3 - I_1(\underset{\approx}{\mathbf{A}})\lambda^2 + I_2(\underset{\approx}{\mathbf{A}})\lambda - I_3(\underset{\approx}{\mathbf{A}}) = 0,$$
$$\lambda^3 - I_1(\underset{\approx}{\mathbf{D}})\lambda^2 + I_2(\underset{\approx}{\mathbf{D}})\lambda - I_3(\underset{\approx}{\mathbf{D}}) = 0. \tag{10.22}$$

So, if the material has a center of symmetry, then in this case for determining the wave velocities we have two cubic equations (10.22), which are easily solved. Generally, based on (10.22) we can conclude that in an arbitrary anisotropic micropolar elastic medium with a center of symmetry no more than six waves appear in each direction.

It should be noted that in the case of a classical medium we will have one cubic equation similar, for example, to the first Eq. (10.22) given that $\underset{\approx}{\mathbf{A}}$ is determined by the first relation (10.14). Consequently, in an arbitrary anisotropic classical elastic medium no more than three waves appear in each direction. We also note that in case of a micropolar medium, it is appropriate to call $\underset{\sim}{\mathbb{M}}$ by the dispersion TBM, in case of the classical medium $\underset{\approx}{\mathbf{A}}$, we can call it the dispersion tensor, since their characteristic equations are the dispersion equations.

Having dispersion equations for the micropolar (classical) medium, it is not difficult to find the wave velocities in the considered media under any anisotropy. Of course, the required velocities can be found both in the traditional representation of material objects (TBM, elastic modulus tensor) and with the help of their eigenvalues, solving in advance the eigenvalue problems of the corresponding object. We note that the eigenvalue problems for the tensor and TBM of any even rank are solved in Nikabadze (2014, 2017d); Nikabadze and Ulukhanyan (2016). We also note that in Nikabadze (2014, 2017d); Nikabadze and Ulukhanyan (2016) the concept of an anisotropy (structure) of a material is introduced and the classification of micropolar and classical media is given. Consequently, for each material included in these classifications, based on the corresponding dispersion equation for the micropolar (classical) medium obtained above, a dispersion equation for the micropolar (classical) medium can be got, and then we can determine the number of waves and their velocities.

Further, before we consider some particular cases of materials and find the wave velocities inside these materials using the eigenvalues of the corresponding tensor objects, we introduce the definition of the anisotropy (structure) symbol of the material (Nikabadze, 2014, 2017d; Nikabadze and Ulukhanyan, 2016).

Definition 10.3. Symbol $\{\alpha_1, \alpha_2, \dots, \alpha_k\}$, where k is the number of different eigenvalues of the TBM (tensor), and α_i is the multiplicity of the eigenvalue λ_i ($i = 1, 2, \dots, k$), is called the anisotropy symbol (structure) of the TBM (tensor).

It should be noted that the anisotropy symbol (structure) is defined for a TBM (tensor) of even rank. In this case, the anisotropy symbol (structure) of the material TBM (material tensor) is also called the anisotropy symbol (structure) of the material. Next, let us consider some particular cases of materials and find the wave velocities inside these materials using the eigenvalues of the corresponding tensor objects.

10.3 Classical Materials with Anisotropy Symbols {1,5} and {5,1}

For the materials with the anisotropy symbols $\{1, 5\}$ and $\{5, 1\}$ the elastic modulus tensor can be presented as (Nikabadze, 2014, 2017d; Nikabadze and Ulukhanyan, 2016)

$$\underset{\approx}{\mathbf{A}} = (\lambda_1 - \lambda_2)\underline{\mathbf{a}}_1\underline{\mathbf{a}}_1 + \lambda_2\underset{\approx}{\mathbf{E}}, \quad \underset{\approx}{\mathbf{B}} = \mu_1\underset{\approx}{\mathbf{E}} - (\mu_1 - \mu_6)\underline{\mathbf{b}}_6\underline{\mathbf{b}}_6. \tag{10.23}$$

Here

$$\underset{\approx}{\mathbf{E}} = \sum_{k=1}^{6} \underline{\mathbf{a}}_k\underline{\mathbf{a}}_k = \sum_{k=1}^{6} \underline{\mathbf{b}}_k\underline{\mathbf{b}}_k = (1/2)(\underset{\approx}{\mathbf{C}}_{(2)} + \underset{\approx}{\mathbf{C}}_{(3)}) \tag{10.24}$$

is the fourth-rank unit tensor, $\underline{\mathbf{a}}_k$, $k = \overline{1,6}$ and $\underline{\mathbf{b}}_k$, $k = \overline{1,6}$, are the complete orthonormal systems of eigentensors for tensors $\underset{\approx}{\mathbf{A}}$ and $\underset{\approx}{\mathbf{B}}$, respectively, $\underset{\approx}{\mathbf{C}}_{(2)}$, $\underset{\approx}{\mathbf{C}}_{(3)}$ are fourth-rank isotropic tensors, λ_1, λ_2 and μ_1, μ_6 are the eigenvalues, $\underline{\mathbf{a}}_1 = \underline{\mathbf{a}}_1^T$ and $\underline{\mathbf{b}}_6 = \underline{\mathbf{b}}_6^T$ are the eigentensors corresponding to eigenvalues λ_1 and μ_6, respectively.

It should be noted that if $\underline{\mathbf{a}}_1$ ($\underline{\mathbf{b}}_6$) is the symmetric tensor and satisfies the orthonormality condition

$$\underline{\mathbf{a}}_1 \overset{2}{\otimes} \underline{\mathbf{a}}_1 = 1 \quad (\underline{\mathbf{b}}_6 \overset{2}{\otimes} \underline{\mathbf{b}}_6 = 1), \tag{10.25}$$

then in the basis constructed using a basis in an arbitrary coordinate system, this tensor is characterized by five components, and in the main basis for $\underline{\mathbf{a}}_1$ ($\underline{\mathbf{b}}_6$) it is characterized by two components. Hence, the tensor $\underset{\approx}{\mathbf{A}}$ ($\underset{\approx}{\mathbf{B}}$) is characterized by seven parameters in the first basis, namely, two eigenvalues and five components of the tensor $\underline{\mathbf{a}}_1$ ($\underline{\mathbf{b}}_6$), and in the basis formed by the canonical basis for $\underline{\mathbf{a}}_1$ ($\underline{\mathbf{b}}_6$) it is characterized by four parameters, namely two eigenvalues and two components of the tensor $\underline{\mathbf{a}}_1$ ($\underline{\mathbf{b}}_6$). We also note that there is a statement that is not difficult to prove, and therefore we shall not dwell on it.

Statement. Let $\underset{\sim}{a} = \underset{\sim}{a}^T$. Then $\underset{\sim}{a}\underset{\sim}{a}$ is the fourth-rang isotropic tensor if and only if $\underset{\sim}{a}$ is the spherical tensor. At the same time if $\underset{\sim}{a} \overset{2}{\otimes} \underset{\sim}{a} = 1$ then $\underset{\sim}{a} = [(\pm\sqrt{3})/3]\underset{\sim}{E}$, and $\underset{\sim}{a}\underset{\sim}{a} = (1/3)\underset{\approx}{C}_{(1)}$. Here $\underset{\approx}{C}_{(1)} = \underset{\sim}{E}\underset{\sim}{E}$ is the first of the three fourth rank isotropic tensors.

According to this statement we can conclude that tensors $\underset{\approx}{A}$ and $\underset{\approx}{B}$ (see (10.23)) will be isotropic if and only if

$$\underset{\sim}{a}_1\underset{\sim}{a}_1 = \underset{\sim}{b}_6\underset{\sim}{b}_6 = (1/3)\underset{\approx}{C}_{(1)}. \qquad (10.26)$$

Obviously, in this case they can be written as

$$\begin{aligned}\underset{\approx}{A} &= (1/3)(\lambda_1 - \lambda_2)\underset{\approx}{C}_{(1)} + \lambda_2\underset{\sim}{E} = \lambda\underset{\approx}{C}_{(1)} + 2\mu\underset{\sim}{E}, \\ \underset{\approx}{B} &= \mu_1\underset{\sim}{E} - (1/3)(\mu_1 - \mu_6)\underset{\approx}{C}_{(1)}.\end{aligned} \qquad (10.27)$$

It should be noted that the isotropic material whose properties is characterized by the $\underset{\approx}{A}$($\underset{\approx}{B}$) (see (10.27)) has a positive (negative) Poisson's ratio (Nikabadze, 2017d; Nikabadze and Ulukhanyan, 2016). We also note that the tensor $\underset{\approx}{A}$ (see the first relation (10.27)) was also presented in the traditional form (via the Lamé parameters $\lambda = 1/3(\lambda_1 - \lambda_2)$, $\mu = 1/2\lambda_2$).

We can see that $\underset{\approx}{A}$ and $\underset{\approx}{B}$ (see (10.23)), as well as the corresponding tensors $\underset{\sim}{A}$ and $\underset{\sim}{B}$ have similar structure and are presented in the form

$$\begin{aligned}\underset{\sim}{A} &= (1/\rho)\mathbf{n} \cdot \underset{\approx}{A} \cdot \mathbf{n} = a(\underset{\sim}{E} + \mathbf{nn}) + b\underset{\sim}{u}, \\ \underset{\sim}{B} &= (1/\rho)\mathbf{n} \cdot \underset{\approx}{B} \cdot \mathbf{n} = f(\underset{\sim}{E} + \mathbf{nn}) + g\underset{\sim}{v},\end{aligned} \qquad (10.28)$$

$$\begin{aligned}&a = \lambda_2/(2\rho) = \mu/\rho > 0, \quad b = (\lambda_1 - \lambda_2)/\rho = (3\lambda)/\rho > 0, \\ &f = \mu_1/(2\rho) > 0, \quad g = (\mu_6 - \mu_1)/\rho < 0, \\ &\underset{\sim}{u} = \mathbf{n} \cdot \underset{\sim}{a}_1\mathbf{n} \cdot \underset{\sim}{a}_1, \quad \underset{\sim}{v} = \mathbf{n} \cdot \underset{\sim}{b}_1\mathbf{n} \cdot \underset{\sim}{b}_1.\end{aligned} \qquad (10.29)$$

Taking $\underset{\sim}{a}_1 = \underset{\sim}{b}_6 = (\pm\sqrt{3}/3)\underset{\sim}{E}$ in (10.28) we get the dispersion tensors $\underset{\sim}{A}$ and $\underset{\sim}{B}$, which correspond to the isotropic tensors (10.27), in the form

$$\underset{\sim}{A} = a\underset{\sim}{E} + b_1\mathbf{nn}, \quad \underset{\sim}{B} = f\underset{\sim}{E} + g_1\mathbf{nn}, \qquad (10.30)$$

$$\begin{aligned}&a = \lambda_2/(2\rho) = \mu/\rho > 0, \quad b_1 = (2\lambda_1 + \lambda_2)/(6\rho) = (\lambda + \mu)/\rho > 0, \\ &f = \mu_1/(2\rho) > 0, \quad g_1 = (\mu_1 + 2\mu_6)/(6\rho) > 0, \\ &\underset{\sim}{u} = \mathbf{n} \cdot \underset{\sim}{a}_1\mathbf{n} \cdot \underset{\sim}{a}_1, \quad \underset{\sim}{v} = \mathbf{n} \cdot \underset{\sim}{b}_1\mathbf{n} \cdot \underset{\sim}{b}_1.\end{aligned} \qquad (10.31)$$

It is seen that the tensors (10.30), which are special cases of the tensors (10.28), also have the same structure. Therefore, let us consider in detail, for example, the first tensor (10.28). Then the corresponding relations for the other tensors (10.28) and (10.30) we can obtain from the derived ones by redefining the coefficients and tensors.

Thus, let us find the wave velocities in material $\{1,5\}$, which dispersion equation has the form (see the first relation (10.28)) and the characteristic equation has the form of the first equation (10.22). For this, first, we will find out $I_1(\mathbf{A}^k)$, $k = 1, 2, 3,$

then using (10.20), which is valid also for the tensor of corresponding rank, we will find $I_k(\underset{\approx}{A})$, $k = 1, 2, 3$. After some calculations we get

$$I_1(\underset{\sim}{A}) = 4a + bI_1(\underset{\sim}{u}), \quad I_1(\underset{\sim}{A}^2) = 6a^2 + 2ab[I_1(\underset{\sim}{u}) + I^2] + b^2 I_1^2(\underset{\sim}{u}),$$

$$I_1(\underset{\sim}{A}^3) = 10a^3 + 3a^2 b[I_1(\underset{\sim}{u}) + 3I^2] + 3ab^2 I_1(\underset{\sim}{u})[I_1(\underset{\sim}{u}) + I^2] + b^3 I_1^3(\underset{\sim}{u}),$$

$$I_2(\underset{\sim}{A}) = 5a^2 + 3abI_1(\underset{\sim}{u}) - abI^2, \quad I_3(\underset{\sim}{A}) = 2a^3 + 2a^2 bI_1(\underset{\sim}{u}) - a^2 bI^2,$$

$$I_1(\underset{\sim}{u}) = \mathbf{n} \cdot \underset{\sim}{\mathbf{a}}_1^2 \cdot \mathbf{n}, \quad I = \underset{\sim}{\mathbf{a}}_1 \overset{2}{\otimes} \mathbf{nn}.$$

$$(10.32)$$

Next, according to the invariants (10.32), constituting the characteristic equation (see the first relation (10.22)) for $\underset{\sim}{A}$ and solving it we get the next relations for the roots

$$\eta_1 = a, \quad \eta_{2,3} = \frac{1}{2}\left[3a + bI_1(\underset{\sim}{u}) \pm \sqrt{[a - bI_1(\underset{\sim}{u})]^2 + 4abI^2}\right]. \quad (10.33)$$

It is not difficult to prove the following theorem.

Theorem 10.1. *The dispersion tensor (or the dispersion TBM) is positive definite.*

According to this theorem, the eigenvalues (the roots of the characteristic equation) of the dispersion tensor and the dispersion TBM are positive. So, it is easy to prove that the eigenvalues (10.33) obtained above are positive, too. Knowing the roots of the characteristic equation of $\underset{\sim}{A}$ (see (10.33)), it is easy to find the wave velocities in an initially stationary medium. In fact, we will have

$$c_1 = \sqrt{\eta_1}, \quad c_2 = \sqrt{\eta_2}, \quad c_3 = \sqrt{\eta_3}. \quad (10.34)$$

Thus, in the initially stationary medium with the structure symbol $\{1,5\}$, using the formulas (10.34) we can find the wave velocities in an arbitrary direction. In general, their number is not more than three. It should be noted that if the medium does not stationary then to determine the wave velocities we will have the formulas

$$c_1(c_1 - v_n^+) = \eta_1, \quad c_2(c_2 - v_n^+) = \eta_2, \quad c_3(c_3 - v_n^+) = \eta_3. \quad (10.35)$$

It is seen that each relation from (10.35) is a square equation with respect to the wave velocity. Solving them we obtain explicit expressions for the wave velocities in the initially nonstationary medium with the structure symbol $\{1,5\}$. It can be assumed that the number of waves arising in such a medium in an arbitrary direction is not more than six, but not less than three. It is not difficult to study this problem. So we will not dwell on these.

Next, let us consider the first dispersion tensor (10.30) corresponding to the first tensor (10.27), characterizing the properties of the isotropic material and being a particular case of the first tensor (10.28). In this case, the relations similar to (10.32) are represented in the form

$$I_1(\underset{\sim}{\mathbf{A}}) = 3a + b_1, \quad I_1(\underset{\sim}{\mathbf{A}}^2) = 2a^2 + (a + b_1)^2, \quad I_1(\underset{\sim}{\mathbf{A}}^3) = 2a^3 + (a + b_1)^3,$$
$$I_2(\underset{\sim}{\mathbf{A}}) = a(3a + 2b_1), \quad I_3(\underset{\sim}{\mathbf{A}}) = a^2(a + b_1).$$

$$(10.36)$$

Taking into account the invariants (10.36), from the first equation (10.22) we get the characteristic equation for the studied dispersion tensor $\underset{\sim}{\mathbf{A}}$. Solving it we get

$$\eta_1 = \eta_2 = a = \frac{\lambda_2}{2\rho} = \frac{\mu}{\rho}, \quad \eta_3 = a + b_1 = \frac{\lambda_1 + 2\lambda_2}{3\rho} = \frac{\lambda + 2\mu}{\rho}. \quad (10.37)$$

According to (10.37) the wave velocities in the infinite stationary isotropic elastic medium are defined by

$$c_1 = \sqrt{\eta_1} = \sqrt{\frac{\lambda_2}{2\rho}} = \sqrt{\frac{\mu}{\rho}}, \quad c_2 = \sqrt{\eta_3} = \sqrt{\frac{\lambda_1 + 2\lambda_2}{3\rho}} = \sqrt{\frac{\lambda + 2\mu}{\rho}}. \quad (10.38)$$

It can be seen that the wave velocities are expressed both in terms of the eigenvalues and the Lamé parameters. In the considered case the relations similar to (10.35) can be written out easily, so we will not dwell on them.

The first tensor (10.30) shows that any vector perpendicular to $\mathbf{n}$ and located in the tangent plane to the wave surface, as well as $\mathbf{n}$, are the eigenvectors of the dispersion tensor $\underset{\sim}{\mathbf{A}}$. Consequently, the system of vectors $(\mathbf{s}, \mathbf{l}, \mathbf{n})$, where $\mathbf{s}$ and $\mathbf{l}$ are mutually perpendicular unit tangent vectors to the wave surface, is a complete orthonormal system of eigenvectors of the considered dispersion tensor $\underset{\sim}{\mathbf{A}}$ (see the first tensor (10.30)). Of course, the eigenvectors of the tensor $\underset{\sim}{\mathbf{A}}$ can be found by solving the system of equations corresponding to these vectors, but in this case, they can be easily identified. Consequently, the canonical representation of the dispersion tensor $\underset{\sim}{\mathbf{A}}$ has the form

$$\underset{\sim}{\mathbf{A}} = \eta_1(\mathbf{ss} + \mathbf{ll}) + \eta_3 \mathbf{nn} = \eta_1 \underset{\sim}{\mathbf{I}} + \eta_3 \mathbf{nn}, \quad \underset{\sim}{\mathbf{I}} = \mathbf{ss} + \mathbf{ll}. \quad (10.39)$$

It is easy to see that according to (10.38) from (10.39) for the initial stationary medium we get

$$\underset{\sim}{\mathbf{A}} = c_1^2(\mathbf{ss} + \mathbf{ll}) + c_2^2 \mathbf{nn} = c_1^2 \underset{\sim}{\mathbf{I}} + c_2^2 \mathbf{nn} = \mathbf{v}_s \mathbf{v}_s + \mathbf{v}_l \mathbf{v}_l + \mathbf{v}_n \mathbf{v}_n, \quad (10.40)$$

and also we will have

$$\underset{\sim}{\mathbf{V}} = \sqrt{\underset{\sim}{\mathbf{A}}} = c_1(\mathbf{ss} + \mathbf{ll}) + c_2 \mathbf{nn} = c_1 \underset{\sim}{\mathbf{I}} + c_2 \mathbf{nn} = \mathbf{v}_s \mathbf{s} + \mathbf{v}_l \mathbf{l} + \mathbf{v}_n \mathbf{n}$$
$$= \mathbf{s} \mathbf{v}_s + \mathbf{l} \mathbf{v}_l + \mathbf{n} \mathbf{v}_n, \quad (10.41)$$

where $\mathbf{v}_s = c_1 \mathbf{s}$, $\mathbf{v}_l = c_1 \mathbf{l}$ and $\mathbf{v}_n = c_2 \mathbf{n}$. From (10.39), (10.40) and (10.41) can be seen that the tensors $\underset{\sim}{\mathbf{A}}$ and $\underset{\sim}{\mathbf{V}}$ are represented as the sum of two or three ortogonal tensors. Besides, from (10.38) it is seen that the wave velocities are the eigenvalues of $\underset{\sim}{\mathbf{V}}$, and the eigenvectors coincide with the eigenvectors of $\underset{\sim}{\mathbf{A}}$. In this connection, $\underset{\sim}{\mathbf{V}}$ can be called the velocity tensors.

We also can see that the vector equation for determining the eigentensors for $\underset{\approx}{\mathbf{A}}$ (see the first relation (10.30)) can be written in the form

$$\{(a - \eta)\underset{\sim}{\mathbf{I}} + (b_1 + a - \eta)\mathbf{nn}\} \cdot [\mathbf{v}] = 0. \tag{10.42}$$

Consequently, for the arbitrary medium motion $[\mathbf{v}]$ can be represented as

$$[\mathbf{v}] = [\mathbf{v}_\tau] + [\mathbf{v}_n], \quad \mathbf{v}_\tau = v_\tau \boldsymbol{\tau}, \quad \mathbf{v}_n = v_n \mathbf{n}, \quad \boldsymbol{\tau} \perp \mathbf{n}. \tag{10.43}$$

Taking into account (10.43) from (10.42) we have

$$\{(a - \eta)\underset{\sim}{\mathbf{I}}\} \cdot [\mathbf{v}_\tau] + \{(b_1 + a - \eta)\mathbf{nn}\} \cdot [\mathbf{v}_n] = 0. \tag{10.44}$$

It is valid the theorem, which we can prove using the kinematic conditions (see the first relation (10.4)).

Theorem 10.2. $\mathrm{rot}\,\mathbf{v} = \nabla \times \mathbf{v} = 0$ ($\mathrm{div}\,\mathbf{v} = \nabla \cdot \mathbf{v} = 0$) *if and only if* $[\mathbf{v}] \parallel \mathbf{n}$ ($[\mathbf{v}] \perp \mathbf{n}$).

Let us also note that if $[\mathbf{v}] \parallel \mathbf{n}$ ($[\mathbf{v}] \perp \mathbf{n}$) than the wave is called longitudinal (transverse) wave. Now, if the motion of the medium is such that $[\mathbf{v}] \parallel \mathbf{n}$ ($[\mathbf{v}] \perp \mathbf{n}$), that is $\mathbf{v} = \mathbf{v}_n = v\mathbf{n}$, ($\mathbf{v} = \mathbf{v}_\tau = v\boldsymbol{\tau}$), then from (10.44) we get

$$\{(b_1 + a - \eta)\mathbf{nn}\} \cdot [\mathbf{v}_n] = 0 \quad (\{(a - \eta)\underset{\sim}{\mathbf{I}}\} \cdot [\mathbf{v}_\tau] = 0) \tag{10.45}$$

and from here we can easily get the velocities of the longitudinal (transverse) waves. They are (10.38) (see also (10.35) (10.37)).

Next, before we discuss the materials of different structure, let us note that we use four-index and two-index views (Nikabadze, 2014, 2017d; Nikabadze and Ulukhanyan, 2016) for the fourth-rank tensor

$$\begin{aligned}
&\underset{\approx}{\mathbf{A}} = A_{ijkl}\mathbf{e}_i\mathbf{e}_j\mathbf{e}_k\mathbf{e}_l = \sum_{m=1}^{9}\sum_{n=1}^{9} A_{mn}\underset{\sim}{\mathbf{e}}_m\underset{\sim}{\mathbf{e}}_n = A_{mn}\underset{\sim}{\mathbf{e}}_m\underset{\sim}{\mathbf{e}}_n, \\
&\mathbf{e}_i \cdot \mathbf{e}_j = \delta_{ij}, \quad i,j = 1,2,3; \quad \underset{\sim}{\mathbf{e}}_1 = \mathbf{e}_1\mathbf{e}_1, \quad \underset{\sim}{\mathbf{e}}_2 = \mathbf{e}_2\mathbf{e}_2, \quad \underset{\sim}{\mathbf{e}}_3 = \mathbf{e}_3\mathbf{e}_3, \\
&\underset{\sim}{\mathbf{e}}_4 = \frac{1}{\sqrt{2}}(\mathbf{e}_1\mathbf{e}_2 + \mathbf{e}_1\mathbf{e}_2), \quad \underset{\sim}{\mathbf{e}}_5 = \frac{1}{\sqrt{2}}(\mathbf{e}_2\mathbf{e}_3 + \mathbf{e}_3\mathbf{e}_2), \\
&\underset{\sim}{\mathbf{e}}_6 = \frac{1}{\sqrt{2}}(\mathbf{e}_3\mathbf{e}_1 + \mathbf{e}_1\mathbf{e}_3), \quad \underset{\sim}{\mathbf{e}}_7 = \frac{1}{\sqrt{2}}(\mathbf{e}_1\mathbf{e}_2 - \mathbf{e}_1\mathbf{e}_2), \\
&\underset{\sim}{\mathbf{e}}_8 = \frac{1}{\sqrt{2}}(\mathbf{e}_2\mathbf{e}_3 - \mathbf{e}_3\mathbf{e}_2), \quad \underset{\sim}{\mathbf{e}}_9 = \frac{1}{\sqrt{2}}(\mathbf{e}_3\mathbf{e}_1 - \mathbf{e}_1\mathbf{e}_3), \\
&\underset{\sim}{\mathbf{e}}_m \overset{2}{\otimes} \underset{\sim}{\mathbf{e}}_m = \delta_{mn}, \quad i,j,k,l = 1,2,3, \quad m,n = \overline{1,9}.
\end{aligned} \tag{10.46}$$

Moreover, if the components of the tensor $\underset{\approx}{\mathbf{A}}$ have symmetries

$$A_{ijkl} = A_{klij} = A_{jikl} \tag{10.47}$$

then we have

$$\underset{\approx}{\mathbf{A}} = A_{ijkl}\mathbf{e}_i\mathbf{e}_j\mathbf{e}_k\mathbf{e}_l = \sum_{m=1}^{6}\sum_{n=1}^{6} A_{mn}\underset{\sim}{\mathbf{e}}_m\underset{\sim}{\mathbf{e}}_n = A_{mn}\underset{\sim}{\mathbf{e}}_m\underset{\sim}{\mathbf{e}}_n,$$

$$i,j,k,l = 1,2,3, \quad m,n = \overline{1,6}.$$
(10.48)

Note that the two-index notation is similar to Voigt's, Nye's, Kelvin's or Mandel's notation.

10.4 Classical Material with an Anisotropy Symbol {1, 2, 3} (Cubic Symmetry)

In this case we have the next representation of the elastic modulus tensor $\underset{\approx}{\mathbf{A}}$

$$\underset{\approx}{\mathbf{A}} = (\lambda_1 - \lambda_4)\underset{\sim}{\mathbf{W}}_1\underset{\sim}{\mathbf{W}}_1 + (\lambda_2 - \lambda_4)(\underset{\sim}{\mathbf{W}}_2\underset{\sim}{\mathbf{W}}_2 + \underset{\sim}{\mathbf{W}}_3\underset{\sim}{\mathbf{W}}_3) + \lambda_4\underset{\approx}{\mathbf{E}}$$
(10.49)

with

$$\underset{\approx}{\mathbf{E}} = \sum_{m=1}^{6}\underset{\sim}{\mathbf{W}}_m\underset{\sim}{\mathbf{W}}_m,$$
(10.50)

where the eigenvalues and the eigentensors have the form

$$\lambda_1 = A_{11} + 2A_{12}, \quad \lambda_2 = \lambda_3 = A_{11} - A_{12}, \quad \lambda_4 = \lambda_5 = \lambda_6 = A_{44},$$

$$\underset{\sim}{\mathbf{W}}_1 = \frac{\pm\sqrt{3}}{3}(\mathbf{e}_1 + \mathbf{e}_2 + \mathbf{e}_3) = \frac{\pm\sqrt{3}}{3}\underset{\sim}{\mathbf{E}}, \quad \underset{\sim}{\mathbf{W}}_2 = \frac{\pm\sqrt{2}}{2}(\mathbf{e}_1 - \mathbf{e}_2),$$
(10.51)

$$\underset{\sim}{\mathbf{W}}_3 = \frac{\pm\sqrt{6}}{6}(\mathbf{e}_1 + \mathbf{e}_2 - 2\mathbf{e}_3), \quad \mathbf{W}_4 = \mathbf{e}_4, \quad \mathbf{W}_5 = \mathbf{e}_5, \quad \mathbf{W}_6 = \mathbf{e}_6.$$

Taking into account the relations for the eigentensors (10.51), from (10.49) we will obtain the representation of the tensor $\underset{\approx}{\mathbf{A}}$

$$\underset{\approx}{\mathbf{A}} = \frac{1}{3}(\lambda_1 - \lambda_2)\underset{\approx}{\mathbf{C}}_1 + (\lambda_2 - \lambda_4)\sum_{k=1}^{3}\mathbf{e}_k\mathbf{e}_k + \lambda_4\underset{\approx}{\mathbf{E}}, \quad \underset{\approx}{\mathbf{E}} = \frac{1}{2}(\underset{\approx}{\mathbf{C}}_2 + \underset{\approx}{\mathbf{C}}_3).$$
(10.52)

According to (10.51), we will have the relations for the dispersive tensor $\underset{\sim}{\mathbf{A}}$

$$\underset{\sim}{\mathbf{A}} = a\mathbf{n}\mathbf{n} + b\underset{\sim}{\mathbf{E}} + d\sum_{k=1}^{3} n_k^2\mathbf{e}_k.$$
(10.53)

Based on (10.53) we can find $I_1(\mathbf{A}^k)$, $k = 1,2,3$, then by (10.20), which are valid also for $\underset{\sim}{\mathbf{A}}$, we ca find $I_k(\underset{\sim}{\mathbf{A}})$, $k = 2,3$. Indeed, after some simple calculations we have

$$I_1(\underset{\sim}{\mathbf{A}}) = a + 3b + d, \quad I_1(\mathbf{A}^2) = a(a+2b) + 2bd + 3b^2 + (2ad + d^2)\sum_{k=1}^{3} n_k^4,$$

$$I_1(\mathbf{A}^3) = a(a^2 + 3ab + 3b^2) + 3b^2(b+d)$$

$$\qquad + d[3a(a+2b) + 3bd]\sum_{k=1}^{3} n_k^4 + d^2(3a+d)\sum_{k=1}^{3} n_k^6,$$

$$I_2(\underset{\sim}{\mathbf{A}}) = \frac{1}{2}[6b^2 + 4ab + 2ad + 4bd + d^2 - d(2a+d]\sum_{k=1}^{3} n_k^4],$$

$$I_3(\underset{\sim}{\mathbf{A}}) = \frac{1}{3!}[6ab^2 + 3ad^2 + 3bd^2 + 6abd + 6b^3 + 6b^2d + d^3$$

$$\qquad -(9ad^2 + 6abd + 3bd^2 + 3d^3)\sum_{k=1}^{3} n_k^4 + (6ad^2 + 2d^3)\sum_{k=1}^{3} n_k^6]. \qquad (10.54)$$

Knowing invariants $I_k(\underset{\sim}{\mathbf{A}})$, $k = 1, 2, 3$, according to, for example, the first equation (10.22) we can construct the characteristic equation for the considered tensor. This equation is the third-degree equation and always has three positive roots, counted every root as many times as its multiplicity. Thus, in this case, it is possible to define the number of waves and find their velocities in any direction. Let us consider three mutually perpendicular wave velocities directions in this medium, where $n_i = \delta_{i1}$, $n_i = \delta_{i2}$ $n_i = \delta_{i3}$, $i = 1, 2, 3$. Note that the first invariant of the tensor $\underset{\sim}{\mathbf{A}}$ does not depend on the direction of $\mathbf{n}$, and the second and third invariants have the similar values. In particular,

$$I_1(\underset{\sim}{\mathbf{A}}) = a + 3b + d, \quad I_2(\underset{\sim}{\mathbf{A}}) = b^2(2a + 3b + 2d), \quad I_3(\underset{\sim}{\mathbf{A}}) = b^2(a + b + d). \qquad (10.55)$$

According to (10.55), we have the same dispersion relation for each direction, which has the roots

$$\mu_1 = \frac{\lambda_1 + 2\lambda_2}{3\rho}, \quad \mu_2 = \mu_3 = \frac{\lambda_4}{2\rho}. \qquad (10.56)$$

Hence, only two waves appear in each direction in the infinite medium under the cubic symmetry and according to (10.56) their velocities are defined by

$$c_1 = \sqrt{(\lambda_1 + 2\lambda_2)/(3\rho)}, \quad c_2 = \sqrt{\lambda_4/(2\rho)}. \qquad (10.57)$$

10.5 Classical Material with an Anisotropy Symbol {1,1,2,2} (Transversal Isotropy)

In this case, the canonical representation of the elastic modulus tensor $\underset{\approx}{\mathbf{A}}$ is

$$\underset{\approx}{\mathbf{A}} = \mu_1\underset{\sim}{\mathbf{w}}_1\underset{\sim}{\mathbf{w}}_1 + \mu_2\underset{\sim}{\mathbf{w}}_2\underset{\sim}{\mathbf{w}}_2 + \mu_3(\underset{\sim}{\mathbf{w}}_3\underset{\sim}{\mathbf{w}}_3 + \underset{\sim}{\mathbf{w}}_4\underset{\sim}{\mathbf{w}}_4) + \mu_5(\underset{\sim}{\mathbf{w}}_5\underset{\sim}{\mathbf{w}}_5 + \underset{\sim}{\mathbf{w}}_6\underset{\sim}{\mathbf{w}}_6), \qquad (10.58)$$

where the eigenvalues are defined by

$$\mu_1 = \frac{1}{2}(A_{11} + A_{12} + A_{33} - \sqrt{(A_{11} + A_{12} - A_{33})^2 + 8A_{13}^2}),$$

$$\mu_2 = \frac{1}{2}(A_{11} + A_{12} + A_{33} + \sqrt{(A_{11} + A_{12} - A_{33})^2 + 8A_{13}^2}),$$

$$\mu_3 = \mu_4 = A_{11} - A_{12}, \quad \mu_5 = \mu_6 = A_{55},$$

and the eigentensors are represented in the form

$$\underset{\sim}{\mathbf{w}}_1 = -\frac{\sqrt{2}}{2}\sin\alpha(\underset{\sim}{\mathbf{e}}_1 + \underset{\sim}{\mathbf{e}}_2) + \cos\alpha\underset{\sim}{\mathbf{e}}_3 = -\frac{\sqrt{2}}{2}\sin\alpha\underset{\sim}{\mathbf{I}} + \cos\alpha\underset{\sim}{\mathbf{e}}_3,$$

$$\underset{\sim}{\mathbf{w}}_2 = \frac{\sqrt{2}}{2}\cos\alpha(\underset{\sim}{\mathbf{e}}_1 + \underset{\sim}{\mathbf{e}}_2) + \sin\alpha\underset{\sim}{\mathbf{e}}_3 = \frac{\sqrt{2}}{2}\cos\alpha\underset{\sim}{\mathbf{I}} + \sin\alpha\underset{\sim}{\mathbf{e}}_3,$$

$$\underset{\sim}{\mathbf{w}}_3 = \frac{\sqrt{2}}{2}(\underset{\sim}{\mathbf{e}}_1 - \underset{\sim}{\mathbf{e}}_2), \quad \underset{\sim}{\mathbf{w}}_4 = \underset{\sim}{\mathbf{e}}_4, \quad \underset{\sim}{\mathbf{w}}_5 = \underset{\sim}{\mathbf{e}}_5, \quad \underset{\sim}{\mathbf{w}}_6 = \underset{\sim}{\mathbf{e}}_6,$$

$$\tan 2\alpha = \frac{2\sqrt{2}A_{13}}{A_{11} + A_{12} - A_{33}}.$$

(10.59)

Note that transversally isotropic materials according to classification (Nikabadze, 2017d; Nikabadze and Ulukhanyan, 2016), can be the following view {1,1,2,2}, {1,2,1,2}, {1,2,2,1}, {2,1,1,2}, {2,1,2,1}, {2,2,1,1}.

Taking into account (10.59), we can represent the tensor (10.58) as

$$\underset{\approx}{\mathbf{A}} = a_2\underset{\approx}{\mathbf{C}}_{(1)} + (a_1 - a_2)\underset{\approx}{\mathbf{E}} + a_3(\underset{\sim}{\mathbf{I}}\mathbf{e}_3 + \mathbf{e}_3\underset{\sim}{\mathbf{I}}) + a_4\mathbf{e}_3\mathbf{e}_3$$
$$+ \frac{1}{2}a_5(\mathbf{e}_3\underset{\sim}{\mathbf{I}}\mathbf{e}_3 + \mathbf{e}_I\mathbf{e}_3\mathbf{e}_I + \mathbf{e}_I\mathbf{e}_3\mathbf{e}_I\mathbf{e}_3 + \mathbf{e}_3\mathbf{e}_I\mathbf{e}_3\mathbf{e}_I),$$

(10.60)

$$\underset{\approx}{\mathbf{C}}_{(1)} = \underset{\sim}{\mathbf{I}}\underset{\sim}{\mathbf{I}}, \quad \underset{\approx}{\mathbf{E}} = \frac{1}{2}(\underset{\approx}{\mathbf{C}}_{(2)} + \underset{\approx}{\mathbf{C}}_{(3)})|1 = \mathbf{e}_1\mathbf{e}_1 + \mathbf{e}_2\mathbf{e}_2 + \mathbf{e}_4\mathbf{e}_4, \quad \underset{\sim}{\mathbf{I}} = \mathbf{e}_1 + \mathbf{e}_2,$$

$$a_1 = \frac{1}{2}(\mu_1 \sin^2\alpha + \mu_2 \cos^2\alpha + \mu_3), \quad a_2 = \frac{1}{2}(\mu_1 \sin^2\alpha + \mu_2 \cos^2\alpha - \mu_3), \quad (10.61)$$

$$a_3 = \frac{\sqrt{2}}{2}(\mu_2 - \mu_1)\sin\alpha\cos\alpha, \quad a_4 = \mu_1\cos^2\alpha + \mu_2\sin^2\alpha, \quad a_5 = \mu_5.$$

After some simple calculations the dispersive tensor for (10.60) will have the form

$$\underset{\sim}{\mathbf{A}} = A_1\mathbf{mm} + A_2\underset{\sim}{\mathbf{I}} + A_3(\mathbf{me}_3 + \mathbf{e}_3\mathbf{m}) + A_4\mathbf{e}_3,$$

$$\mathbf{m} = n_I\mathbf{e}_I, \quad A_1 = \frac{1}{\rho}(a_1 + \frac{1}{2}a_2), \quad A_2 = \frac{1}{\rho}(\frac{1}{2}a_2m^2 + a_5n_3^2),$$

$$A_3 = \frac{1}{\rho}(a_3 + a_5)n_3, \quad A_4 = \frac{1}{\rho}(a_4n_3^2 + a_5m^2).$$

(10.62)

It is easy to find $\underset{\sim}{\mathbf{A}}^2$ and $\underset{\sim}{\mathbf{A}}^3$. Indeed, according to (10.62) we will have

$$\underset{\sim}{\mathbf{A}}^2 = B_1\mathbf{mm} + B_2\underset{\sim}{\mathbf{I}} + B_3(\mathbf{me}_3 + \mathbf{e}_3\mathbf{m}) + B_4\mathbf{e}_3,$$
$$\underset{\sim}{\mathbf{A}}^3 = C_1\mathbf{mm} + C_2\underset{\sim}{\mathbf{I}} + C_3(\mathbf{me}_3 + \mathbf{e}_3\mathbf{m}) + C_4\mathbf{e}_3,$$

(10.63)

$$B_1 = A_1^2 m^2 + 2A_1 A_2 + A_3^2, \quad B_2 = A_2^2, \quad C_2 = A_2^3,$$
$$B_3 = A_1 A_3 m^2 + A_2 A_3 + A_3 A_4, \quad B_4 = A_3^2 m^2 + A_4^2,$$
$$C_1 = A_1^3 m^4 + (3A_1^2 A_2 + 2A_1 A_3^2)m^2 + 3A_1 A_2^2 + 2A_2 A_3^2 + A_3^2 A_4, \quad (10.64)$$
$$C_3 = A_1^2 A_3 m^4 + (2A_1 A_2 A_3 + A_1 A_3 A_4 + A_3^3)m^2 + A_2^2 A_3$$
$$+ A_2 A_3 A_4 + A_3 A_4^2, \quad C_4 = A_1 A_3^2 m^2 + (2A_3^2 A_4 + A_2 A_3^2)m^2 + A_4^3.$$

Knowing (10.62) and (10.63), easy to find $I_1(\underset{\sim}{\mathbf{A}}^k)$, $k = 1, 2, 3$, and then by (10.20) we can find $I_m(\underset{\sim}{\mathbf{A}})$, $m = 2, 3$. Indeed,

$$I_1(\underset{\sim}{\mathbf{A}}) = A_1 m^2 + 2A_2 + A_4, \quad I_1(\underset{\sim}{\mathbf{A}}^2) = B_1 m^2 + 2B_2 + B_4,$$
$$I_1(\underset{\sim}{\mathbf{A}}^3) = C_1 m^2 + 2C_2 + C_4, \quad I_2(\underset{\sim}{\mathbf{A}}) = \frac{1}{2}\{A_1^2 m^4 + [2A_1(2A_2 + A_4)$$
$$- B_1]m^2 + (2A_2 + A_4)^2 - 2B_2 - B_4\},$$
$$I_3(\underset{\sim}{\mathbf{A}}) = \frac{1}{3!}\{A_1^3 m^6 + 3[A_1^2(2A_2 + A_4) - A_1 B_1]m^4$$
$$+ [3A_1(2A_2 + A_4)^2 - 3A_1(2B_2 + B_4) - 3B_1(2A_2 + A_4) + 2C_1]m^2$$
$$+ (2A_2 + A_4)^3 - 3(2A_2 + A_4)(2B_2 + B_4) + 2(2C_2 + C_4)\}.$$

(10.65)

Knowing $I_m(\underset{\sim}{\mathbf{A}})$, $m = 1, 2, 3$ (see (10.65)), from the dispersion equation we can find the eigenvalues for $\underset{\sim}{\mathbf{A}}$ in any direction $\mathbf{n}$, and then we can define the wave velocities.

Let us find, for example, the wave velocities in directions $n_i = \delta_{i1}$, $n_i = \delta_{i2}$ and $n_i = \delta_{i3}$, $i = 1, 2, 3$. Note that from $I_m(\underset{\sim}{\mathbf{A}})$, $m = 1, 2, 3$ (see (10.65)) it is following that they have the equal values in the directions $n_i = \delta_{i1}$ and $n_i = \delta_{i2}$, $i = 1, 2, 3$. Indeed, according to the corresponding relations (10.62) and (10.65) for these directions we will have

$$I_1(\underset{\sim}{\mathbf{A}}) = \frac{a_1 + a_2}{\rho} + \frac{a_2}{2\rho} + \frac{a_5}{\rho},$$
$$I_2(\underset{\sim}{\mathbf{A}}) = \frac{a_1 + a_2}{\rho}\frac{a_2}{2\rho} + \frac{a_1 + a_2}{\rho}\frac{a_5}{\rho} + \frac{a_2}{2\rho}\frac{a_5}{\rho}, \quad (10.66)$$
$$I_3(\underset{\sim}{\mathbf{A}}) = \frac{a_1 + a_2}{\rho}\frac{a_2}{2\rho}\frac{a_5}{\rho}.$$

It is seen from (10.66) that the characteristic equation of the tensor $\underset{\sim}{\mathbf{A}}$ has roots

$$\mu_1 = \frac{a_1 + a_2}{\rho}, \quad \mu_2 = \frac{a_2}{2\rho}, \quad \mu_3 = \frac{a_5}{\rho}. \quad (10.67)$$

From here we have the next values for the wave velocities in the initial stationary medium

$$c_1 = \sqrt{\frac{a_1 + a_2}{\rho}}, \quad c_2 = \sqrt{\frac{a_2}{2\rho}}, \quad c_3 = \sqrt{\frac{a_5}{\rho}}. \quad (10.68)$$

For the direction $n_i = \delta_{i3}$, $i = 1, 2, 3$, the relations similar to (10.66)–(10.68) have the form

$$I_1(\underset{\approx}{\mathbf{A}}) = \frac{a_4}{\rho} + 2\frac{a_5}{\rho}, \quad I_2(\underset{\approx}{\mathbf{A}}) = 2\frac{a_4}{\rho}\frac{a_5}{\rho} + \left(\frac{a_5}{\rho}\right)^2, \quad I_3(\underset{\approx}{\mathbf{A}}) = \frac{a_4}{\rho}\left(\frac{a_5}{\rho}\right)^2,$$

$$\mu_1 = \frac{a_4}{\rho}, \quad \mu_2 = \mu_3 = \frac{a_5}{\rho}, \quad c_1 = \sqrt{\frac{a_4}{\rho}}, \quad c_2 = \sqrt{\frac{a_5}{\rho}}.$$

Note that the consideration of the trigonal syngony with the symbol of structure $\{1,1,1,2,1\}$ (which has 6 essential components) reduces to the previous case, therefore we will not dwell on this.

10.6 Micropolar Material with a Center of Symmetry and the Anisotropy Symbol {1,5,3}

In this case, the properties of the medium are characterized by two fourth-rank tensors $\underset{\approx}{\mathbf{A}}$ and $\underset{\approx}{\mathbf{D}}$, which have the same structure. Thus, it is enough to consider only one of them (see, for example, (10.22)). Let us consider, for instance, the tensor $\underset{\approx}{\mathbf{A}}$. Its canonical representation has a form

$$\underset{\approx}{\mathbf{A}} = \lambda_1 \underset{\sim}{\mathbf{u}}_1 \underset{\sim}{\mathbf{u}}_1 + \lambda_2 \sum_{m=2}^{6} \underset{\sim}{\mathbf{u}}_m \underset{\sim}{\mathbf{u}}_m + \lambda_7 \sum_{m=7}^{9} \underset{\sim}{\mathbf{u}}_m \underset{\sim}{\mathbf{u}}_m$$

$$= (\lambda_1 - \lambda_2)\underset{\sim}{\mathbf{u}}_1 \underset{\sim}{\mathbf{u}}_1 + \lambda_2 \underset{\approx}{\mathbf{E}} + (\lambda_7 - \lambda_2) \sum_{m=7}^{9} \underset{\sim}{\mathbf{u}}_m \underset{\sim}{\mathbf{u}}_m, \qquad (10.69)$$

$$\underset{\approx}{\mathbf{E}} = \underset{\approx}{\mathbf{C}}_{(2)} = \sum_{m=1}^{9} \underset{\sim}{\mathbf{u}}_m \underset{\sim}{\mathbf{u}}_m.$$

Now, based on (10.69) it is easy to find $I_1(\mathbf{A}^k)$, $k = 1, 2, 3$, and then according to (10.20) to find $I_m(\underset{\approx}{\mathbf{A}})$, $m = 2, 3$. Knowing $I_m(\underset{\approx}{\mathbf{A}})$, $m = 1, 2, 3$, we can get the characteristic equation and roots of $\underset{\approx}{\mathbf{A}}$, and then we can determine the required wave velocities. It is seen that despite big formulas they can be presented in an explicit form. It should be noted that cases $\{5,1,3\}$ and $\{1,5,3\}$ are considered in the same way. Let us consider the micropolar isotropic elastic medium with the centre of symmetry and the symbol of the structure $\{1,5,3\}$. In this case $\underset{\approx}{\mathbf{A}}$ and $\underset{\approx}{\mathbf{D}}$ have the representations

$$\underset{\approx}{\mathbf{A}} = a_1 \underset{\approx}{\mathbf{C}}_{(1)} + a_2 \underset{\approx}{\mathbf{C}}_{(2)} + a_1 \underset{\approx}{\mathbf{C}}_{(3)}, \quad \underset{\approx}{\mathbf{D}} = d_1 \underset{\approx}{\mathbf{C}}_{(1)} + d_2 \underset{\approx}{\mathbf{C}}_{(2)} + d_1 \underset{\approx}{\mathbf{C}}_{(3)}, \quad (10.70)$$

where the coefficients in the right side of (10.70) are defined by (Nikabadze, 2017d; Nikabadze and Ulukhanyan, 2016)

$$a_1 = \frac{1}{3}(\lambda_1 - \lambda_2), \quad a_2 = \frac{1}{2}(\lambda_2 + \lambda_7), \quad a_3 = \frac{1}{2}(\lambda_2 - \lambda_7),$$

$$d_1 = \frac{1}{3}(\mu_1 - \mu_2), \quad d_2 = \frac{1}{2}(\mu_2 + \mu_7), \quad d_3 = \frac{1}{2}(\mu_2 - \mu_7).$$

It is seen that the dispersion tensors $\underset{\approx}{A}$ and $\underset{\approx}{D}$ corresponding to $\underset{\approx}{A}$ and $\underset{\approx}{D}$ (see (10.70)), respectively, have the form

$$\underset{\sim}{A} = \frac{1}{\rho}[(a_1 + a_3)\mathbf{nn} + a_2\underset{\sim}{E}, \quad \underset{\sim}{D} = \frac{1}{J}[(d_1 + d_3)\mathbf{nn} + d_2\underset{\sim}{E}. \qquad (10.71)$$

From (10.71) it is followed that $\underset{\sim}{A}$ and $\underset{\sim}{D}$ are similar to the tensors considered above (see, for instance, (10.30)). So, introducing

$$a = \frac{a_2}{\rho}, \quad b = \frac{a_1 + a_3}{\rho}, \quad f = \frac{d_2}{J}, \quad g = \frac{d_1 + d_3}{J},$$

we can get the tensors (10.71) in the form

$$\underset{\sim}{A} = a\underset{\sim}{E} + b\mathbf{nn}, \quad \underset{\sim}{D} = f\underset{\sim}{E} + g\mathbf{nn}. \qquad (10.72)$$

Then, similar to (10.37) and (10.38) the roots of the dispersion equation of the tensors $\underset{\sim}{A}$ and $\underset{\sim}{D}$ and the wave velocities in the initial stationary medium are defined by

$$\eta_1 = \eta_2 = a = \frac{\lambda_2 + \lambda_7}{2\rho}, \quad \eta_3 = a + b = \frac{\lambda_1 + 2\lambda_2}{3\rho} = \frac{\lambda + 2\mu}{\rho},$$
$$\eta_4 = \eta_5 = f = \frac{\mu_2 + \mu_7}{2\rho}, \quad \eta_6 = f + g = \frac{\mu_1 + 2\mu_2}{3\rho}. \qquad (10.73)$$

$$c_1 = \sqrt{\frac{\lambda_2 + \lambda_7}{2\rho}}, \quad c_2 = \sqrt{\frac{\lambda_1 + 2\lambda_2}{3\rho}},$$
$$c_3 = \sqrt{\frac{\mu_2 + \mu_7}{2\rho}}, \quad c_4 = \sqrt{\frac{\mu_1 + 2\mu_2}{3\rho}}. \qquad (10.74)$$

Thus, in an infinite micropolar isotropic initially stationary medium four waves with the velocities (10.74) arise in each direction.

Note that in order to determine the directions of propagation of these waves, it is necessary to find a complete system of eigenvectors and give a canonical representation of the dispersion tensors, and then to study them in the same way as it was done above in the case of a classical isotropic material. Consequently, such a study can always be carried out in the more general case of anisotropy, but we will not dwell on this. It should be noted that some questions about the application of eigenvalue problems for tensor objects are set out in the papers (Nikabadze, 2016, 2017a,b,c).

10.7 Conclusion

The formulation of the eigenvalue problem for TBM of any order and any even rank is given. Using the canonical representations of material tensor objects and the obtained kinematic and dynamic conditions, the problems of finding of the wave propagation velocities are reduced to the eigenvalue problems of the corresponding dispersion tensor objects. In particular, the materials with the anisotropy symbol {1.5} and {5.1} are considered, as well as isotropic materials, and for them the expressions for the velocities of wave propagation are obtained. In addition, we obtained expressions for the velocities of wave propagation for different anisotropic materials. We also obtained the expressions for the velocities of wave propagation for a micro-polar medium with the anisotropy symbol {1.5.3} and {5.1.3}, and for an isotropic micropolar material.

Acknowledgements This work was supported by the Shota Rustaveli National Science Foundation (project no. DI-2016-41) and the Russian Foundation for Basic Research (project no. 18–29–10085-mk).

References

Altenbach H, Eremeyev VA, Lebedev LP, Rendón LA (2010) Acceleration waves and ellipticity in thermoelastic micropolar media. Archive of Applied Mechanics 80(3):217–227

Baskakov VA (1991) Analysis of the propagation and dynamic influence of shock waves on a deformable solid (in Russ.). PhD thesis, Ulyanov Chuvash State University

Baskakov VA, Bestuzheva NP, Konchakova NA (2001) Linear Dynamic Theory of Thermoelastic Media with a Microstructure (in Russ.)

Eringen AC (1999) Microcontinuum Field Theories, vol I. Foundations and Solids. Springer, New York

Konchakova NA (1998) Investigation of wave processes in the thermoelastic Cosserat medium (in russ.). PhD thesis, Voronezh State Technical University

Kupradze VD (ed) (1979) Three-dimensional Problems of the Mathematical Theory of Elasticity and Thermoelasticity, Applied Mathematics and Mechanics, vol 25. North-Holland

Nikabadze MU (2007) Some issues concerning a version of the theory of thin solids based on expansions in a system of Chebyshev polynomials of the second kind. Mechanics of Solids 42(3):391–421

Nikabadze MU (2014) Development of the Method of Orthogonal Polynomials in the Classical and Micropolar Mechanics of Elastic Thin Bodies. Moscow Univ. Press, (in Russian)

Nikabadze MU (2016) Eigenvalue Problems of a Tensor and a Tensor-Block Matrix (TMB) of Any Even Rank with Some Applications in Mechanics. In: Altenbach H, Forest S (eds) Generalized Continua as Models for Classical and Advanced Materials, Advanced Structured Materials, vol 42, Springer, pp 279–317

Nikabadze MU (2017a) An eigenvalue problem for tensors used in mechanics and the number of independent Saint-Venant strain compatibility conditions. Moscow University Mechanics Bulletin 72(3):66–69

Nikabadze MU (2017b) Eigenvalue problem for tensors of even rank and its applications in mechanics. Journal of Mathematical Sciences 221(2):174–204

Nikabadze MU (2017c) To the problem of decomposition of the initial boundary value problems in mechanics. Journal of Physics: Conference Series 936(1):1–12

Nikabadze MU (2017d) Topics on tensor calculus with applications to mechanics. J Math Sci 225(1):1–194

Nikabadze MU, Ulukhanyan AR (2016) Analytical Solutions in the Theory of Thin Bodies. In: Altenbach H, Forest S (eds) Generalized Continua as Models for Classical and Advanced Materials, Advanced Structured Materials, vol 42, pp 319–361

Nowacki W (1970) Teoria Sprezystosci. Panstwowe Wydawnictwo Naukowe, Warsaw

Petrashen GI (1978) Fundamentals of the mathematical theory of elastic waves propagation (in Russ.). In: Issues of the Dynamic Theory of Seismic Wave Propagation., vol 18, Nauka, Leningrad

Petrashen GI (1980) Wave Propagation in Anisotropic Elastic Media (in Russ.). Nauka, Leningrad

Poruchikov VB (1986) Methods of Dynamic Theory of Elasticity (in Russ.). Nauka, Moscow

Sagomonyan AY (1985) Stress Waves in Continuous Media (in Russ.). Izd. Moscov. Univ., Moscow

Chapter 11
Theoretical Estimation of the Strength of Thin-film Coatings

Sergey N. Romashin, Victoria Yu. Presnetsova, Larisa Yu. Frolenkova, Vladimir S. Shorkin, and Svetlana I. Yakushina

Abstract The theoretical strength of a perfect crystal lattice, which corresponds to the simultaneous breaking of intermolecular bonds, is very high − only ten times less than the Young's modulus. Strength of real solids is several orders of magnitude smaller. This is linked with the existence of lattice defects. In the work of the various types of defects are considered only crack. Although really brittle materials is very small. Question of cracks in brittle solids is of great practical importance because many plastic materials (metals) are destroyed "brittle" manner. Problem of brittle fracture paid much attention. Trying to explain the discrepancy between the actual and theoretical values of strength in the presence of cracks was made by Griffith in his theory of brittle fracture of amorphous materials. He suggested that the real materials have a large number of small cracks, which can act as stress concentrators, increasing their value to the theoretical strength. The process reduces the gap while increasing the length of cracks until complete separation of the sample into two parts. In this paper we propose a simplified model of the phenomenon of destruction of elastic material, which allows to establish a connection between the theoretical tensile strength and the actual breaking stress. The model is based on the idea that each representative elementary particle continuum elastic medium incorporates the micro-crack. So that the material, as in Griffith's theory, there is a whole network of micro-cracks. Fracturing process occurs due to the merger with the apparent micro-cracks (assuming no cracks) voltage equal to the actual tensile strength, and tensile material reaching peaks between neighboring cracks voltage theoretical tensile strength. Evaluation carried out calculations of the magnitude of micro-cracks and damage some materials. Installed in the connection between the theoretical tensile strength and the real destructive voltage calculation results correspond to known concepts and reference data.

S.N. Romashin, V.Yu. Presnetsova, L.Yu. Frolenkova, V.S. Shorkin, S.I. Yakushina
Orel State University named after. I.S. Turgenev, 29 Naugorskoe Shosse, 302020 Orel, Russia
e-mail: sromashin@yandex.ru; alluvian@mail.ru; Larafrolenkova@yandex.ru
e-mail: vshorkin@yandex.ru; jakushina@rambler.ru

© Springer Nature Switzerland AG 2019
H. Altenbach et al. (eds.), *Higher Gradient Materials and Related Generalized Continua*, Advanced Structured Materials 120,
https://doi.org/10.1007/978-3-030-30406-5_11

Keywords: Elementary representative area · Theoretical tensile strength · Actual breaking stress · Size of micro-cracks damage

11.1 Introduction

Theoretical and real strength differ significantly. This is explained by the Griffith's theory of cracks (Griffith, 1920), which admits the presence of micro-cracks in the network material.

For example, as in Petch (1968), breaking the interatomic bonds by stretching requires stress $\sigma_t \approx E/5$, where E denotes Young's modulus. However, brittle fracture occurs at a stress $\sigma_t \approx E/500$. In particular, in the framework of this theory (Petch, 1968) that the breaking stress for a material (Griffith's estimate), in which there are cracks of length l_{cr}, is determined by the expression

$$\sigma_u = (4EW_{saf}/\pi l_{cr})^{1/2}, \tag{11.1}$$

where W_{saf} denotes surface free energy.

Orowan's estimate, which was obtained under the assumption that the radius of curvature at the crack tip is equal to the atomic size, has the form

$$\sigma_u = (EW_{saf}/2l_{cr})^{1/2}. \tag{11.2}$$

It is also noted that the commonly observed values of strength $\sigma_t \approx E/500$ can be explained by the presence of Griffith's cracks of length $l_{cr} \approx 10^{-6}$ m.

It should be noted that the size of micro-cracks was estimated by other authors. In Rybin (1986), an estimate is given $l_{cr} \approx (3 \div 5) \cdot 10^{-6}$ m. Busov (2007) presents an estimate of the size of micro-cracks corresponding to the beginning of their unification under uniaxial tension. Their length must be within $l_{cr} \approx (2 \div 20) \cdot 10^{-6}$ m.

The Griffith's theory considers the development of a single crack in an infinitely extended medium, which is subject to stretching by forces applied at the edges infinitely distant from the crack. During the operation of thin-film coatings, a situation is possible when the base is subjected to stretching, drawing the coating into the process leading it to destruction. The stretching of the film using the base from which it is bonded by adhesive forces occurs through tangential forces applied along the contact surface. The destruction of the stretchable film (without base) occurs due to the loss of stability and the catastrophic development of a single crack. For a coating on a stretchable basis, the appearance of a network of cracks parallel to each other (as in Volynskii et al, 2006) is characteristic with the distance between the nearest micro-cracks is of order $h = 6 \cdot 10^{-6}$ m, which is commensurate with $l_{cr} = 10^{-6}$ m.

The emergence of a network of micro-cracks, which appeared due to the fragmentation of a crystal in the fields of external influences due to a violation of the translational invariance of the crystal structure, was theoretically and experimentally considered and substantiated in Panin and Egorushkin (2009). In varying degrees, the influence of the fragmentation of the structure of a material on its strength was taken into account in a number of works, for example, in Zhurkov et al (1977); Sarafanov and Pereverzentsev (2010); Hild et al (2005); Zhu et al (2008).

The paper proposes a simplified model of the phenomenon, as well as a destructive stress estimate, made on its basis. This result will make it possible to carry out estimates of the actual strength of thin-film coatings, subject to the availability of information on the average size of their micro-cracks.

11.2 Theoretical Position

It is assumed that the material has a cellular structure. In each of the cells there is a micro-crack. The shape and orientation of the micro-cracks in the film material is random. However, when it is stretched by means of a substrate, those which are perpendicular to the direction of stretching, are activated.

The body B, which simulates the coating, is represented as a union of elementary parts ΔB_n, $n = 1, ..., N : \bigcup_{n=1}^{N} \Delta B_n$, $\Delta B_n \cap \Delta B_m = \varnothing$, $m \neq n$. This ratio is transferred to the areas that are occupied by the whole body and its parts. Bodies ΔB_n are the elementary structural cells of the studied body B. In each of them, the characteristics of the material used for macro-description are repeated..

To take into account the microinhomogeneity of the material in the continual description, internal parameters are introduced for the structural cell (Sedov, 1972). To exclude a detailed account of the heterogeneity of the state in the microregion, the characteristics of the material averaged over it are introduced into the description. The concept of a representative region is used when the diameter of the area ΔB_n has (for a given accuracy of averaging) a non-zero lower bound.

It is assumed that a micro-crack has formed in each structural cell ΔB_n. So the cell ΔB_n is the area Ω_D of influence of the presence of a crack on the stress-strain state of the material surrounding the crack. The area Ω_D is bounded by the surface A_Ω. At the time of crack formation, it is a mathematical cut. The crack opening, if it should be under the conditions considered, is assumed to be small. The surface of the coast and their area are denoted as $A_{cr}^{\pm} = A_{cr}$. It is bounded by a closed line L_{cr}. The diameter $A_{cr}^{\pm}$ of the areas is l_{cr}.

In accordance with Griffith (1920), when a crack appears, a part of the energy of elastic deformations of its environment turns into the surface energy F_{saf} of its coasts, which is distributed through them evenly with intensity W_{saf}:

$$F_{saf} = W_{saf} \left(A_{cr}^{\pm} + A_{cr}^{-} \right) = 2W_{saf} A_{cr}. \tag{1.3}$$

The area Ω_D of influence of a crack on the state of the material around it has a characteristic size D, commensurate with its counterpart l_{cr} for the crack. The assumption of commensurability of the characteristic dimensions of the crack and the area of its influence coincides with a similar assumption in the Griffith's theory in Griffith (1920). In this theory, it is believed that the line L_{cr} is the edge of a crack that lies on the surface A_Ω. Therefore, it is accepted $D \cong l_{cr}$. In this case, D exceeds l_{cr}.

The area Ω_D in turn is conventionally divided into two non-intersecting areas Ω_α and $\Omega_{1-\alpha}$. Let be A_{Gr} – the surface separating these regions. It is assumed that a area $\Omega_{1-\alpha}$ has a property endowed with a theory (Griffith, 1920) - the line L_{cr} belongs to it: $L_{cr} \subset A_{Gr}$. In this case, as in the theory of Griffith (1920), the areas farthest from the crack are located, where perturbations of the stress and deformation fields can be neglected. Hence, the surfaces A_{Gr} and A_Ω have common areas.

The flat shape of the crack introduces a certain orientation to the space Ω_D. Given this, it is assumed that the area $\Omega_{1-\alpha}$ has a cylindrical shape with a forming – a straight line parallel to the normal to the crack plane, and directing – a line L_{cr}. The ends of the cylinder for simplicity are assumed to be flat, parallel to the plane of the crack, which are separated at a distance h from it. The area Ω_D is also cylindrical. Its lateral surface is parallel to the lateral surface of the area $\Omega_{1-\alpha}$, the ends are parts of the same planes as the ends of the area $\Omega_{1-\alpha}$.

Considering that all three areas Ω_D, Ω_α and $\Omega_{1-\alpha}$ have the same height $2h$, it can be argued that the volumes of these areas V_D, V_α and $V_{1-\alpha}$ are exactly the same as the areas A_D, A_α and $A_{1-\alpha}$, that is, they are in relation to

$$V_D : V_\alpha : V_{1-\alpha} = A_D : A_\alpha : A_{1-\alpha} = 1 : \alpha : 1 - \alpha. \tag{11.3}$$

Here, α is the relative area of a part of the section of the area Ω_D by the plane of the crack, which has not been destroyed. In this case, $\beta = 1 - \alpha$ - the damage of this section by a crack. The parameter α is entered as a parameter determining the state of the elastic medium.

The introduction of a quantitative assessment of the discontinuity of the material - its damage, which depends on the objectives of the research and the tasks solved at the same time, can be carried out in different ways (Kachanov, 1974; Rabotnov, 1970; Aptukov, 2007; Kashtanov and Petrov, 2006; Astaf'ev et al, 2001). In this situation, the simplest option is chosen. It is believed that damage is the relative area of a micro-crack in the cross section of a characteristic unit cell of the medium by its plane: $\beta = A_{1-\alpha}/A_D$. The parameter $\alpha = A_\alpha/A_D$ is to be determined—in Vitkovsky et al (2007), this definition of damage was used to quantify the damage to the adhesive contact of solids under the assumption that A_D, A_α and $A_{1-\alpha}$ are infinitesimal.

Before the formation of a crack, the whole medium and the regions Ω_D, Ω_α and $\Omega_{1-\alpha}$, mentally allocated in it, were in a uniform deformed state. After its formation, this state inside Ω_D is distorted, ceases to be homogeneous. Since Ω_D is the area occupied by the elementary structural cell ΔB of the body B, that is, it is a representative elementary area.

It is assumed that the material properties are described on the basis of the linear theory of elasticity. The deformed state in this case is characterized by the Cauchy deformation tensor (Nowacki, 1975):

$$\mathbf{g} = (\mathbf{e}_i \mathbf{e}_j) g_{ij} = (1/2)(\mathbf{e}_i \mathbf{e}_j)(u_{i,j} + u_{j,i}). \tag{11.4}$$

We use the Cartesian coordinate system x_i, $(i = 1, 2, 3)$ with an orthonormal basis $\mathbf{e}_i$. $\mathbf{u} = \mathbf{u(r)}$ — the displacement vector, $\mathbf{r}$ — the radius vector of a point in the reference configuration of the body B, $\mathbf{R} = \mathbf{r} + \mathbf{u(r)}$ — the radius vector of the same point in the current configuration of the same body, $\nabla u = (\mathbf{e}_i \mathbf{e}_j) u_{i,j}$ — the gradient of the displacement vector, the index j after the comma means differentiation in x_j.

The volume density of changes in the free energy caused by the deformations and the corresponding stress tensor are determined by the equality (Nowacki, 1975):

$$w = \mu g_{ij} g_{ij} + \frac{\lambda}{2} g_{kk} g_{ll},$$

$$\mathbf{P} = (\mathbf{e}_i \mathbf{e}_j) P_{ij} = (\mathbf{e}_i \mathbf{e}_j) \frac{\partial w}{\partial g_{ij}} = (\mathbf{e}_i \mathbf{e}_j)(2\mu g_{ij} + \lambda \delta_{ij} g_{kk}). \tag{11.5}$$

The characteristics of the deformed state of the material particles ΔB inside Ω_D are considered to be constant, equal to the average of Ω_D values. Therefore, the deformed state of the particle ΔB is characterized by $\langle \nabla \mathbf{u} \rangle$.

$$\langle \nabla \mathbf{u} \rangle = \frac{1}{V_D} \int\limits_{V_D} (\nabla \mathbf{u}\,(\mathbf{r}))\, dV_D = \frac{1}{V_D} \int\limits_{V_\alpha} (\nabla \mathbf{u}\,(\mathbf{r}))\, dV_\alpha +$$

$$+ \frac{1}{V_D} \int\limits_{V_{(1-\alpha)}} (\nabla \mathbf{u}\,(\mathbf{r}))\, dV_{(1-\alpha)} = \frac{V_\alpha}{V_D} \langle \nabla \mathbf{u} \rangle_\alpha + \frac{V_{(1-\alpha)}}{V_D} \langle \nabla \mathbf{u} \rangle_{(1-\alpha)} = \tag{11.6}$$

$$= \alpha \langle \nabla \mathbf{u} \rangle_\alpha + (1 - \alpha) \langle \nabla \mathbf{u} \rangle_{(1-\alpha)}.$$

In this expression, the values of $\langle \nabla \mathbf{u} \rangle_\alpha$, $\langle \nabla \mathbf{u} \rangle_{1-\alpha}$ are average over the areas Ω_α and $\Omega_{1-\alpha}$ by the gradients of the vector of displacements of the points of these areas. They do not coincide with each other.

The stress tensor (11.5) depends on the gradient of displacements linearly. Therefore, it is true that the expression coincides in form with (11.6).

$$\langle \mathbf{P} \rangle = \alpha \langle \mathbf{P} \rangle_\alpha + (1 - \alpha) \langle \mathbf{P} \rangle_{1-\alpha}. \tag{11.7}$$

The bulk density of the free energy change caused by the deformation of the material inside V_D is $[w]$.

$$[w] = \int_0^{\langle \nabla \mathbf{u} \rangle} \langle \mathbf{P} \rangle\,(\langle \nabla \mathbf{u} \rangle)\, d\,\langle \nabla \mathbf{u} \rangle = \mu\,\langle g_{ij} \rangle\,\langle g_{ij} \rangle + \frac{\lambda}{2}\,\langle g_{kk} \rangle\,\langle g_{ll} \rangle,$$

$$\langle \mathbf{g} \rangle = \alpha\,\langle \mathbf{g} \rangle_\alpha + (1 + \alpha)\,\langle \mathbf{g} \rangle_{1-\alpha}, \tag{11.8}$$

$$[w] = \alpha^2\,[w]_{\alpha\alpha} + 2\alpha(1 - \alpha)\,[w]_{\alpha(1-\alpha)} + (1 - \alpha)^2\,[w]_{(1-\alpha)(1-\alpha)}.$$

This value does not coincide with the average value of the energy density of elastic deformations over the area V_D: $[w] \geq \langle w \rangle$. When $\alpha \to 1$, $D \to 0$ the area of the crack decreases and the perturbation area tends to zero, also when the norm $\left\| \nabla^2 \mathbf{u} = (\mathbf{e}_i \mathbf{e}_j \mathbf{e}_k) u_{i,j,k} \right\|$ is limited, condition $[w] \to w$ is satisfied.

The condition for determining the parameter α is proposed to consider the condition of stationarity of the free energy of the material inside Ω_D.

$$\frac{\partial [w]}{\partial \alpha} = 0. \tag{11.9}$$

The consequence of this expression is the formula

$$\alpha = \kappa \left[1 + \frac{[w]_{\alpha\alpha} - [w]_{\alpha(1-\alpha)}}{[w]_{(1-\alpha)(1-\alpha)} - [w]_{\alpha(1-\alpha)}} \right]^{-1}. \tag{11.10}$$

The fragmentation of the body B into elementary cells ΔB with the described structure has a translational periodicity. In this case, in the plane of the crack A_{cr}, there is a whole network of periodically repeating cracks. Between the neighboring edges of cracks there is an intact isthmus. The stress-strain state of such structures was considered in works Kashtanov and Petrov (2006); Zhou and Wang (2006). If we assume that the area Ω_D is a representative elementary area, which is assumed to be infinitesimal when constructing the model of mechanical behavior of the body B, then we can construct one of the variants of the model of a damaged elastic medium. The parameter α (or $\beta = 1 - \alpha$) is one of the parameters of the thermodynamic state of the medium, along with the temperature (which is considered constant in this work) and the strain tensor. Expression (11.10) indicates that as soon as a stress state arises in a material, micro-cracks can arise in it, since the necessary condition for their existence $- 0 < \alpha < 1$ is to be fulfilled. The tendency of the parameter α to zero ($\alpha \to 0$) means the tendency of the material to fracture. Considering the well-known fact, indicating that the value of a real tensile strength (breaking stress) is less than the theoretical value due to the presence of micro-cracks, further the mechanism of destruction of the particle ΔB under the action of tensile stress, connecting the breaking stress σ_u with the theoretical tensile strength σ_t, the value of parameter α and the size of the crack l_{cr} is proposed.

11.3 Basic Assumptions

The following is assumed.

1. At any moment of time, the tension stretching the particle ΔB is calculated under the assumption that there is no micro-crack in it, although it may be in it (if evaluated by expression (11.10)). At the moment when this stress has reached the value of the breaking stress σ_u, it is actually on the isthmus between the cracks it reaches the theoretical tensile strength σ_t. Because of this, there is an effect of destruction of the material with a lower strain compared to it. In this case, the elastic energy of a solid cell corresponding to the stress is converted

into the surface energy of the crack shores corresponding to the same stress σ_u, and the elastic energy in the part Ω_α corresponding to the theoretical strength limit σ_t.

2. It is assumed that the crack opened, although the width of the disclosure is neglected. The opening, therefore, means the disappearance of the influence of the crack coasts on each other. The stress state in the areas $\Omega_{1-\alpha}^{\pm}$ is absent due to crack opening. So there are no internal forces in these areas. All the energy of the elastic deformations of the areas $\Omega_{1-\alpha}^{\pm}$ has been transformed into the surface energy of the coasts $\Omega_{1-\alpha}^{\pm}$ of the cracks that limit the areas $\Omega_{1-\alpha}^{+}$ and $\Omega_{1-\alpha}^{-}$. The work of the internal forces developing in areas $\Omega_{1-\alpha}^{\pm}$ on the displacements of areas $\Omega_{\alpha}^{\pm}$, as well as the forces of areas $\Omega_{\alpha}^{\pm}$ on the displacements of points $\Omega_{1-\alpha}^{\pm}$ is zero. This assumption is very rough, but it significantly simplifies the task of establishing a connection between the stress state of the medium and its damage.

3. The cells Ω_D^{+} and Ω_D^{-} (for definiteness of the form) are considered cubes with the side length equal to l. So the diameter of the cross section A_D is equal to $D = l\sqrt{2}$. The crack section is located in the left front corner of the middle section of the cell, see Fig. 11.1.

4. Considering the stated hypotheses, it is possible to obtain two relations between the cell size, the stress values in it, and the crack size inside it.

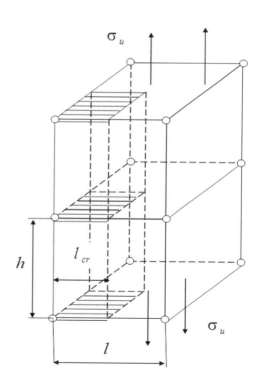

Fig. 11.1 Diagram of the structure of the material before destruction by the forces that stretch it.

The first of them reflects the equality of energy $[w] = \sigma_u^2/E$ in the cell before the formation of a micro-crack in it under the action of destructive strain (real tensile strength) of energy after its formation, but on the eve of complete destruction. In part $(1 - \alpha)$ between parallel cracks, all elastic energy is converted into surface energy of the two coasts of the crack. In this case, in expression (11.8):

$$[w]_{(1-\alpha)(1-\alpha)} = \frac{4W_{saf}}{2h}, \quad [w]_{\alpha(1-\alpha)} = 0. \tag{11.11}$$

In part α average strain equal to the theoretical tensile strength σ_t. Therefore:

$$[w]_{\alpha\alpha} = \frac{\sigma_t^2}{2E}. \tag{11.12}$$

Balance equality (taking into account (11.8)) reads

$$\frac{\sigma_u^2}{2E} = (1 - \alpha)^2 \frac{2W_{saf}}{h} + \alpha^2 \frac{\sigma_t^2}{2E}. \tag{11.13}$$

In accordance with the definition

$$\alpha = \frac{l^2 - l_{cr}^2}{l^2}. \tag{11.14}$$

Taking into account the assumptions made, expression (11.10) takes the form:

$$\alpha = \frac{1}{1 + \dfrac{\sigma_t^2 h}{2W_{saf}E}}. \tag{11.15}$$

The system of relations (11.13)–(11.15) contains as unknown quantities σ_u, α, l_{cr}, l. The values of σ_t and W_p can be calculated, for example, by the methods described in Frolenkova and Shorkin (2013); Vitkovsky et al (2012). The uncertainty of the system of equations gives grounds for two of the unknowns to be determined experimentally or hypothetically, for example, by the assumption that $\sigma_u \approx 0.01\sigma_t$, $l \approx h$.

From the expression (11.15) follows the relationship:

$$\frac{2W_{saf}}{l} = \frac{\alpha}{1 - \alpha} \left(\frac{\sigma_t^2}{2E} \right). \tag{11.16}$$

With its account from (11.13) follows the dependence:

$$\sigma_u = \sigma_t \sqrt{\alpha} = \sigma_t \sqrt{1 - \beta}. \tag{11.17}$$

Given that $\sigma_u \approx 0.01\sigma_t$, it can be argued that the parameter α has a small value: $\alpha \approx 10^{-4}$. Taking it as a basis and considering (11.14) we can get that $l \approx l_{cr}$. Then

$$l_{cr} = \frac{4W_{saf}E}{\alpha\sigma_t^2} = \frac{4W_{saf}E}{\sigma_u^2}, \tag{11.18}$$

or

$$\sigma_u = \left(\frac{4W_{saf}E}{l_{cr}}\right)^{1/2}. \tag{11.19}$$

Expression (11.17) establishes the relationship between the theoretical tensile strength σ_t, breaking stress σ_u and damage to a material of the type considered – a network of parallel micro-cracks.

11.4 Comparison of Calculation Results with Known Data

Comparing the obtained results, first, we can be convinced of the commensurability of the estimate of the value of breaking stress, determined on the basis of expression (11.19) with the existing estimates of Griffiths in (11.1) and Orowan (11.2). Secondly, the calculation and the length of the micro-crack l_{cr} under the assumption that $\sigma_t \approx E/5$, as well as the parameter α, corresponding to the well-known (Busov, 2007) reference data on the values σ_u in Table 11.1 indicates the commensurability of the calculated values l_{cr} with the value $l_{cr} \approx 10^{-6}$ m (Petch, 1968). The value α, estimated based on real values σ_u and known estimate $\sigma_t \approx E/5$ also corresponds to the conventional representation $\sigma_u \approx 0.01\sigma_t$. It is characteristic that the size estimate $l \approx l_{cr} \approx 10^{-6}$ m is approximately equal, as expected, to the distance between the cracks of the stretched thin pellicles having a value $\approx 5 \cdot 10^{-6}$ m.

11.5 Conclusion

In this paper, studies of the relationship of the theoretical tensile strength for stretching with real breaking stress and damage of the studied material have been carried

Table 11.1 Results of the calculation of the length of micro-cracks and the minimum relative area of not damaged by a crack unit cell cross section (accepted $\sigma_u \approx 0.01\sigma_t$)

Material	$\sigma_t, 10^{11}$ Pa calculated	$\sigma_u, 10^9$ Pa calculated	$\sigma_u, 10^9$ Pa (Babichev et al, 1991)	$l_{cr}, 10^6$ m	$\alpha, 10^4$
Al	0.07	0.07	0.05	15	0.84
Cr	0.37	0.37	0.41	17	1.05
Ti	0.21	0.21	0.24	17	1.07
Mo	0.53	0.53	0.80	7	1.23

out. Based on the idea of an elementary representative particle of a material studied by the methods of mechanics of a deformable solid body, it is assumed that each such particle has a micro-crack, the presence of which is characterized by its real diameter and relative (with respect to the sectional area of the particle) area - damage. A mechanism of destruction has been proposed, based on taking into account the energy changes during the formation of a micro-crack, the stationarity of the energy of elastic deformations that exist after its formation. The connections between the theoretical tensile strength, the real breaking stress, the damage of the material and the size of micro-cracks have been established. The disadvantage of the model is the non-closure of the system of algebraic equations for the unknown of real breaking stress, the damage, the size of micro-cracks. This required the use of data on the real breaking stress or its connection with the theoretical tensile strength, the value of which, as well as the surface energy and the Young's modulus are considered known—they can be found by the methods proposed in Petch (1968). Estimated calculations of the value of micro-cracks and damage of some materials have been carried out. The calculation results correspond to known ideas and reference data. In addition to practical significance, the result of the work (relation (11.10)) indicates that as soon as a stress-strain state arises in an elastic medium, a network of micro-cracks can develop in it.

Acknowledgements The work was performed within the framework of the basic part of the State task for 2017 - 2019, project code 1.5265.2017 / BCh.

References

Aptukov VN (2007) Model' uprugo-povrezhdennoi ortotropnoi sredy [model of the elastic damaged orthotropic environment]. Vestnik Permskogo universiteta Seriia: Matematika Mekhanika Informatika 7(12):84–90

Astaf'ev VI, Radaev YN, Stepanova LV (2001) Nelineinaia mekhanika razrusheniia [Nonlinear Mechanics of Damage]. Izdatel'stvo "Samarskii universitet"

Busov VL (2007) Rasseianie ul'trazvukovykh voln na mikrotreshchinakh v fragmentirovannykh polikristallakh [ultrasonic wave scattering by microcracks in the fragmented polycrystals]. Akustichnii vestnik Acoustic Bulletin 10(3):19–24

Frolenkova LY, Shorkin VS (2013) Metod vychisleniia poverkhnostnoi energii i energii adgezii uprugikh tel [method of calculating the surface and adhesion energies of elastic bodies]. Vestnik Permskogo natsional'nogo issledovatel'skogo politekhnicheskogo universiteta Mekhanika PNRPU Mechanics Bulletin 1:235–259

Griffith AA (1920) The phenomena of rupture and flow in solids. Phil Trans Roy Soc London Ser A 221:163–198

Hild F, Forquin P, Denoual C, Brajer X (2005) Probalistic-deterministic transition involved in a fragmentation process of brittle materials: Application to high performance concrete. Latin American Journal of Solids and Structures 2(1):41–56

Kachanov LM (1974) Osnovy mekhaniki razrusheniia [Fundamentals of Mechanics of Damage]. Moskva: Nauka

Kashtanov AV, Petrov YV (2006) Energy approach to determination of the instantaneous damage level. Technical Physics The Russian Journal of Applied Physics 51(5):604–608

Nowacki W (1975) Teoriia uprugosti [Elasticity Theory]. Mir, Moskva

Panin VE, Egorushkin VE (2009) Physical mesomechanics and nonequilibrium thermodynamics as a methodological basis for nanomaterials science. Physical Mesomechanics 12(5):204–220

Petch NJ (1968) Metallographic aspects of fracture. In: Liebowitz H (ed) Fracture: An Advanced Treatise, Academic Press, New York, vol 1: Microscopic and Macroscopic Fundamentals, pp 376–420

Rabotnov Y (1970) O razrushenii tverdykh tel [On the Damage of Solid Bodies]. Sudostroenie, Leningrad

Rybin VV (1986) Bol'shie plasticheskie deformatsii i razrushenie materialov [Large Plastic Deformations and Damage of Materials]. Moskva: Metallurgiia

Sarafanov GF, Pereverzentsev VN (2010) Zarozhdenie mikrotreshchin v fragmentirovannoi strukture [nucleation of microcracks in the fragmented structure]. Vestnik Nizhegorodskogo universiteta im N I Lobachevskogo 5(2):90–94

Sedov LI (1972) A Course in Continuum Mechanics, vol 1-4. Wolters-Noordhoff Publ., Groningen

Vitkovsky IV, Konev AN, Shorkin VS, Yakushina SI (2007) Theoretical estimation of discontinuity flaw of adhesive contacts between multilayer elements of the liquid metal blanket in a fusion reactor. Technical Physics The Russian Journal of Applied Physics 52(6):705–710

Vitkovsky IV, Frolenkova LY, Shorkin VS (2012) Adhesion-diffusion formation of a multilayer wall for the liquid metal flow channel of a fusion reactor blanket. Technical Physics The Russian Journal of Applied Physics 57(7):1013–1018

Volynskii AL, Iarysheva LM, Moiseeva SV, Bazhenov SM, Bakeev NF (2006) Novyi podkhod k otsenke mekhanicheskikh svoistv tverdykh tel ekstremal'no malykh i bol'shikh razmerov [new approach to an assessment of mechanical properties of solid bodies of extremely small and big sizes]. Rossiiskii khimicheskii zhurnal Zhurnal Rossiiskogo khimicheskogo obshchestva im D I Mendeleeva Russian Journal of General Chemistry 50(5):126–133

Zhou ZG, Wang B (2006) Nonlocal theory solution of two collinear cracs in the functionally graded materials. International Journal of Solids and Structures 43(5):887–898

Zhu TT, Bushby AJ, Dunstan DJ (2008) Materials mechanical size effects: a review. Materials Technology 23(4):193–209

Zhurkov SN, Kuksenko VS, Petrov VA, Savel'ev VN, Sulgonov U (1977) O prognozirovanii razrusheniia gornykh porod [about forecasting of destruction of rocks]. Izvestiia AN SSSR Fizika Zemli Izvestiya - Academy of Sciences of the USSR Physics of the Solid Earth 6:11–18

Printed in the United States
By Bookmasters